Impounded Rivers

ENVIRONMENTAL MONOGRAPHS AND SYMPOSIA

A series in the Environmental Sciences

Convener and General Editor
NICHOLAS POLUNIN, CBE
Geneva, Switzerland

MODERNIZATION OF AGRICULTURE IN DEVELOPING COUNTRIES: Resources, Potentials, and Problems

I. ARNON, *Hebrew University Jerusalem, Agricultural Research Service, Bet Dagan, and Settlement Study Centre, Rehovot, Israel*

STRESS EFFECTS ON NATURAL ECOSYSTEMS
Edited by
G. W. BARRETT, *Institute of Environmental Sciences and Department of Zoology, Miami University, Oxford, Ohio, USA*
and
R. ROSENBERG, *Fishery Board of Sweden, Institute of Marine Research, Lysekil, Sweden*

AIR POLLUTION AND PLANT LIFE
Edited by
MICHAEL TRESHOW, *Department of Biology, University of Utah, Salt Lake City, Utah, USA*

IMPOUNDED RIVERS: Perspectives for Ecological Management
GEOFFREY E. PETTS, *Department of Geography, University of Technology, Loughborough, Leicestershire, UK.*

Impounded Rivers

PERSPECTIVES FOR ECOLOGICAL MANAGEMENT

GEOFFREY E. PETTS
Department of Geography, University of Technology, Loughborough, UK

A Wiley–Interscience Publication

JOHN WILEY & SONS
Chichester · New York · Brisbane · Toronto · Singapore

Library of Congress Cataloguing in Publication Data:

Petts, Geoffrey E.
  Impounded rivers.

  (Environmental monographs and symposia)
  'A Wiley–Interscience publication.'
  Bibliography: p.
  Includes indexes.
  1, Dams—Environmental aspects. 2, Stream ecology.
  3, Stream conservation. I. Title.   II. Series.
QH545.D35P47 1984   574.5'26323   84–7538
ISBN 0 471 10306 3

British Library Cataloguing in Publication Data:

Petts, Geoffrey E.
  Impounded rivers.—(Environmental monographs and symposia)
  I. Dams—Environmental aspects
  I. Title   II. Series
  304.2   TD195.H93

  ISBN 471 10306 3

Phototypeset by Input Typesetting Ltd, London SW19 8DR.
Printed by Page Brothers., (Norwich) Ltd.

For my Family

Contents

viii

Series Preface

For civilization to survive in anything like its present form, the world's human population will need continuingly to widen its knowledge of the environment. Moreover, this knowledge will need to be closely followed by concomitant action to safeguard the biosphere and so maintain the framework and chief structures of our life support system. The *increase* of knowledge and awareness must come through observation, research and applicational testing, its *widening* through environmental education, and the necessary concerted *action* through duly organized application of the knowledge that has been thus acquired and disseminated.

The environmental movement has long been an undefined but widely effective vehicle for increasing appreciation of the vital nature and fundamental importance of man's and nature's environment. It is hoped that the *World Campaign for The Biosphere 1982*—and its adopting World Campaign for The Biosphere/International Society for Environmental Conservation—will focus attention on the fragility of this 'peripheral envelope of Earth together with its surrounding atmosphere in which living things exist naturally', on our utter dependence on its health as it constitutes our only life-support system, and on the necessity to foster it in every possible way.

To help to encourage such ideals and guide appropriate actions which in many cases are imperatives for man and nature, as well as to distil and widen knowledge in component fields of scientific and allied environmental endeavour, we founded and are now fostering an open-ended series of *Environmental Monographs and Symposia*. This emanated from an invitation by the international publishers John Wiley & Sons, and consists of authoritative volumes of two main kinds: monographs in the full sense of being detailed treatments of particular subjects by from one to three leading specialists, and symposia by more than three specialist authors covering a particular subject between them under the guidance and editorship of a suitable specialist or up to three specialists (whether such a volume results in part or wholly from an actual 'live' symposium or consists entirely of 'contributed' papers conforming to an agreed plan).

There seems to be virtually no end to the possibilities for our series; we are constantly getting or being given new ideas, and now have very many to think about and, in chosen cases, to work on. At the same time, we hope to complement the existing SCOPE reports, emanating from what in a sense is the world's environmental 'summit'. In addition to the present work and the already published *Modernization of Agriculture in Developing Countries*, by Professor I. Arnon, *Stress Effects on Natural Ecosystems*, edited by Professor Gary W. Barrett and Dr Rutger Rosenberg, *Air Pollution and Plant Life*, edited by Professor Michael Treshow, near-future volumes of Environmental Monographs and Symposia are expected to include *The Stratospheric Ozone Shield*, by Drs Byron W. Boville and Rumen Bojkov and *Ecosystem Theory and Application*, by Professor George A. Knox and the undersigned.

Whether or not we shall in time come to cover, in however general a manner, the entire vast realm of environmental scientific endeavour, must remain to be seen, though this was the gist of the distinguished publishers' original invitation and poses a challenge that we can scarcely forget. Meanwhile we believe we have decided on a constructive compromise with this propitious and promising series, in which we look forward to the effective participation of more and more of the world's leading environmentalists.

Nicholas Polunin
Convenor and General Editor of the Series
Geneva, Switzerland

Preface

Rivers are complex physical, chemical, and biological systems. Every river is to some extent unique, but there are some general characteristics which make it possible to classify riverine environments. These environments are determined by processes operating not just within the confines of the channel but, more importantly, on and within the hill-slopes which direct water, sediment, organic matter, and nutrients, into the channel network. It is the headwater drainage basins that provide the source areas for the downstream river system—including the floodplain and, at least to some extent, the delta, estuarine, and near-shore zones. Dams interrupt this pattern of downstream transfers, and the Impounded River has its discharges, sediment and organic loads, and water quality, determined by releases from the reservoir.

Without doubt the damming of rivers has been one of the most dramatic and widespread, deliberate impacts of Man on the natural environment. Yet, paradoxically, the nature of the impact—the range and magnitude of the induced environmental changes—has become widely appreciated only during recent years. We may not, even now, be aware of the full extent of the biotic species extinctions and other changes that can result, directly or indirectly, from river impoundment. The lack of a positive response to these problems during project evaluations, reflects not only the relative weights of economic and environmental argument but also three levels of failure by scientists, planners, and decision-makers, when concerned with rivers and their management. These three levels are, first, the adoption of an isolationist attitude by engineers, earth scientists, and biologists, during the 1960s in particular, which resulted in a failure to appreciate the full extent of observed and potential changes. Secondly, environmental scientists failed, until recently, to demonstrate the time-scale (measured in years or even tens of years) needed before some changes could be detected. The lack of effects demonstrated by some short-term post-project surveys created, not surprisingly, a general complacency towards the management of rivers downstream from dams. Certainly the more immediate effects of creating the man-made lake have attracted greatest concern. Thirdly, environmental scientists have generally failed to communicate their findings convincingly to planners and politicians concerned with the decision-making process. It is encouraging that, as I complete this monograph, an improved public awareness, demonstrated by extensive media coverage, has been generated in response to the new Itaipu barrage, Brazil–Paraguay, and to the Stage 2 (Lower) Gordon River Power Development, Tasmania, Australia. Ironically, a survey designed to provide background environmental data for this latter project revealed unexpected and far-reaching consequences of the Stage 1 dams, located in the headwaters (King & Tyler, 1982).

Gilbert F. White has advocated the geographical appraisal of water management problems, an approach which emphasizes that future management must be based upon a historical point of view—utilizing more critical post-audits of what in fact has happened in the trail of large-scale development, and giving due consideration to the integrity of natural systems (White, 1977a). This book presents a geographical analysis of impounded rivers; it seeks to generate an improved awareness of the fate of rivers, and of the scale of change in space and time, after dam construction; and it is intended for a wide readership. The focus is placed initially on the physical and chemical alterations to river systems imposed by impoundment, and then on the implications of these alterations for the biological components.

Chapter 1 of the book describes the global distribution of impounded rivers, the history of related scientific inquiry, and the spatial variability of natural river-systems, leading to a preliminary appraisal of the different effects of impoundment. This is followed by consideration of the primary impacts upon hydrological characteristics (2), water quality (3), and seston transport (4). In each chapter attention is directed to discussion of the type and magnitude of change, and to the patterns of change which may be observed along and between rivers. The effects of these primary impacts upon channel morphology (5) and aquatic, riparian, and floodplain vegetation (6), are examined before considering how these changes affect the benthic macroinvertebrates (7) and fish and fisheries (8). In the final chapter the environmental considerations that are most pertinent to successful river management are discussed, namely the problems of forecasting ecological change prior to impoundment, and the potential effectiveness of responsive approaches (9).

Current knowledge enables us to manipulate some components of river systems to minimize impacts resulting from upstream impoundment. Generally, such 'environmental engineering' is, as yet, of only limited success. Yet, our perceptions of environmental problems are often biased by short-term socio-economic attractions. The common view is of hydroelectric power as a 'clean' alternative; of river-flows as a freely exploitable resource; and of flood-control as a necessary convenience to allow urban, industrial, or agricultural, development. This monograph does not seek to show that such views are mere folly. However, it does aim to demonstrate the dramatic and far-reaching environmental consequences of damming rivers, and to identify considerations that are necessary at the planning stage for the effective evaluation of alternative schemes.

If there is one plea, it is for acceptance of the need for a long-term perspective in river management. Although many changes become apparent only after a period of years, much of the river system, including its floodplain, estuary or delta, and even the near-shore coastal zone—will, over time, be altered more or less drastically by any impoundment. The fundamental issue relates to the value which we give to these inevitable changes.

GEOFFREY E. PETTS

University of Technology, Loughborough, Leicestershire, UK
30 August 1983

Acknowledgements

This book is the outcome of seven years of research and discussion with friends and professional colleagues. Many people have helped me, knowingly or inadvertently, and to all I offer my sincere thanks. I owe a great dept to Professor K. J. Gregory for his guidance and encouragement during the early years. For conversations which inspired renewed enthusiasm, I thank Dr R. F. Warner and E. E. Serr. For constructive comment on chapter drafts I am indebted to Dr R. Jones, Dr M. Wade, M. Greenwood and G. White. J. D. Pratts, T. R. Foulger, and D. J. Gilvear helped in many ways; Mrs Jenny Jarvis typed the manuscript; and Mrs Anne Tarver drew the figures. My wife, Judy, helped me throughout, not least with the compilation of the bibliography; without her strong support this monograph would probably never have been completed.

I wish to express my gratitude also for permission to use and reproduce material to: The Institute of British Geographers; John Wiley & Sons Ltd; *Science*, The American Association for the Advancement of Science, and R. J. Gibbs; The Royal Society of Victoria; The American Society of Limnology and Oceanography, Inc.; The American Society of Civil Engineers; The International Association of Hydraulic Research; Elsevier Scientific Publishing Company; The American Geophysical Union; The United States Geological Survey; The Foundation for Environmental Conservation and Nicholas Polunin; Edward Arnold (Publishers) Ltd; Dr W. Junk Publishers; the *Journal of the Fisheries Research Board of Canada*; *Water, Power and Dam Construction*; Blackwell Scientific Publications Ltd; B. A. Whitton; The Board of Regents of the University of Wisconsin System; G. E. Hollis; J. D. Pratts; Sigma Xi and the Scientific Research Society; J. Cotillon and the Commission Internationale des Grands Barrages; the Pennsylvania Academy of Science; Academic Publishing House of the Czechoslovak Academy of Science; E. Schweizerbart'sche Verlagsbuchlandlung (Nägele u. Obermiller); the Entomological Society of America; the Severn–Trent Water Authority; the Department of Water Resources, Red Bluff, California; Plenum Publishing Corporation; R. N. Winget; J. V. Ward; H. Décamps; J. Henricson; *Annales de Limnologie*; and J. Lewin.

G. E. PETTS

Loughborough, UK

Inside: Grand Dixence Dam, Switzerland, completed in 1962, the world's largest dam today is 285 m high. The reservoir has a total capacity of about 400 million cubic meters and receives water, primarily by inter-basin pipe-lines, from a total catchment area of 357 km², of which 180 km² are glaciers. The scheme has markedly altered the streams of the Val d'Herens in the west and the Visp valley in the east. The total gross hydro-electric power production is 1680 GWL.

CHAPTER 1

The Impounded River

'. . . human operations . . . do act in the ways ascribed to them though our limited faculties are at present, perhaps forever, incapable of weighing their immediate, still more their ultimate, consequences. But our inability to assign definite values to these causes of the disturbance of natural arrangements is not a reason for ignoring the existence of such causes in any general view of the relations between man and nature, and we are never justified in assuming a force to be insignificant because its measure is unknown, or even because no physical effect can now be traced to it as its origin.'

George P. Marsh, 1864

In: Lowenthal (1965, p. 465).

The total volume of water on Earth has been estimated as 1.4×10^9 km^3, but less than 0.0087% exists as surface water in rivers, streams, and natural lakes; over 90% is ocean water. Nevertheless, the surface water, if distributed evenly in space and time, could meet existing and future needs. The problem for management is that water is not equitably distributed: in some areas excessive precipitation produces perennial flood problems, whereas in others, arid conditions may not provide enough water to sustain all demands. In the developed countries the demand for water, its efficiency of use, and its quality, were generally secondary issues until the early 20th century. This situation started to change with the advent of the industrial revolution. Today, river exploitation for hydroelectric power has been added to the needs for water-supply and flood control; the industrialized river is a man-made waterway almost from source to mouth. Its headwaters are impounded and its flows are determined by the controlled releases from the dam; the quality of the waters released is determined by the processes that are operative within the impounded reservoir; it receives effluent from industry and from urban areas; it may be canalized for navigation; and its floodplain is used for agricultural, industrial, or urban development. It may yet prove to our cost that technology has advanced far more quickly than our knowledge of the environmental consequences of such impoundment.

For more than 2500 years, man has sought to understand, to predict, and also to control, nature in order to realize the maximum benefits from available resources. Too often, developments have been evaluated on structural or architectural grounds, with little or no regard for the geomorphological, ecological, and aesthetic, harmony of the result. In environmental management, decisions are most often made in a context in which the outcome is in doubt and the consequences of a given choice cannot be fully predicted. The costs and benefits of alternative actions are rarely absolute, and are frequently impossible to state in monetary terms. Indifference to, or ignorance of, the intricate relationships maintaining stability within the environment, has allowed changes to take place. However, our failure to predict precisely the environmental costs of human activity cannot provide justification to ignore the probable or possible consequences. Nor can it remove our responsibility to our successors to ensure that we manage our rivers within a long-term perspective.

George P. Marsh has been called 'the mighty prophet of modern conservation' (Lowenthal, 1965 p. ix), and certainly his *Man and Nature* provided a graphic and comprehensive exposé of the damage which Man has done to the Earth, and an awareness of the potential long-term consequences of human activity. Many of Man's activities have a direct or indirect effect upon river systems, and in extreme cases the consequent environmental changes have been widespread, affecting the in-channel, riparian, floodplain, and delta or estuarine, fauna and flora. Many long-term and irreversible effects cause changes in the genetic structure, or even in the extinction, of certain species. Gregory (1976a) cites the work of several research workers who, prior to 1920, described changes in river hydrology, sediment transport, and channel characteristics, as a consequence of Man's activities. Important symposia, such as 'Man's role in Changing the Face of the Earth', held at Princeton, New Jersey, USA, in 1955 (Thomas, 1956), sought to provide an integrated basis for the development of interdisciplinary studies; but a paucity of field data inhibited the detailed investigation of human impacts upon river systems.

Nevertheless, during the past two decades, the number of studies of rivers has increased rapidly and numerous examples now exist of Man's unintentional and deliberate alterations to, or destruction of, natural systems (e.g. Oglesby *et al.*, 1972; Gregory, 1977; Hollis, 1979; Goudie, 1981). A major reason for this recent upsurge of interest stems from the need to utilize knowledge of processes in the interpretation and prediction of temporal change. It is only now, in this second half of a century in which Man probably has done more to alter his environment than in any previous millennium of his history, that we are coming to appreciate some of the bad consequences of our actions (Hynes, 1970a). However, ecological data for many of the world's major rivers are scant and, for logistical reasons, are likely to remain so for some time. In the case of Australia's most important river, the Murray, it is scarcely possible to list the floral and faunal species present, let alone describe their interactions with each other and the physical and chemical environment (Croome *et al.*, 1976).

The aim of environmental management should be to optimize development and resource-use by minimizing both costs and impacts (Clark, 1978). For this goal to be achieved, three developments are required: (1) an improved awareness of the complex interrelationships between the components within environmental systems; (2) an improved knowledge of the mechanisms of environmental response to human activity, and (3) an improved perspective of time. To these may be added a fourth, namely the need for planners, engineers, politicians, and any decision-maker involved with any development (but particularly with river impoundment and reservoir operations), to gain an adequate appreciation of the long-term environmental effects of their actions.

The Spread of Dams

At the turn of the century, Voeïkov (1901) believed that the control and mastery of water-supply was one of the main tasks which Man had yet to accomplish. Technical advance during the twentieth century has been such that Mermel (1981) was able to conclude that the number and distribution of dams 'reflects favourably on the challenges confronting engineers in developing water resources' (p. 58).

Dams were first constructed for the purpose of river regulation over 5000 years ago in Egypt (N. Smith, 1971); they had become popular in the Mediterranean area by Roman times, and were introduced into western Europe with the overshot wheel during the late Middle Ages (Beckinsale, 1972). One of the earliest large dams was the 27-m-high Tashahyan Dam on the Abang Xi, China, completed in AD 833, which is still being used for irrigation today. But in Great Britain the first large structure (over 15 m high) was Coombs Dam, which was not completed until 1787. The era of major dam-building activity did not begin until the early 1900s and was coincident with changes in earth-moving and concrete technology; but during the past 50 years, most of the major rivers of the world have been impounded at least to some degree.During the 1930s the construction of large dams and the grouping of multipurpose projects within entire river basins became symbols of the efficient application of engineering techniques to water management. First achieved in the Tennessee Valley, USA, the fully integrated development of entire basins was subsequently pursued in many basins throughout the world, notably within the Volga, USSR, and in the Snowy Mountains scheme of Australia (White, 1977b). Important new information, generated particularly by the International Hydrological Decade and the International Biological Programme, led, however, to an improved insight into the nature of river systems that had been perturbed by water development. By the mid-1970s White (1977b p. 5) noted that

> 'The great multipurpose dam which in mid-century was a symbol of social advancement and technological powers, came into bad odour . . . and was often attacked as destructive and poorly conceived.'

The *World Register of Dams*, published by the International Commission on Large Dams in 1973, catalogues all structures more than 15 m in height. By 1971 more than 12 000 had been built, impounding 4000 km³ of water and inundating an area of 800 000 km² (Fels & Keller, 1973). Beaumont (1978) used the *Register* to identify three periods of dam building prior to 1970 (Fig. 1): before 1900, dam construction was increasing but numbers were small; between 1900 and 1945 there was a period of moderate building activity separated by troughs associated with wars and economic depression; and between 1945 and 1971 there was a period of accelerating impoundment, with the completion of 8140 large dams world-wide. Western Europe experienced a relatively high rate of dam construction between 1840 and 1880, and an increase in building activity occurred in North America, Western Europe, and South-east Asia, between 1900 and 1940; but all regions experienced an extreme acceleration after 1950. The data exclude many dams in the USSR, but the most important omission is the People's (Democratic) Republic of China. The world picture presented by Beaumont has been dominated since 1950 by North America, where over 200 major dams were completed each year between 1962 and 1968. Beaumont (1978) identified the peak of dam-

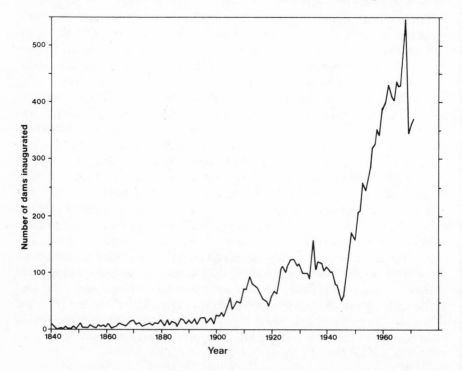

Fig. 1. World dam construction. (From Beaumont, 1978. Reproduced by permission of the Institute of British Geographers.) Since 1971 the annual rate of dam-building has not declined but, for the countries represented in the 1973 *Register*, has been maintained at about 400 per year; this figure is increased, however, to 700 per year if all countries are considered (Mermel, 1981).

building activity as being in 1968, and suggested that, on a world scale, the peak of dam-building activity may well have passed. However, the data of the *Register* support the conclusion that on a conservative estimate more than two dams per day are being added to the world's rivers in attempts to enhance world resources (Mermel, 1976).

At the world-scale, intensive dam construction has continued since 1970, and in fact, for some countries the rate of dam-building is increasing. An update of the *Register* to cover the years 1971–74 demonstrated that the number of dams was being added to at a rate per year of 125 in USA, 30 in Japan, 25 in China, 20 in India, 15 to 20 in Spain, 10 to 15 in Mexico, 10 in Australia, 8 to 10 in France, 5 to 10 in Korea, and 5 to 8 in Turkey. A second update reported increased dam-building activity during the last half of the 1970s in several countries: Brazil, Argentina, Canada, India, Japan, Turkey, Spain, and the People's Republic of China (Mermel, 1981). The major dams of the world are listed in Table I, and major dam-building projects that are due for completion by 1990 are given in Table II. In the 1930s the world's highest dam was the 221-m Hoover Dam, on the Colorado River, USA, and the largest reservoir was Lake Mead ($37 \times 10^9 m^3$), which it impounded; but today they rank 18th and 20th, respectively, in the list of dams completed or actually under construction!

The distribution of the large dams (Table III) obviously relates to geographical factors, particularly those influencing site-availability. Large numbers of dams are currently under construction in Central and South America, particularly Brazil, and southern Asia, especially China and India. Certainly, dams are as common in many Third World countries as in industrialized nations. Virtually all rivers on the African continent which are capable of generating power have been dammed, and those which are not large enough to produce power have been impounded anyway for flood control, water-supply, or fisheries (Obeng, 1981). Although tropical rivers were largely unaffected by dams until relatively recently, considerable efforts have been made to exploit the great hydroelectric power-potential which exists in these areas (Jassby, 1980). Kariba Dam on the Zambezi, closed in 1958, was the first large dam to be constructed in the underdeveloped tropical countries; but it was quickly followed during the 1960s and 1970s by five more in Africa (on the Volta, Zambezi, Nile, Niger, and White Bandama), one on the Surinam River in South America, and one on the Ord in northern Australia (Petr, 1978).

As a result of river impoundment and the artificial storage of flood-water, the proportion of stable runoff has been augmented on every continent (Lvovitch, 1973). In Africa and North America about 20% of the stable runoff is contributed by impoundments, in Europe and Asia the figures are 15% and 14%, respectively, while rivers in South America (4.1%) and Australasia (6.1%) are least affected. However, on every continent the regulatory effects of man-made lakes upon the stream-flows exceed that of natural lakes by more than three times. Croome *et al.* (1976) suggest that by the year 2000, about 66% of the world's total stream-flow will be controlled by dams.

Table I   Major world dams constructed before 1983. (Data from Mermel, 1982. Reproduced by permission of *Water, Power and Dam Construction*.)

A. Largest Reservoirs (threshold $65 \times 10^9 m^3$)

Name	Completion date	Location	Reservoir capacity ($\times 10^9 m^3$)
Owen Falls	1954	Lake Victoria/River Nile, Uganda	204.8
Bratsk	1964	River Angara, USSR	169.3
High Aswan	1970	River Nile, Egypt	164.0
Kariba	1959	River Zambezi, Zimbabwe	160.4
Akosombo	1965	River Volta, Ghana	148.0
Daniel Johnson	1968	River Maniconagan, Canada	141.9
Bennett W.A.C.	1967	River Peace, Canada	74.3
Krasnoyarsk	1972	River Uenisei, USSR	73.3
Zeya	1975	River Zeya, USSR	68.4

B. Highest Dams (threshold 225 m)

Name	Completion date	Location	Dam height (m)
Grand Dixence	1962	River Dixence, Switzerland	285
Vaiont	1961	River Vaiont, Italy	262
Guavio	1982	R. Orinoco, Columbia	250
Mica	1973	River Columbia, Canada	245
Chicoasén	1981	River Grijalva, Mexico	245
Sayan-Shushensk	1980	River Yenisei, USSR	242
Mauvoisin	1957	Drange de Bagnes, Switzerland	237
Chivor	1975	River Bata, Columbia	237
Oroville	1968	River Feather, USA	235
Chirkei	1977	River Sulak, USSR	233
Bhakra	1963	River Sutlej, India	226

C. Largest Hydroelectric Power Dams (threshold 4000 MW)

Name	Completion date	Location	Planned power capacity (MW)
Grand Coulee	1942	River Columbia, USA	10 830*
Tucurui	1982	River Tocantins, Brazil	6480*
Sayano-Shushensk	1980	River Yenisei, USSR	6400
Krasnoyarsk	1972	River Yenisei, USSR	6000
La Grande 2	1982	River La Grande, Canada	5328
Churchill Falls	1971	River Churchill, Canada	5225
Bratsk	1964	River Angara, USSR	4600
Ust-Ilim	1980	River Angara, USSR	4500
Cabora Bassa	1974	RiverZambezi, Mozambique	4000

* The latest updatings have given corrected figures for the Planned rated capacity of Grand Coulee Dam as 5494 MW and of Tucurui as 8000 MW (Mermel, 1983).

Table II Major world dams under construction and due for completion by 1990.
(Data from Mermel, 1982. Reproduced by permission of *Water, Power and Dam
Construction*.)

A. Large Reservoirs (threshold $65 \times 10^9 m^3$)

Name	Completion date	Location	Reservoir capacity ($\times 10^9 m^3$)
Guri	1985	River Caroni, Venezuela	136

B. High Dams (threshold 225m)*

Name	Completion date	Location	Dam height (m)
Rogun	1985	River Vakhsh, USSR	325
Nurek	1985	River Vakhsh, USSR	300
Inguri	1985	River Inguri, USSR	272
Tehri	1990	River Bhagirathi, India	261
Kishaw	1985	River Tons, India	253
El Cajon	1985	River Humuya, Honduras	226

C. Hydroelectric Power Dams (threshold 4000 MW)

Name	Completion date	Year of initial operation	Location	Planned power capacity (MW)
Itaipu	1985	1983	River Paraná, Brazil/ Paraguay	12 600
Guri	1985	1968	River Caroni, Venezuela	10 000
Corpus Posadas	1988	1990	River Paraná, Argentina/Paraguay	6000
Yacreta-Apipe	1988	1986	River Paraná, Argentina/Paraguay	4050

* In the latest update, Mermel (1983) recorded that the Borocua Dam, Costa Rica, will be the
fifth-highest dam when completed, at 267 m.

Investigations on Impounded Rivers

The development of research on impounded rivers obviously reflects the
graph of dam-building activity. An examination of the citation structure of
major publications (Fig. 2) reveals that relatively few learned papers were
published on impounded rivers prior to 1960, but a dramatic rise in the
number of citations during the subsequent two decades is clearly shown, and
this reflects the sudden realization that deleterious environmental effects can
arise from river impoundment. A post-1950 growth of citations is common
in science as a whole, and this relates to the general growth of scientific
knowledge and the expansion of research groups and outlets for publication
(Stoddart, 1967; Park, 1980). However, until the 1950s, in hydrological terms
at least, river impoundments were few and generally of a small size when
producing *man-modified* rivers: floods were regulated and sediment was
trapped, but in many cases the river retained at least some of its natural
character; seasonal flow-variations were less marked than formerly but were
nonetheless apparent, and the relatively small reservoirs affected the channel
for only a limited distance below the dam.

Table III   Global distribution of major dams and reservoirs.

A. Distribution by number (after Mermel, 1981. Reproduced by permission of *Water, Power and Dam Construction*):

	Completed to 1981			Dams under construction (109)*
	Large reservoirs (25)*	High Dams (24)*	Hydroelectric Power Dams (77)*	
NORTH AMERICA	4	6	21	14
USA	2	4	10	5
Canada	2	2	11	9
CENTRAL AND SOUTH AMERICA	1	2	11	25
AUSTRALIA AND NEW ZEALAND	0	0	0	1
SOUTH-EAST ASIA	1	0	6	10
Japan	0	0	4	4
China	1	0	2	6
SOUTH-WEST ASIA	0	1	0	12
India	0	1	0	5
AFRICA	5	0	1	5
EUROPE	0	5	2	6
USSR	9	3	9	8

B. Distribution of Large Dams in 1971 by storage capacity (%) Data from the *World Register of Dams* (International Commission on Large Dams, 1973):

	Storage capacity ($\times 10^6 m^3$)					Proportion of total (%)
	<1	1–10	10–100	100–1000	>1000	
NORTH AMERICA	24	34	23	14	5	38
CENTRAL AND SOUTH AMERICA	7	37	30	18	8	7
AUSTRALIA AND NEW ZEALAND	17	28	33	15	6	3
SOUTH-EAST ASIA†	61	23	12	3	1	18
SOUTH-WEST ASIA	9	40	34	14	3	10
AFRICA	26	34	26	11	3	3
EUROPE	24	37	27	11	1	20
USSR‡	40	24	35	1	–	1
Weighted averages§	28	34	24	11	3	
Weighted averages for reservoirs larger than 1×10^6 m^3		47	33	15	4	

* The number of cases considered from a listing by rank order.
† No data included for China.
‡ Limited data available.
§ Excluding the USSR.

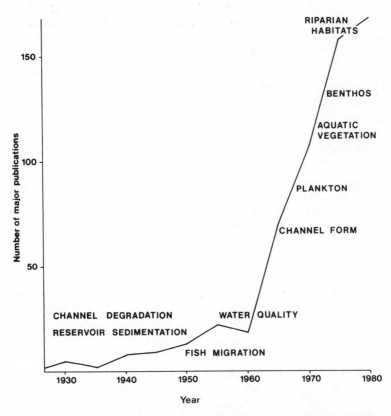

Fig. 2. The growth of major publications on impounded rivers. (Compiled by D. J. Gilvear.) The number of publications—excluding departmental, agency, or similar, reports—during each 5-years' period is shown, together with an indication of when specific themes first became important within this literature.

Subsequently, however, the tendency has been to build bigger and bigger dams of larger and larger capacity, and increasingly these have been designed to exert a total control upon the flow-regime of the river downstream to provide hydroelectric power, flood-control, and water-supply. As a consequence, truly *man-made* rivers have been created: out-of-season high-flows are no longer correlated with the seasonal rains, and natural extremes of discharge, water temperature, and sediment transport, are eliminated while unnaturally rapid and repeated flow fluctuations or sudden periodic flow changes may be superimposed upon a constant background discharge.

The growth of concern for the effects of an impoundment upon the downstream river in part reflects the recognition of this change to the creation of man-made rivers. Indeed, during the 1960s, a strengthening vocal movement of conservationists questioned the need for large dams, and scientists began to find that irreparable changes could be caused by river impoundment (Turner, 1971). However, it is only since about 1970 that the general desecration of landscape by dams has attracted the attention of the public; before this time an awareness was confined to academics or the most esoteric government research laboratories (Hynes, 1970a).

Conferences on man-made lakes in London in 1965 (Lowe-McConnell, 1976), Accra in 1966 (Obeng, 1969), and Knoxville in 1971 (Ackermann et al., 1973), directed attention to the environmental and socio-economic changes associated with the construction of large dams, and international concern was further exemplified by the report (SCOPE 2) on *Man-made Lakes as Modified Ecosystems*, Scientific Committee on Problems of the Environment (SCOPE, 1972) published in 1972 by the International Council of Scientific Unions. However, the impacts of the impoundments upon the physical, chemical, and biological, character of the rivers downstream of the dams, received only limited attention. An important review of the impact of reservoirs upon rivers downstream had been published by Neel (1963) and one of the most comprehensive studies of an impounded river was reported by Peňáz et al. (1968) for the River Svratka, Czechoslovakia. But, the first major international symposium addressing the question of the environmental impacts of large dams upon the downstream river was not held until 1973 (11th International Congress on Large Dams, Madrid, cf. Cheret, 1973).

Subsequently, several general reviews have been published (Ridley & Steel, 1975; Baxter, 1977; Brooker, 1981), and these have been supplemented by reviews of specific problems (Efford, 1975; Petts, 1979; Welcomme, 1979), reviews with a national perspective (Baxter & Glaude, 1980), and river-specific reports. These last have been, for example, of the Nile (Rzóska, 1976), of the Peace River, Canada (Townsend, 1975), of the Susquehanna River, USA (El-Shamy, 1977), of the River Murray, Australia (Walker et al., 1979), and of the Rivers Wye and Tees, UK (Edwards & Crisp, 1982). The International Geographical Union's Commission on Man and Environment symposium on Complex River Development, held on the Volga–Don in July 1976, emphasized that the prediction of the environmental consequences of water development remained a major problem (White, 1977a). In 1979 a symposium on 'Regulated Streams', held at Erie, USA, reflected the magnitude of world-wide stream regulation and the growth of international awareness of the ecological effects induced (Ward & Stanford, 1979a). The 'Second International Symposium on Regulated Streams', held in Oslo in 1982, attracted participants from several countries that were not represented at Erie—including USSR, Spain, and New Zealand.

In detail, the graph of citations reflects the type of effects induced by river impoundment. Prior to 1960, the literature was dominated by engineering studies of reservoir sedimention and channel degradation, which were related to questions of reservoir safety. More than one hundred catastrophic dam failures have been reported since 1930 for the USA alone (Greenhalgh, 1980). During the late 1950s and early 1960s, research concerned with the effects of large reservoirs upon river-water quality, and the consequences of dams as barriers to fish migrations, became important themes. It is only since the late 1960s that data have been presented to demonstrate the full ramifications of building a large dam upon channel morphology, aquatic plants, planktonic and benthic invertebrate communities, indigenous fish, and riparian, wetland, and floodplain, habitats.

Dams can, of course, be of virtually any size, and there is no limit to man-made lakes. Even in economic terms, a number of small reservoirs on headwater streams can have a resource potential of equal significance to a single large dam, and many countries, particularly Japan and Spain, depend upon large numbers of small reservoirs for water resource management, and for sustaining inland fisheries. No matter whether a large river or a small stream is impounded, the river downstream will be changed. Considerable attention has been directed to the effects of large dams, such as Glen Canyon Dam on the Colorado River (e.g. Turner & Karpiscak, 1980), Kariba Dam on the Zambezi River (e.g. Begg, 1973), Aswan High Dam on the River Nile (e.g. Din, 1977; Abu-Zeid, 1983), W. A. C. Bennett Dam on the Peace River (e.g. Kellerhals, 1982), and the V. I. Lenin Dam on the Volga River (e.g. Eliseev & Chikova, 1974). However, large reservoirs having a storage capacity of more than $1 \times 10^9 m^3$ represent less than 5% of the large dams recorded in the *Register* (Table IIIB); more than 80% of the dams impound reservoirs of less than $0.1 \times 10^9 m^3$, and probably more than 50% of all large dams impound a storage capacity of less than $0.01 \times 10^9 m^3$. Indeed, important storage within the Huang-Huai-Hai Plain, China, is provided by gates creating in-channel storage of over $3 \times 10^9 m^3$ (Fenglan & Wenkai, 1983); individual gates on large rivers can store up to $0.05 \times 10^9 m^3$, with a maximum backwater length of 50 km. Yet the effects of these very numerous but relatively small reservoirs upon downstream rivers have attracted less concern, even though such small impoundments, located on tributary streams, can exert an important impact upon the physical, chemical, and biological, characteristics of the main river.*

Problems Arising from River Impoundment

River impoundment and the resultant man-made lake generate a complex web of impacts which affect the human, biological, and physical, components of the environment. The sudden transformation of a river and its adjacent terrestrial environment into a lake, will directly affect the social and economic welfare of people—most importantly the relocatees. Impoundment of the Zambezi at Caborra Bassa, Mozambique, required the resettlement of 25 000 people in new villages, with due respect for family and tribal groupings; but the concentration of people away from their traditional scattered communities led to many problems, not least of emotional adaptation to community living and uprooting from traditional homelands (Jackson, 1975). Large dams, and especially those with associated irrigation schemes, may increase opportunities for the transmission of infections, such as hepatitis, poliomyel-

* A crash programme of dam construction has recently been approved in Spain, with 111 small dams—each one impounding only 23 000 m³ on average—being built now or to be started before 1985. Opposition to the construction of these dams was expected to be slight; to date, ecological protests have mainly been directed towards nuclear plants (Anon, 1983).

itis, typhoid, dysentery, and cholera. The creation of suitable habitats for the survival and proliferation of a wide range of vectors and intermediate hosts of diseases, has posed the greatest threat. Mosquitoes (especially *Anopheles funestus*),vectors of malaria and numerous virus infections, snails, the intermediate hosts of several worm diseases such as bilharzia, and the Buffalo Gnat or Blackfly (*Simulium damnosum*), which carries the disease onchocerciasis, are common problems, for example in South Africa (McIntosh *et al.*, 1973).

Laying aside these direct socio-economic problems that are not included in the mandate of this monograph on impounded rivers, the physical and biological effects of a dam will extend far beyond the limits of the man-made lake. Downstream changes to the physical and biological characteristics of the river, floodplain, estuary, or delta, and near-shore zone, testify to the deleterious ramifications of river impoundment. The Aswan High Dam has produced 7 thousand million kWh of electricity annually, brought 900,000 acres (364 500 ha) of land under cultivation, and increased the national income by more than $500 millions each year, but reservoir creation and irrigation development have caused an increase in the infection rate of bilharzia from zero to 80%. Moreover, impoundment of the River Nile has contributed to a 95% loss in the sardine catch (*Sardinella aurita* and *S. madarensis*) of the eastern Mediterranean, and, by trapping over 100 million tonnes of silt annually, the dam has prevented beach replenishment within the delta—thus allowing coastal erosion to threaten the recreation industry, particularly around Alexandria, and eliminating the annual silt enrichment of the Nile floodplains (Turner, 1971). Moreover, the increasing use in lowland areas of rivers for water abstraction and recreation, means that the study of the physical, chemical, and biological, changes induced by upstream impoundment is of more than purely academic interest.

Many benefits have accrued as a result of river impoundment— hydroelectric power, irrigation, navigation, flood-control, fisheries, urban water-supply, and various forms of recreation. Dams are symbols of economic advancement; they are an accepted tool for harnessing the resource potential of rivers, and for controlling the hazards to human activity that they impose; and it could well be argued that some reservoirs exemplify the successful enhancement of environmental quality of Man's modification of river systems. Indeed, SCOPE (1972 p. 17) proposed the view of:

> 'Man as the dominant species in ecosystem earth, who by building dams adds to his long-demonstrated capacity for technological change, and who seeks beneficially to integrate that skill with an ongoing process of interaction among human and other elements of the ecosystem. With inspiration and skill, he may mould a better world for his descendants.'

It is unfortunate that this admirable sentiment has been put into practice often only through the development of 'improved' responsive approaches—for the control of impacts after dam construction, rather than through anticipatory evaluations. A long-range planning approach is commonly rejected or ignored because of the constraints of uncertainty which lie between the present and the future. Furthermore, many of the changes experienced by impounded rivers are caused not directly by the dam itself, but by agricultural and industrial/urban development stimulated by the provision of a reliable water or power supply, or of flood control.

A Classification of Natural Rivers

A wide range of disciplines have accumulated a rich descriptive and functional literature on the particular effects of dams upon the downstream river system. However, the interests of scientists have usually been quite narrow—in part a reflection of the history of each separate discipline. Thus, individual scientific studies may provide neither information that is generally applicable to all situations, nor information concerning all the components of a single river system. Similar physical and biological processes are at work in large and small reservoirs, although they show profound differences from rivers in limnology and management. However, the effects of an impoundment of particular dimensions will depend upon its geographical location, and upon the characteristics of the natural river. Reservoirs have been built on rivers in subarctic, temperate, and tropical regions, and in arid areas traversed by exogenous streams, such as the Nile. The different energy conditions of these zones create some fundamental differences in the water-balance, temperature, regime, and biological processes. Certainly, a clear distinction can be made between the effects of river impoundment in areas of high and constant flow—where discharges are naturally regular, continuous, and dependable—and those in semi-arid or high-seasonal-rainfall areas, where river flow is strikingly variable; also between natural alluvial rivers having large sediment loads, and those with few sediment sources that transport little material

Catchment Ecosystems

'The problem of classifying natural areas, especially aquatic ecosystems, would be greatly simplified if the watershed were the base for classifying, since it is the watershed, functioning in response to external forces, that controls aquatic systems.'

Lotspeich (1980 p. 582)

Each catchment (= watershed) is delineated by a topographic divide, that isolates it from adjacent watersheds; the only interactions being the migration of mobile biotic populations, and some ground-water transfers. Within a catchment, the climate and geology are the primary factors controlling ecosystem production and the integrated development of landforms, soils, and vegetation. Climate provides energy and water to establish and drive ecosystems, but the geology—lithology and structure—gives each catchment a distinctive appearance. Watersheds are the products of geomorphological processes, including chemical and physical weathering, and the processes of sediment transport which ultimately remove debris from slopes to channels, and, finally, to oceans, or to inland basins of deposition. The soil characteristics (texture and structure), which are derived from the parent materials through rock-weathering and soil-forming processes, control runoff, infiltration, percolation to ground-water, and the susceptibility of the surface to erosion.

Vegetation provides the primary production, through photosynthetic activity, of catchment ecosystems. The composition of the plant communities is influenced by macro- and micro-climate, and by the soil's texture, structure, and chemical characteristics; and plants influence runoff-rate, water-loss through evapotranspiration, and soil erosion. This scenario supports the view of the river, characterized by its hydrology, sediments, morphology, and biotic community, as the result of all the processes operative within the catchment. The importance of the catchment in investigations of rivers was stressed by Thompson & Hunt (1930) and, in geomorphology, the catchment (= drainage basin) has become the fundamental unit of study (Chorley, 1969). Subsequently the catchment ecosystem approach has become accepted for a variety of river studies (Bormann & Likens, 1969; Black, 1970; Curry, 1972; Hynes, 1975; Schumm, 1977).

Fundamentally, rivers are a part of the hydrological cycle, and it is the stochastic nature of the runoff processes which gives rivers their dynamic characteristics. These relate, in part, to the magnitude of low-flows during the dry season, and to the predictability of seasonal flow variations, which are related directly to climate. However, they will also reflect the relative importance of the different streamflow sources during and between storm events, not least because these relate to the sediment-load and quality of the discharges. Ground-water flows will dominate if the soil is deep, or the rock is permeable, to produce relatively stable discharges.

Most rivers have an important flood component, which relates directly to the morphology, soils, and vegetation, of the catchment (e.g. Fig. 3). *Horton(ian) overland flow* (Horton, 1945) is generated if the rate of rainfall or melting is greater than the capacity of the soil to absorb water, and is uncommon in well-vegetated catchments. In these catchments, percolating water will be diverted laterally by a relatively impermeable soil horizon, and the water will reach the stream as *subsurface stormflow*. At some slope locations, vertical and lateral percolation may cause the soil to become saturated throughout its depth, and some subsurface stormflow may emerge from the soil surface as *return flow*, which may be supplemented by direct precipitation on the saturated area to produce *saturated overland flow*. In humid temperate areas the latter two processes dominate storm runoff, and the yielding proportion of the catchment, in headwater areas, expands, or shrinks—depending upon the rainfall amount and antecedent wetness of the soil (Hewlett & Hibbert, 1967). However, the processes which control the availability of sediment and organic matter are also of primary importance. Indeed, to understand any part of the river system, something must be known about the quantity and type of material, and the manner in which it is supplied from the source area.

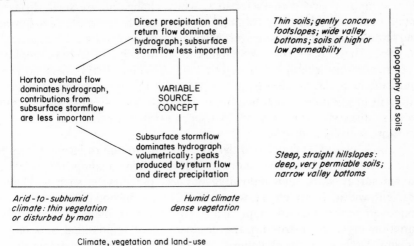

Fig. 3. Schematic illustration of the relationship between runoff processes and geographical controls. (From Dunne, 1978. Reproduced by permission of John Wiley & Sons Ltd.)

Streams are basically heterotrophic; almost all the biotic energy that drives stream communities is dead organic matter, derived from outside the channel. Allochthonous material is leached, and also decays in the water, making its products available as food for the benthic invertebrates. Most of the organic matter goes into solution in, or on, the soils of the catchment, which is where most of the litter falls, rather than in the stream itself. Annually, the litterfall from trees in forested catchments can amount to several tonnes per hectare, in the form of branches, stems, bark, flowers, leaves, etc. Although large quantities of litter may fall, or be washed, or blown into streams, the removal in water of leachates from the soil litter produces streamflow that is heavily loaded with dissolved organic matter, and this may provide nearly half of the total energy input (Fisher & Likens, 1973). Large river systems are composed of a hierarchical arrangement of catchments; small headwater catchments, because of their size, tend to be more homogeneous in geology, soils, and vegetation, than large ones; but even in large catchments the unidirectional flows makes the river dependent upon its headwater streams.

Spatial Variation of World Rivers

The world's rivers show a diversity of structure and behaviour that reflects the complex interaction between precipitation and the catchment ecosystems. With the increasing concern for conservation, it is important that the observed, and expected, consequences of river impoundment are viewed within the context of the natural, spatial differences in river character, and a system of river classification is invaluable in any assessment of the impact of dams. Classifications are rendered problematical because of the influence of scale upon a river's characteristics, which change progressively from the small, headwater, finger-tip tributary, downstream to the wide, slow-flowing, and often sinuous, floodplain river. Many workers (e.g. Abell, 1961; Harrel *et al.*, 1967; Harrel & Dorris, 1968; Whiteside & McNutt, 1972) have employed stream ordering techniques (Horton, 1945; Strahler, 1957) as an objective and widely applicable classification system for the examination of rivers of different magnitudes. Small headwater channels, as first-order streams, join to form a second-order stream; two second-order streams join to create a third-order river; and so on.

Barila *et al.* (1981) concluded that the use of a stream ordering system for lotic classification may prove more useful—providing a composite index—in stream survey work than individual physical parameters, such as channel gradient, width, depth, etc. However, Petts & Greenwood (1981) successfully employed a morphometric approach (*see* p. 20), describing the downstream variation of channel-size, for the comparison of impounded and natural streams. Drainage-network volume, which integrates the downstream change of channel-size and channel network length (Gregory & Ovenden, 1979), may come to provide an improved methodology. However, morphometric channel properties will reflect major climatic differences, modified by lithological and gross topographical variations (inherently related to palaeoclimate) which are expressed by runoff and sediment yield (Park, 1977).

It is fundamental to any scientific classification that the classes are based upon the causes of the class differences, rather than upon the effects that the differences produce (Strahler, 1975). Thus, Lotspeich (1980) proposed the use of climate and geology as the essential controls, with soil and vegetation as secondary factors, for the classification of catchments and their drainage systems. For the assessment of the effects of impoundment upon rivers, four factors particularly influence river characteristics; these are the river discharge and its seasonal variability, the sediment load, and the water chemistry.

Spatial variations in water quality are conditioned by the overall framework within which the hydrological system functions but, within climatic regions, catchment characteristics—relief, vegetation, and/or soil—became more significant (Gower, 1980). Altitude (Vitousek, 1977) can influence the water quality due to variations in the water-balance, with precipitation increasing and evaporation decreasing at higher altitudes. Vegetation also plays a major role in regulating nutrient concentrations in stream-flow (Johnson *et al.*, 1969), and a general relationship appears to exist between nutrient output and changes in net ecosystem production through successional time (Vitousek & Reiners, 1975). However, geology has been shown to be the most important control (Douglas, 1968; Webb & Walling, 1974), and river waters have been classified on the basis of calcium concentrations (Ohle, 1937) and the relative concentration of calcium and magnesium (Dittmar, 1955).

Gibbs (1970) identified three geographical groups by reference to both geology and climate, each of them being related to the mechanisms that control the composition of the major dissolved salts (Fig. 4). (i) Tropical rivers of Africa and South America, having sources in thoroughly-leached areas of low relief, in which the rate of supply of dissolved salts to the rivers is very low and the amount of rainfall high, are characterized by low-salinity waters having a chemical composition that is dominated by the amount of dissolved salts which are furnished by precipitation. The primary constituents are sodium and chloride, and ions derived from the rocks may be limited to silica with only traces of other components, of which potassium is commonly the most important. (ii) In hot, arid regions exogenous rivers evolve a chemistry determined by processes of evaporation and reflecting the precipitation of calcium carbonate ($CaCo_s$) from solution to leave sodium (Na^+) and chloride (Cl^-) as the dominant ionic constituents. (iii) Rivers in humid temperate and high-latitude areas, on the other hand, generally have moderate solute loads, comprised of dissolved salts derived from the rocks and soils of their drainage basin, and the end-member of the series is calcium-dominated. Within these three primary groups, Gibbs's data describe a large range in the relative proportions of the ionic constituents for a given total

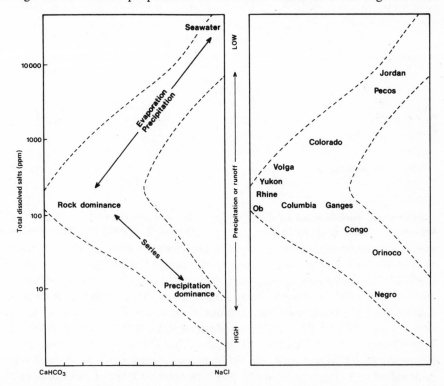

Fig. 4. The primary controls of river water chemistry. (From Gibbs, 1970, *Science*, **170**, 1088–1091. Reproduced by permission; copyright 1970 by the American Association for the Advancement of Science.)

18

dissolved-solids concentration, and in reality the groups are points on a continuum of water types. Furthermore, the solute load and its changing composition reflect the variation of precipitation or runoff.

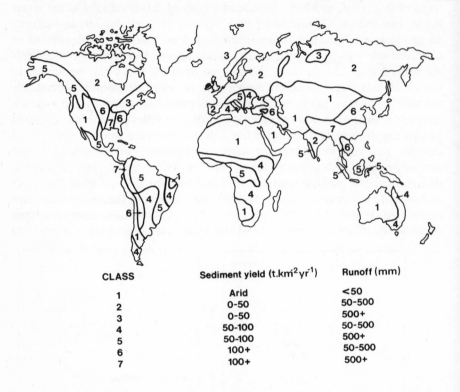

CLASS	Sediment yield (t.km^{2}yr^{-1})	Runoff (mm)
1	Arid	<50
2	0-50	50-500
3	0-50	500+
4	50-100	50-500
5	50-100	500+
6	100+	50-500
7	100+	500+

Fig. 5. A general classification of world rivers. (Sediment yield is indicated in tonnes per square kilometre per year.)

Using the runoff data of Lvovitch (1958) and sediment-load data from Strakhov (1967), seven geographical regions may be identified (Fig. 5). The world average annual yield of sediment by rivers is estimated to be between 14 and 64 thousand million tonnes (Livingstone, 1963). Strakhov (1967), using suspended sediment data, suggests that the highest annual sediment loads, of over 200t per km^2, occur in south-east Asia, followed in turn by south-east USA, the tropics, and temperate latitudes, with the lowest yields in arctic areas. This reflects the influence of vegetation (Langbein & Schumm, 1958; Douglas, 1967; Fleming, 1969) and relief (Hadley & Schumm, 1961; Schumm, 1963) as controls on the pattern of sediment yield, although removal of vegetation by Man can produce extreme rates of soil erosion.

The average annual runoff of the land surface of the Earth is 267 mm (Barry, 1969), and varies spatially from less than 50 mm to more than 1000 mm. The distribution of average annual runoff closely reflects that of the world's climates, which control the influential processes of precipitation and evapotranspiration. However, such averages can conceal considerable contrasts in the pattern of catchment runoff, and the analysis of river regimes, as proposed by Pardé (1955), illustrated by Beckinsale (1969) for the global scale, and by Ledger (1964) for the regional scale, provides an improved insight into the characterization of rivers. In some rivers, runoff will be fairly evenly distributed throughout the year, whereas in others the runoff may be concentrated into one or two periods of high flows.

Throughout the lowland, high-latitude areas of northern Canada and Siberia, moderate runoff and low sediment yields reflect the influence of the permafrost and forest cover; the rivers may freeze in winter and remain frozen in early spring, but the thaw of ice and snow produce a strong late-spring peak of runoff and turbidity after an equally marked period of winter low-flow. In north-east Siberia and eastern Asia, the thaw is followed by summer rains to produce strong summer maxima.

In humid temperate mid-latitudes, between about 55° and 30° north and south, the vegetation cover can effectively prevent erosion in all but the high-relief areas. In the Mediterranean area, western Europe, south-west Australia, and small areas of South Africa and south-east coastal South America, most, or even all, of the rainfall occurs in winter, producing a winter–spring maximum and summer low-flow or drought. Areas having a marked dry season are associated with higher sediment yields than others, because the desiccation of grasslands leads to intense erosion during the early part of the wet season. Very high sediment yields may also be found in areas of loess deposits, particularly in eastern Asia.

A subtropical arid zone merges with the zone of strong summer maximum runoff, and the rainy season lengthens in the regions of truly tropical climate. In the tropical and subtropical areas of Africa, South America, monsoonal south-east Asia, and north-east Australia, heavy rains produce peak discharges in August and September, but the period of low-flow may last for up to 7 months. Sediment loads are characteristically moderate or high. However, truly tropical streams, in low-relief areas, furnish almost no solids or dissolved loads, because of the massive plant cover or deeply leached soils. Also, the all-year-round rainfall in the main valley of the Congo, parts of Indonesia, and the upper Amazon, are characterized by high runoff with a duality of maxima in spring and autumn.

Mountainous areas are characterized by flow regimes dominated by glacier ice or snow-melt, and this may produce a typical diurnal flow variation throughout the summer months. Superficial deposits may provide a readily available sediment supply, giving rise to high-sediment loads. Moreover, a feature of some large rivers is that they have a headwater area in a region of high precipitation, and often high sediment yield, but then flow through a region of relatively low-runoff, or a low-sediment-load zone: the Nile and Amazon, respectively, provide examples of such large rivers.

The River Continuum Concept

'To maintain our rivers in ecological conditions as natural as is compatible with Man's legitimate uses, recognition must be given to the existence of different river zones, and in determining river management policies, account should be taken of their different ecological characteristics.'

Hawkes (1975 p. 324)

From headwaters to mouth, the physical variables within a river system progressively change, and the gradient from a shallow, turbulent, mountain stream to a deep, meandering, lowland river results in a consistent pattern of community structure and function. The physical structure of the stream-channel, coupled with the hydrological cycle and energy inputs, produce a series of responses within the constituent populations—resulting in a continuum of biotic adjustments, and consistent patterns of loading, transport, utilization, and storage, of organic matter along the length of a river (Vannote et al., 1980). The 'River Continuum Concept' proposes that the gradient of physical factors, formed by the drainage network, exerts a direct control upon the biological strategies and dynamics of river systems.

The downstream variation of the physical factors, describing the characteristics of the channel with increasing discharge, was demonstrated by Leopold & Maddock (1953). Their *hydraulic-geometry* approach uses power-function relationships to characterize the downstream variations of flow width, depth, and velocity, relative to discharge as it increases along a river at a constant flow-frequency. In order to ensure that the selected discharge-frequency is of consistent significance, for sediment transport and channel morphology at each point along the river, the downstream adjustment of the 'bankfull' channel dimensions is analysed in relation to the 'bankfull' discharge. Basin area, or total upstream channel length, have been used as surrogates for discharge in perennial rivers (Hack, 1957), and Park (1978) has proposed a *morphometric approach*. This is based upon numerical descriptions of covariation between elements of landscape geometry, utilizing power functions, to relate measures of channel size and form to measures of spatial and scale location within the channel network, or drainage basin.

Characteristically, channel dimensions increase downstream, and within the Midwest American rivers, for example, the more rapid downstream increase of width compared with that of depth suggests an adjustment of channel shape, with large rivers being slightly wider in relation to their depth than small streams (Leopold & Maddock, 1953). The gradient of the long profile decreases downstream, but velocity usually increases, or at least remains constant, because of the increasing cross-section efficiency (larger hydraulic radius), and decreasing boundary resistance as the bed materials are reduced in size, with boulders being replaced successively by cobbles, gravel, sand, and finally silt. Both the hydraulic-geometry and morphometric approaches assume smooth, continuous adjustments of channel morphology as functions of discharge or drainage area. However, major adjustments occur at tributary junctions, and downstream trends of discharge, drainage area, and channel geometry, are step functions (Richards, 1980), reflecting the organization of the drainage-network. Furthermore, the discharges and sediment loads delivered by tributaries, the materials comprising the channel bed and bank, channel slope, and riparian vegetation, can all vary locally, so that the downstream adjustment of channel geometry must be interpreted in relation to these multivariate controls.

Stream classifications for biological studies have used various physical parameters: the types of stream-bed (Shelford, 1911), size of catchment (Thompson & Hunt, 1930), channel slope (Huet, 1959; Hocutt & Stauffer, 1975), flow regime (Horwitz, 1978), and water temperature (Burton & Odum, 1945; Illies, 1952; Vannote & Sweeney, 1980). The zonation of rivers according to the longitudinal succession of fish species, benthic invertebrate taxa, and Algae, has been reviewed by Hawkes (1975), and five major fish-groups were identified (Table IV). Recently, Schlosser (1982) demonstrated that the patterns of fish community structure and function support the qualitative aspect of the stream continuum concept. Cool-water fish populations of low diversity in headwater streams, change progressively to a more diverse, warm-water community in the middle and lower reaches, and become characterized by piscivorous or invertivorous, and then planktivorous, species respectively.

Vannote et al. (1980) proposed a grouping of lotic communities into headwater streams, middle-order streams, and large rivers, on the basis of invertebrate functional feeding groups. Primary consumption involves a detritivory and a herbivory (McIntire & Colby, 1978). Within the former, shredders utilize coarse particulate organic matter, and collectors either filter the fine particles from suspension or gather particles from the channel substrate. In both cases, the microbial biomass and products of microbial metabolism are important for their nutrition. The herbivory is dominated by the grazers, which rely primarily upon periphyton for their food-supply. Energy transfer from primary to secondary consumers, or from secondary to tertiary consumers by predation, generally has the same relative dominance along a river.

The three ecological units described by Vannote et al. (1980) are comparable to the three geomorphological zones recognized by Schumm (1977). The headwater, Zone 1, is the primary sediment producer, and strong links with the terrestrial system are reflected by the marked influence of riparian vegetation, which reduces autotrophic production in the water by shading, and contributes large amounts of allochthonous detritus. The substrate is composed of coarse gravels, boulders, and rock outcrops, and the water temperature shows a low seasonal range and only weak daily temperature-variation, because of the proximity of sources and shading effects. Species diversity may be low because of the restricted temperature-range and nutritional base. The invertebrate community will be dominated by shredders utilizing the coarse particulate matter derived from riparian sources, and by collectors.

The intermediate, Zone 2, is the zone of transfer or predominant transport,

Table IV Major fluvial zones (Adapted from Hawkes, 1975. Reproduced by permission of Blackwell Scientific Publications Ltd.)

Zone	Key physical characteristics	Invertebrate community (Vannote et al., 1980)	Zone (Illies & Botosaneau, 1963)	Illies (1961)	Dominant and sub-dominant fish in Germany (Haut, 1954) and U. K. (*, Carpenter, 1928)
1	Small channel; turbulent flow; erosional zone; coarse substrate; stable temperature	Low species-diversity; dominated by shredders and collectors	I	Eucrenon	Salmo trutta (Trout) and S. salar parr (Atlantic Salmon)*
			II	Hypocrenon	ditto
			III	Epirhithron	ditto
			IV	Metarhithron	ditto
2	Variable flow and temperature; stable channel; heterogeneous substrate	High species-diversity; dominated by collectors and grazers	V	Hyporhithron	Thymallus thymallus (Grayling); Phoximus phoximus (Minnow)*; S. trutta, S. salar,
			VI	Epipotamon	Cottus gobio (Bullhead); Barbus barbus (Barbel); Leuciscus cephalus (Chub)*; Gobbio gobbio (Gudgeon)*
3	Stable flow and temperature; depositional zone; turbid water; fine substrate; reduced levels of dissolved oxygen	Low species-diversity; dominated by collectors	VII	Metapotamon	Abramis brama (Bream); Cyprinus carpio (Carp); Tinca tinca (Tench); Rutilus rutilus (Roach); Leuciscus leuciscus (Dace); Esox lucius (Pike); Perca fluviatilis (Perch)
				Hypopotamon	Anguilla anguilla (Eel)

where, with a stable channel, the input of sediment can equal the output. A coarse substrate, broad seasonal temperature regime, wide maximum diel temperature-range, and variable discharge, favour a diverse fauna. The predictably variable physical characteristics of many middle-order rivers encompass optimum conditions for a large number of species, so that the system may have high species-diversity, or, at least, high complexity in species function. Terrestrial organic inputs are of reduced importance, but this coincides with the enhanced significance of autochthonous primary production. The dominant invertebrates—collectors and grazers—are reliant on Algae, rooted aquatic plants, and organic transport from upstream.

Sediment is deposited, in Zone 3, on an alluvial plain or delta, or in an estuary. The fine substrate sediments, large numbers of degree days*, relatively stable discharge, and low maximum diel temperature-range—due to the buffering effect of the large volume of water—are associated with a low degree of biotic diversity and domination by collectors. Large rivers receive quantities of fine particulate organic matter, derived from the upstream processing of dead leaves and woody debris; the effect of riparian vegetation is insignificant, but primary production may be limited by water depth and turbidity.

The fish zones described by Illies (1961) are dependent upon temperature, so that the spatial extent of the zones varies with geographic location: the rhithron extends to lower altitudes at higher latitudes than at lower ones, and the potamon extends to higher altitudes towards the Equator. Nevertheless, the general predictions of the river continuum concept for lotic invertebrates have been confirmed (e.g. Hawkins & Sedell, 1981), although modifications have been suggested in order to give due consideration to the varying riparian influences, and tributary inputs from different biomes, which large rivers must traverse.

The location of the zone through which the stream passes when its predominant metabolism shifts from heterotrophic to autotrophic, is dependent upon the degree of shading (Minshall, 1978) provided by the riparian vegegation, channel banks, and valley sides (Vannote et al., 1980). Thus, for the South Saskatchewan River, Culp & Davies (1982) confirmed that the longitudinal trends in functional feeding groups, within the benthic macroinvertebrate communities, generally followed the predictions of the river continuum concept. The zonation reflected differences in water temperature, geomorphic features, terrestrial vegetation, soil type, and water quality. A strong longitudinal zonation, observed during the summer–autumn period, was related to three geographical zones through which the river flows—namely subalpine forest, fescue prairie, and mixed prairie—each having distinct botanical, valley form, and channel morphology, characteristics.

EFFECTS OF RIVER IMPOUNDMENT

'We have been startled into profound realizations, no less profound for having become commonplace; that Nature's complex web is easily rent by

* 'degree days' refers to the sum of mean daily temperatures.

our clumsy passage; . . . and that what we do today will affect not only ourselves, and not only our children, but our descendents for many generations to come.'

Lowrance (1976 p. 6)

River impoundment has given rise to dramatic changes in the characteristics of rivers. In general terms, the impacts have been much the same as mild organic pollution (Spence & Hynes, 1971), which cancels out the diversity-enhancing effects of increasing physical heterogeneity along the river, so that species diversity may remain at levels normally associated with lower-order streams (Tramer & Rogers, 1973). Thus, the impact of Hungry Horse Dam on the sixth-order Flathead River was to create conditions similar to those in fourth-order streams (Hauer & Stanford, 1982).

Many factors interact to produce the characteristics of a river, and it is difficult to discuss individual factors in isolation. Indeed, physical, chemical, and biological, factors nearly always operate in conjunction with others to produce their effects. Nevertheless, the consideration of individual factors is an expedient which, although fundamentally deplorable (Macan, 1963), is justified here by treating the effects of impoundment within a hierarchical framework (Fig. 6). This provides for an improved understanding of the complex interactions that may take place between factors, and of the time-scale required before the full effects may be recognized. Biotic interactions, such as competition and predation, are known to play an important role, but this discussion is directed primarily to the changes of abiotic factors, and the consequences of these changes for the biological components of the system.

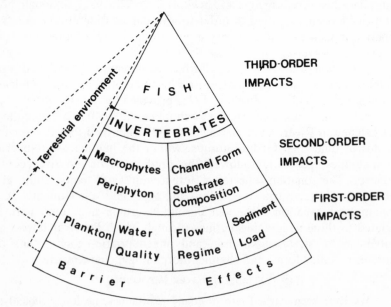

Fig. 6. A framework for the examination of impounded rivers.

A large number of factors have been used by different research workers in order to quantify river habitat (Pennak, 1971; McIntire & Colby, 1978; Binns & Eiserman, 1979), but in the highest level of investigation, i.e. studies of fish productivity, seven primary groups of factors may be identified, and, with specific reference to the effects of impoundments, an eighth (the barrier effect imposed by the dam and man-made lake) may be added. Within this framework, the effects of impoundment upon the downstream river may be viewed in terms of three orders of impact. (i) A first-order impact occurs simultaneously with dam closure, and effects the transfer of energy, and material, into and within the downstream river. (ii) Second-order impacts are the changes of channel structure and primary production, which result from the modification of first-order impacts by local conditions, and depend upon the characteristics of the river prior to dam closure. These impacts may require a time-period of between 1 and 100 years, or more, to achieve a new 'equilibrium' state. (iii) The third-order impacts will reflect all the changes of first- and second-order ones, and the fish population will also be influenced by changes of the invertebrate community, which provides the major food-supply for many species. These impacts may occur with a considerable time-lag in relation to the first-order impacts, but several phases of population adjustment and readjustment may take place in response, particularly, to the changes of second-order factors. It is the interaction of the different components of the system involved within the third-order time-scale, that will have implications for the aesthetic and recreational potentials of rivers. Furthermore, it is only through the due appreciation of changes that may be induced within third-order time, and then by a recognition of the need to consider conservation measures, that we will achieve the efficient management of the world's rivers.

CHAPTER II

The Hydrology of Dammed Rivers

The hydrological characteristics of a river exert the fundamental controls on its lotic systems. The rate at which water is conveyed, discharge variability over a range of time-scales from minutes to months, and the frequency of flow extremes, exert important controls upon every physical, chemical, and biological, attribute of a river. Discharge control resulting from river impoundment involves the redistribution of discharge in time. The reservoir reduces short- and long-term flow variability within the river downstream, filling when the inflow exceeds the outflow, and emptying when the dam releases exceed the inflow. Many factors determine the precise effects of an impoundment upon the flow-regime of a river (Fig. 7). Some of these, particularly basin morphometry and surface area, can produce effects that are common to all reservoirs, whereas the dam design, and related operational procedures, can generate a wide range of discharge patterns depending upon

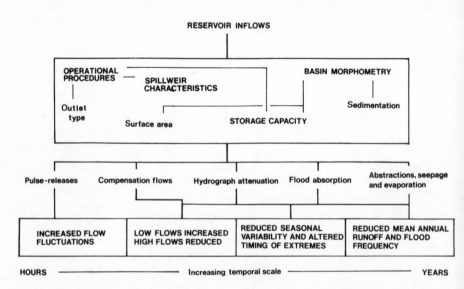

Fig. 7. Factors affecting the hydrological characteristics of dammed rivers.

the reservoir function. An increase in low-flows is associated with navigation- and irrigation-supply releases; reduced high-flows are produced by flood-control dams; and extreme short-term flow-fluctuations can result from hydroelectric power demands. However, dams are being increasingly operated for multiple purposes, so that a unique flow-regime can be generated within each impounded river.

Dams represent perhaps the greatest point-source of hydrological interference by Man (Table V). The mean annual discharge has been reduced by up to 80% as a result of river impoundment, due to increased evaporation, abstraction for industrial, domestic, or other supply purposes, and seepage to ground-water storage. High-flows have been reduced, and the timing of annual extremes of flow has been altered. The changes induced by river impoundment may be transmitted over considerable distances downstream from the dam, so that even the delta or estuary can be affected. Thus, following the regulation of flows on the Peace River, the 2560 km² of the Peace–Athabaska delta—more than 1000 km below Bennett Dam—were transformed from a rich floodplain and wetland environment, to a series of isolated mud-flats (Blench, 1972). Concern for the fate of the Aral Sea has arisen as a result of the fall of water level by 3 m between 1966 and 1976, caused by the impoundment and irrigation use of runoff from the main feeding rivers, the Amu Darya and Syr-Darya (Gerasimov & Gindin, 1977). Moreover, the shoreline has receded by over 50 km in the last 20 years; a major fishing industry has been lost; salinization has reduced the fertility of the surrounding land; and climatic changes are becoming apparent.

These general effects, identified in Table V, are common to all impounded rivers, and differences will occur only in terms of the magnitude of the impact—itself a function of the design and operation of the dam. The biological components of lotic ecosystems are adapted to the natural flow-regime, and depend not only upon average flows, but also upon high-flows and periods of low- or even zero-flow, to satisfy the requirements of their life-cycles. Periodic high-flows may be especially important, for example to clean spawning gravels, and a predictable seasonal flow-pattern may be required to facilitate fish spawning-migrations, or to recharge riparian habitats and floodplain wetlands. River regulation can cause considerable changes, not only within the lotic, riparian, and floodplain, habitats, but also indirect effects, through associated changes of channel morphology and substrate stability, can be influential.

RESERVOIR DESIGN AND FUNCTION

Project design in this context is based upon a knowledge of the hydrological characteristics of the river to be impounded and of its catchment: specifically, this means the quantity of stream-flow and its occurrence with respect to both area and time. Hydrological considerations include the storage requirements for conservation and/or flood-control, the discharge capacity of

Table V   Some examples of the hydrological effects of river impoundment.

River/Reservoir, Country	Reported hydrological changes	Source
	REDUCED AVERAGE ANNUAL RUNOFF	
River Nile/Lake Nasser, Egypt	Average annual water-yield reduced: losses of nearly 500 × 10^6m³ yr through evaporation	Waffa & Labib (1973)
Hawkesbury–Nepean River Warragamba Dam, Australia	Average annual runoff reduced by between 19% and 28%	Warner (1981)
River Churchill, Canada	Average discharge reduced from 1000 m³ per s to 200 m³ per s	Dickson (1975)
River Zambezi, Mozambique	Saltwater incursion in the coastal floodplain and delta area induced by reduced freshwater discharges	Hall et al. (1977)
River Dnieper, Kakhovka Power-station, USSR	Average annual runoff reduced by 20%	Zalumi (1970)
	REDUCED SEASONAL FLOW VARIABILITY	
River Columbia, Dalles Dam, USA	Maximum average monthly flow reduced by 50%	Trefethen (1972)
River Yarmouk/R. Jordan, Israel	Flow will be reduced to a minimal level throughout the year	Ortal & Por (1978)
River Damodar reservoirs, India	Impoundments designed to tranform the seasonal river into a perennial one	Jain et al. (1973)
River Peace, Bennett Dam, Canada	Flow-range compressed from 150 to 9000 m³ per s to 500–2000 m³ per s	Kellerhals (1971)
River Gordon, Tasmania	Minimum regulated flows three times natural flow	King & Tyler (1982)
	ALTERED TIMING OF ANNUAL EXTREMES	
River Jordan, Israel	50 years of dam construction for power and irrigation supply purposes have transformed the flow-regime from a winter to a summer high-flow	Ortal & Por (1978)
Murray–Darling rivers, Australia	Impoundments have effectively reversed the natural pattern of runoff: the former summer low-flow period has been replaced by 'peak' flow in response to irrigation demand	Cadwallader (1978)
River Vir, Czechoslovakia	Natural flow-regime characterized by two periods of high-flow in March and July, but regulated river has only one (in May)	Peňáz et al. (1968).
River Columbia, Dalles Dam, USA	Month of minimum average flow is now 1 month earlier, and high-flow season has been delayed by several months	Trefethen (1972)
	REDUCED FLOOD MAGNITUDES	
TVA Scheme, USA	1957 flood at Chattanooga; flood stage reduced from 7.3 m to 0.7 m	Elliot & Engstrom (1959)

River/Reservoir, Country	Reported hydrological changes	Source
Central European Rivers	50 years' flood reduced by 20%	Lauterbach & Leder (1969)
River Colorado, Glen Canyon Dam, USA	10 years' flood reduced by 75%	Dolan et al. (1974)
River Nile, High Aswan Dam, Egypt	'Flood peaks' reduced by 75%	Kinawy et al. (1973)
River Damodar, Panchet Reservoir, India	Design flood of 28 300 m³ per s will be reduced to 5660m³ per s	Jain et al. (1973)
River Peace, Bennett Dam, Canada	Flow stabilization reduced flood stages for 1200 km below the dam	Geen (1974)
River Zambezi, Kariba Reservoir, Africa	Flood controls effective for 130 km downstream	Guy (1981)
	IMPOSITION OF UNNATURAL PULSES	
River Vyrnwy, Lake Vyrnwy, UK	Regular monthly releases of 1.16 m³ per s for 5 days during summer (5 times normal controlled low-flows)	Severn Trent Water Authority (pers. comm.)
Upper Kennebec River, Maine, USA	Periodic instantaneous, flow-fluctuations from 7.8 m³ per s to 170 m³ per s	Trotzky & Gregory (1974)
Colorado River, Glen Canyon Dam, USA	Daily fluctuations of water-depth by about 1.5 m	Turner & Karpiscak (1980)
Zambezi River, Kariba Reservoir, Africa	Sudden release increased flow from 700 m³ per s to 4531 m³ per s	Begg (1973)
Zambezi River, Cabora Bassa Dam, Mozambique	Flow cut from 3000 m³ per s to 60 m³ per s during four months of filling: recommended minimum discharge of 400–500 m³ per s ignored	Davies (1975)

spillways and other outlets to release the highest expected floods, and conservation-storage with due regard for the channel and floodplain characteristics, and the water needs of users downstream. Furthermore, the day-to-day operation of most gated reservoirs utilizes pre-reservoir hydrometric data in addition to current, and forecast, stream-flow and precipitation data. The operational schedule is devised to achieve the maximum benefit from the storage capacity, and is based upon the purpose, or purposes, of the impoundment in relation to the flow characteristics of the river. The primary aim of river impoundment is to regulate discharge fluctuations within the river downstream of the impoundment. Seasonal fluctuations can be controlled by storage or wet-season runoff, for subsequent release during periods of low-flow or drought. Short-term discharge variations can be regulated by the storage, and subsequent gradual release, of flood-water. Thus, impoundment can provide regular water-supplies and relief from flood-damage caused by floodplain inundation. These two roles often conflict, however, because:

'The concept of flood-control storage is empty space; that of conservation storage is stored water for later use.'

(Rutter & Engstrom, 1964 p. 61)

A reservoir may be viewed as being composed of one, or a number of, storage area(s), associated with one or more gated, or ungated, outlet facilities located at different elevations in the dam (Fig. 8). Conservation-storage provides a more or less permanent volume of water for later supply purposes. A fixed, year-round allocation may be made for specific purposes, or variable allocations can provide for a number of different uses depending on specified seasons or times of the year. For most supply purposes (hydroelectric power, navigation, irrigation, etc.), the conservation-storage should be maintained as fully as possible in order to have sufficient reserve to maintain supply during periods of drought. The provision of flood-control storage, on the other hand, requires that a storage volume is reserved solely for the absorption of flood discharge. The water volume retained within the flood-control storage should be kept as low as possible, so that unexpected high runoff can be contained.

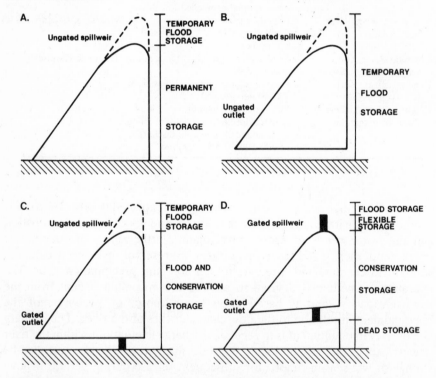

Fig. 8. Major dam types: a simple conservation dam (A), flood-detention dam (B), and structures for multipurpose operation (C and D).

Many reservoirs are also allocated a 'dead' storage volume below the lowest outlet level. This is unavailable to supply use, but can provide storage for sediment yielded by inflowing streams and trapped behind the dam. For the simplest dams, any water flowing into the reservoir is stored, and releases occur only if the storage capacity is exceeded, when water will be discharged over the emergency spillway. More complex structures can have gated spillways, to allow for the more flexible management of the flood-control and conservation-storage, together with one or more gated or ungated low-level outlets. Ungated outlets allow the passage of normal flows, but are designed to limit the outflow rate to safe channel capacities, so that ponding behind the dam occurs during high inflows. Gated conduits allow the controlled release of water, as required for flood-control, navigation, downstream supply purposes, or hydroelectric power generation.

Dams with gated outlets will have a variable impact upon the flow regime of the downstream river, depending upon their individual operating schedules. Knappen *et al.* (1952) describe three main operational procedures, as follows:

(1) 'The Insurance Method' is concerned with the maintenance of minimum flows. Based upon knowledge of the driest probable year, only enough water is released to maintain a predetermined minimum flow at the point of regulation. All other conservation-storage is held in reserve for the next dry period.

(2) 'The Annual Use Method' aims to use practically all available storage-water each year, without special regard for maintaining a downstream minimum or regulated flow. During dry years the reservoir may be emptied, allowing extreme low-flows to occur within the river downstream, although floods may be effectively controlled.

(3) 'The Hydropower Method' combines the key features of (1) and (2) to provide a dependable minimum flow and, at the same time, to use the total volume of stored water to the best advantage. However, short-term release-schedules for flow regulation and to provide for peak-power demand, together with long-sustained flows at abnormally high levels, must be considered in relation to the habitat structure of the channel and floodplain downstream.

The primary attraction of mainstream impoundment is that stored water can be released at any time, providing that controlled outlets have been built into the design. In order to realize the maximum benefits from impoundment, a river will normally be regulated for a variety of purposes. Indeed, many of the existing simple water-conservation structures are being enlarged, and fitted with gated spillways and outlets, to improve their flow-regulation capability. From the operational viewpoint, small, upstream reservoirs and large, downstream impoundments cannot be substituted for one another in basin-wide flood-control schemes, because they provide protection for different parts of a valley (Leopold & Maddock, 1954). However, to be effective for combined purposes, a reservoir must have a large capacity (Knappen *et al.*, 1952).

Many major rivers are regulated by a chain of reservoirs, and the Missouri and Tennessee Rivers, USA, for example, clearly fall into this category. The viability of managing the Missouri reservoirs has been examined by Neel

(1963). Main-stream reservoirs on the Missouri River provide for power production, navigation, and irrigation, and their total capacity amounts to 3 years' water-yield from the catchment down to Sioux City, nearly 90 km below Gavins Point, the dam farthest downstream. Within each reservoir, storage volumes are allocated for both specific and flexible use (Table VI). Only 4.4% of the total storage is held exclusively in reserve for flood control, and nearly 23% is dead storage. Neel demonstrated that such schemes have several important advantages over a single reservoir: the control of water levels in each dam can maximize protection from anticipated floods without a loss of reserves; power production is possible at a number of sites; and upstream reservoirs allow firm commitments for demands on storage from their downstream counterparts. However, they require a greater land-area, and can augment losses through evaporation.

Table VI   The main-stream reservoirs of the Missouri River. (From Neel, 1963. Reproduced by permission of The Board of Regents of the University of Wisconsin System.)

Reservoir	Fort Peck	Garrison	Oahe	Big Bend	Ft Randall	Gavins Point
Closure Date	1937	1953	1958	ca 1965	1952	1956
Total capacity (km³)	23.9	30.2	29.1	2.3	7.5	0.7
Exclusive flood-control reserve	5%	6.8%	5%	9%	15%	12%
Annual flood-control and multiple-use capacity	14%	18%	14%	0	23%	19%
Carry-over multiple-use capacity	58%	56%	58%	14%	39%	40%
Dead storage	23%	20%	23%	77%	23%	0.29%

Flood Regulation

Rutter & Engstrom (1964) stated that the basic concept of flood regulation was 'empty-space'. That is, the effect of a reservoir upon individual flood discharges is related to the content of the reservoir prior to the arrival of the flood-wave. If the reservoir is empty, the magnitude and timing of flood-peaks downstream will be reduced and lagged, respectively, as the flood discharges are contained within the reservoir's storage volume. The amount of storage available within the reservoir will be dependent not only upon the time-interval between successive high-flows, and on the volume of those flows in relation to the flood-storage capacity of the impoundment, but also upon the rate of water release (if any) and the operational procedures, which can vary the release-rate and, therefore, the volume stored. Reservoirs having a large flood-storage capacity in relation to the annual runoff can exert complete control upon the annual hydrograph of the river downstream. The 476×10^6 m³ capacity Canyon Reservoir, USA, for example, controls the flow within the Guadalupe River, so that the flow variation is considerably reduced (Fig. 9): the discharge is controlled by releases from a low-level outlet, which maintains flows at between 4.5 m³ per s and 20 m³ per s for most of the year (Hannan *et al.*, 1979).

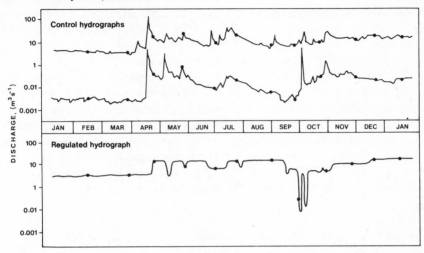

Fig. 9. Flow regulation on the Guadalupe River, USA. (From Hannan *et al.*, 1979. Reproduced by permission of Dr W. Junk Publishers.)

The flood-storage capacity of a reservoir is chosen during the design stage, and the desired degree of protection determines the magnitude of the discharge adopted as the 'design-flood'. Flood control reservoirs are often designed to regulate the 'maximum probable flood'. However, it is impossible to specify the 'maximum possible flood', because of the stochastic nature of river-flows, so that the reservoir storage requirements, dam height, and discharge capacity through spillways and outlets, must be based upon the greatest flood that may reasonably be expected, taking into account all pertinent conditions of location, meteorology, hydrology, and topography (Rutter & Engstrom, 1964). However, the benefit derived from a flood-protection scheme may not justify the cost of protection against such extremely large and rare floods, and the 'design-flood' may be smaller than the 'maximum probable flood'. In such circumstances, the 'design-flood' will depend upon the purpose of flood-control, and the potential costs of flood damages. This will apply, particularly, to areas where floods can occur at any time of the year, because storage must be provided all the time. For rivers having a marked and predictable flood season, storage to control these floods needs to be available only at these times, and the reservoir could assume a multipurpose role, so increasing the benefits realized from the scheme.

Flood-control reservoirs are of two basic types: namely, retarding reservoirs and detention basins (Knappen *et al.*, 1952). Retarding reservoirs are simple flood-storage basins which release water through uncontrolled outlets. The flow-rate through these outlets is proportional to $H^{0.5}$, where H is the head of water above the outlet. Spillway discharge may augment these releases, but the flood peak will be reduced, and the duration of moderately high flows will be increased, within the channel downstream (Fig. 10, A). Detention basins have one or more controlled outlets to allow the flexible operation of the reservoir storage volume. The more effective use of reservoir storage capacity is made possible by using the maximum acceptable capacity of the river channel downstream to convey stored water, thereby increasing the available reservoir capacity prior to the arrival of a flood discharge. The storage volume of Lake Nasser, Egypt, is larger than that of the maximum flood for which control is required, so that the full control of downstream discharge is achieved (Jansen *et al.*, 1979). Such large reservoirs can produce a constant flow within the downstream river (Fig. 10, B), but even small-capacity detention basins can achieve comparable flow regulation by careful operation and flood forecasting. Effective flow regulation can be achieved by the prior evacuation of storage *via* outlet works, and by controlled releases as the reservoir fills on arrival of the flood discharge (Fig. 10, C).

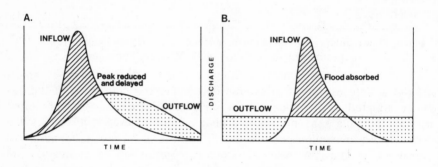

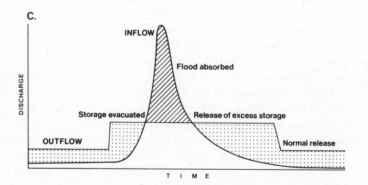

Fig. 10. Primary types of flow regulation: reservoir attenuation (A), reservoir storage (B), and release manipulation (C).

Reservoir Lag

Changes will occur in the rate and timing of runoff as the flood-wave passes through a reservoir, even when the flood-storage capacity is full to emergency spillweir level. As a flood discharge moves along a channel, it is subjected to two processes: namely, translation, whereby the wave moves downstream without changing its shape, and attenuation, which is effected by the temporary storage of water within the channel or floodplain. The pulse of water entering a reservoir does not flow out, nor does it displace a similar water-volume, immediately—most of the water is stored behind the dam. Thus, when the permanent storage volume is full, flood-waves will be attentuated by storage above the spillweir:

> 'The balancing effect of the temporary storage of water in the reservoir above the crest level of the waste weir when the reservoir is full and overflowing, referred to as the reservoir "lag", plays an important role in reducing the maximum rate of outflow from the reservoir.'
>
> I.C.E. (1933 p. 10)

As water flows into a reservoir it is stored: the water-level of the lake is raised and the outflow increases; but, because of storage changes, the peak discharge will not be as high as that for the inflow. When once the inflow ceases, or declines significantly, the stored water drains slowly, maintaining an outflow in excess of the inflow rate, so that the flood-duration is increased. The effect of storage provided by the rise in water-level above the spillweir-crest is related to the surface area of the reservoir, the hydraulic characteristic of the spillweir, and the form of the inflow hydrograph. Reservoir surface-area at emergency spillweir elevation, and the morphometry of the reservoir basin, control the volume of storage available for a given change of water-level. This storage–head relationship is normally expressed as the 'retention factor' (Lauterbach & Leder, 1969), which describes the attenuation of flood peaks routed through a reservoir. It is a function of the quotient of the total reservoir surface-area to catchment area: as the retention factor increases, so the 'lag' effect becomes more marked. The Institute of Civil Engineers (I.C.E., 1933) report on *Floods in Relation to Reservoir Practice* considered (p. 18) that:

> 'Where the water area of a reservoir is 2% or upwards of the catchment area, the storage provided by the rise in water level above the weir has an appreciable effect in reducing the rate and head of overflow.'

Thus, in Fig. 11, A, peak flow is reduced by 40% with a retention factor of 0.02, and by 80% when the retention factor is increased to 0.08.

The head–outflow relationship is controlled by the hydraulic characteristics of the spillweir. The capacity of the spillweir increases rapidly with rising reservoir water-level, and for simple spillweirs the discharge will be proportional to $H^{1.5}$. Therefore, the characteristics of the spillweir will have a marked influence on reservoir lag—the greater the head is, the greater will

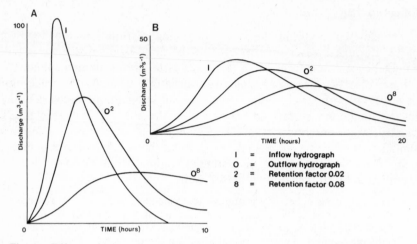

Fig. 11. Effects of lake area, and inflow hydrograph form, upon reservoir lag.

be the outflow control—and a narrow spillweir will impart the greatest control upon flood peaks, although the effect of the weir is less significant than the retention factor.

Reservoir attenuation will vary also in response to the rate of natural runoff, which is particular to individual catchments; certainly, the ratio between peak inflow and outflow will be a function of hydrograph shape (cf. Fig. 11, A and B). Although a function of the magnitude, duration, and location, of the storm event, the shape of the inflow hydrograph characterizes the rate of runoff in response to the rainfall of a given storm. Rates of storm runoff from permeable catchments, from naturally regulated rivers interrupted by lakes or large areas of floodplain storage, may be affected only negligibly by impoundment. Minor variations in the outflow peak may occur as a consequence of persistent high-velocity winds, causing the pile-up of water, and increasing the head above the spillweir crest. Winds may be of particular importance within open, treeless valleys. Under such conditions, Makkaveyev (1970) observed that wind-speeds of 18–20 m per s produced waves 3 m in height within the Kuybyshev Reservoir, and 1-m-high waves can be generated by 10 m per s winds on the Rybinsk Reservoir on the Upper Volga, USSR (Butorin et al., 1973).

Variable Effectiveness of Flood-control

Several controls on the hydrological characteristics of a river are changed by river impoundment, so that the direct comparison of inflow and outflow hydrographs may be rendered problematical. Precipitation directly on to the reservoir surface may be important where the reservoir inundates more than 5% of the catchment area (I.C.E., 1975). The effective shortening of stream lengths, following the inundation of tributaries by the reservoir, may also be important. Both factors would serve to increase the rate of water-level rise within the reservoir, in comparison with that expected from hydrological records obtained for the natural river at the dam-site.

For major floodplain rivers, impoundments may actually increase flood-peaks. Under natural flood conditions, water storage occurs temporarily within the channel and floodplain, and this storage reduces the peak outflow from a reach. Reservoirs may 'fill' the storage previously provided by the channel–floodplain system, and under these conditions of increased flow-depth, the passage of the peak flow is accelerated through the reservoir (Rutter & Engstrom, 1964).

Nevertheless, the magnitude and timing of flood peaks will normally be regulated: the 50 years' return-period flood has been reduced by over 20% (Lauterbach & Leder, 1969; Huggins & Griek, 1974), and the mean annual flood, having a return period of 2.33 years, has commonly been reduced by more than 25% (Table VII). The reduction of the magnitude of peak discharges may be relatively great for smaller, more frequent events, and least for the larger, less frequent floods. Where prolonged intense rainfalls generate persistent, high runoff of more than 50 mm per day, the lag-effect may be neutralized, and spillway discharge could become roughly equal to the inflow—even in reservoirs having relatively large surface-areas (I.C.E., 1933). Furthermore, the containing effect of reservoir storage is independent of the size of the inflow events. Therefore, downstream from a dam, floods of moderate frequency may be considerably lower in magnitude than the inflow flood, but rarer floods may be altered only to a negligible degree (Lauterbach & Leder, 1969). Thus, Warner (1981) reported that subsequent to the closure of the Warragamba Dam, Australia, floods with a recurrence interval of less than 2.33 years were reduced in magnitude by more than half, but rarer events were less significantly affected, and the magnitude of the 50 years' flood was insignificantly different from that for the pre-dam period.

Table VII   Effect of reservoirs on the mean annual-flood.

Location	Reduction of mean annual-flood (%)	Source
Clatworthy Reservoir, River Tone, England, UK	60	Gregory & Park (1974)
Hawkesbury–Nepean system, Australia	50	Warner (1981)
Sutton Bingham Reservoir, River Yeo, England, UK	39	Petts & Lewin (1979)
Blue River, Colorado, USA	36	Huggins & Griek (1974)
John Martin Dam, Colorado, USA	25	Wolman (1967)
Elephant Butte Reservoir, New Mexico, USA	25	Wolman (1967)

The effects of impoundment will become less and less significant downstream from a reservoir as the proportion of uncontrolled catchment increases. Downstream of a dam, the 'lag' of main-stream peak discharges routed through a reservoir may desynchronize the main-stream and tributary peaks although, under certain conditions, and for particular situations, the opposite effect can occur. The superposition of hydrographs may result where alteration of the timing of an event causes the main-stream and tributary peaks to coincide. Nevertheless, the frequency of tributary confluences below the dam, and the relative magnitude of the tributary streams, will determine the length of river affected by impoundment. A comparison of discharges with specified frequencies for the River Tone below Clatworthy Reservoir, UK, with the regional pattern (Gregory & Park, 1974), suggested that peak discharges immediately below the dam (a drainage area of 18.2 km²) are 34%–40% of those expected. Downstream, at a catchment area of 57.8 km², discharges are still less than half the expected values; but with a catchment area of 202 km², peak discharges approximate to the regional pattern. Thus the hydrological effects were transmitted along the river to the point where the area impounded was reduced to less than 10% of the total drainage.

Flood regulation may be achieved for considerable distances below dams by the planned use of outlet valves to reduce the reservoir storage-levels. As an example, Smith (1972) has described the effect of the January 1960 floods on the Derbyshire Derwent, England, UK. Three flood-peaks, produced by heavy rainfalls, were absorbed by the reservoirs before overflow occurred. Immediately below the dam, the two largest flows were reduced by a total of 94.1 m³ per s, whilst at Matlock, 30 km downstream, the reduction was only 57.5 m³ per s, and the town was flooded to a depth of 1.8 m. Had the proposed flood drawdown curve for the main Ladybower Reservoir been operated, the flooding would have been prevented (Richards & Wood, 1977). Thus, for major regulating reservoirs, their control of the river flow-regime may be exerted for a considerable distance. Within the Peace River, Canada, for example—a river with little natural floodplain storage—hydrological changes have been transmitted over the entire 1200 km between Bennett Dam and the delta wetlands around Lake Athabaska (Geen, 1974).

HYDROLOGICAL EFFECTS OF SMALL STRUCTURES

Rivers are impounded for two objectives—flood-control and water conservation—and both of these have been widely achieved in the past, although less frequently today, by the construction of relatively small, simple, single-purpose dams. Despite their apparently small size, they are usually located in upland areas, have small catchments, and, as a result of flood-absorption and attenuation, can exert an important control upon the flow-regime within the river downstream. The essential difference between flood-control and water-conservation dams is the type of outlet. Flood-retarding dams have an uncontrolled low-level principal outlet and a high-level emergency spillweir; a flood-storage pool is created between them, and a sediment-storage pool is provided below the level of the principal outlet. Conservation dams have an emergency high-level spillweir and a controlled low-level outlet, although in some cases this lower outlet may be omitted.

Floodwater-retarding Structures

In the United States, small flood-control structures have been widely employed for the reduction of flood-water, erosion, and sediment, damage. Flood Control Acts of 1936 and 1944, and the Watershed Protection and Flood Prevention Act of 1954, provided the main stimuli for dam construction in small catchments, of between 1.3 km^2 and 25 km^2, in order to reduce damage in and alongside downstream reaches (Sauer & Masch, 1969). Between 1935 and 1962, 450 catchments totalling an area of more than 200 000 km^2—slightly over 20% of the country's land area—became 'protected' by 2000 flood-water-retarding structures, 3000 stabilization and sediment-control structures, and 500 silt and debris basins (Gottschalk, 1962). At that time, Gottschalk identified more than 8000 other catchments with a total area of more than 4×10^6 km^2, or nearly 40% of the total land-areas, which were suitable for protection by flood-water-retarding structures.

The reservoirs are designed so that the flood storage-capacity is exceeded, and the emergency spillway operates, on an average of from more than once in 25 years to once in 100 years. Flow from these structures is ungated, and controlled only by the elevation of water in the reservoir. The flood-water detention-capacity of most structures is less than 2×10^6 m^3, but many reservoirs also provide storage for either 50 or 100 years of sediment accumulation below the elevation of the principal spillweir. For free-release dams, the level of the principal spillweir is determined by the frequency of the required control discharge, and by the delay to releases required to allow the peak discharge from the downstream, uncontrolled area to pass before the peak dam-release. Many structures have a two-stage principal outlet, with a low-level conduit for the first stage, and a higher weir as the second stage (Moore, 1969). These single- and double-stage principal spillweirs generally have capacities of between 0.1 and 0.16 m^3 per s per km^2 (Decoursey, 1975).

Within the reservoirs, consumption losses—attributable to evaporation from the free water surface and to seepage—can markedly reduce the catchment yield. For example, six flood-water-retarding structures on Mukewater Creek, Texas, controlling nearly 40% (70 km^2) of the catchment, in two consecutive years reduced the runoff yield from 102.4 mm to 97.0 mm, and from 24.4 mm to 17.8 mm (Coskun et al., 1969). Evaporative losses dominate consumption, and depend, primarily, upon the mean surface-area of the reservoir and the potential evaporation. However, the hydrological significance of consumption losses will also relate to the annual runoff: as a percentage of the annual runoff, consumption losses will decrease as annual runoff increases. Extreme losses from reservoirs in the southern Great Plains,

USA, of up to 42%, have been reported during years of low rainfall (Decoursey, 1975). Consumption losses can increase the flood-control ability of a structure, however, by ensuring that flood storage is rapidly reduced. Thus, within the southern Great Plains, where annual evaporation is considerably greater than annual rainfall, the effects of flood-water-retarding structures are greater than in any other region of the United States (Decoursey, 1975). Twenty-five structures on Sugar Creek, Oklahoma, reduced by 23% the area of floodplain inundated during one extreme flood, but the duration of the flood was lengthened to 20 days as a result of temporary storage within the reservoirs (Hartman et al., 1969).

The effect of small structures on peak-flows is related to the number of structures, the area of catchment controlled, and the size of the flood in relation to the storage-volume available before the storm (Table VIII). Thus, a single flood-retarding structure on Six Mile Creek, Arkansas, could reduce the peak-flow by 98% if the storage capacity was not exceeded, but the same inflow would still be reduced by about 40% as a result of routing and spillway flow, when the storage capacity is full (Moore, 1969). When the reservoirs are empty, the full capacity can be used for flood storage, and the water released slowly, so that floods from small storms, at least, can be eliminated—provided that the storm frequency does not exceed the time-period required for the storage to drain. However, even the small structures can effectively attenuate floods by reservoir lag; peaks will be reduced and delayed, and the flood duration will be lengthened. The combined effect will be to reduce runoff yield to the river below the dam, and to redistribute the flow, with a lower percentage occurring at high-flows and a higher percentage occurring at intermediate flows.

Water-conservation Structures

Solid water-conservation dams have been constructed for several thousand years, and in western Europe they became particularly common during the eighteenth and nineteenth centuries. A large number of simple storage reservoirs with a single, controlled, low-level outlet and ungated spillweir have been built in the UK. The earliest reservoirs were built by landscape architects, such as Capability Brown, to beautify the estates of large landowners (Kennard, 1972), or by canal engineers. Since the late nineteenth century, many simple dams have been built by public or private water authorities in upland areas, e.g. for direct supply purposes. Today, over 60% of the dams in Britain impound lakes with a surface area of less than 1.0 km², and 80% have areas of less than 4 km². These simple dams are designed to provide a reliable water-supply to meet industrial and domestic demand; excess water is allowed to discharge freely over a high-level spillweir, while the controlled outlet is used only to provide a compensation flow, maintaining minimum water-levels within the river downstream. Thus, these reservoirs are maintained as full as possible throughout the year.

Table VIII   Flood-peak reduction by small structures.

Number of structures	Area of catchment controled (%)	Discharge (m³ per s)		Flow reduction(%)	
		With structures	Natural flow	Reservoir empty	Reservoir full
A.* Sugar Creek, Oklahoma (Hartman *et al.*, 1969)					
4	—	26.9	36.8	27	
8	—	11.3	38.8	71	
25	—	234.9	464.1	51	
B. Mukewater Creek, Texas (Coskun *et al.*, 1969)					
6	39.3			63	
		48.1	73.6		34
		15.6	35.4	56†	
		25.5	26.9		5†
C. Southern Great Plains (Decoursey, 1975)					
—	100	—	—	99	50
—	50–60	—	—	56	25
—	30–45	—	—	75	15

* In A the initial storage capacity of the reservoirs was not cited.
† Data for mean daily flows.

Consequent upon dam closure, the structures exert an immediate effect upon the river discharges downstream. During the period that the reservoir is filling, the entire runoff volume will be impounded by the dam, and only compensation flows will be released (Fig. 12, A and B). When once the permanent storage-pool has filled, the volume and pattern of reservoir discharge will be related to the storage level within the reservoir, which reflects the balance between inflows and losses. These latter include releases for compensation flow, spillweir flow,direct abstraction for supply, evaporation, and seepage. Thus, the seasonal variation of runoff yield is less apparent for the upland Avon Dam than for the lowland Sutton Bingham Reservoir: the catchment of the former has nearly twice the mean annual rainfall and a considerably lower mean annual soil-moisture deficit than the latter.

Reservoirs which are markedly drawn-down during summer can provide a considerable storage volume for flood absorption. This is particularly significant in many areas of Britain where extreme floods are produced by intense cyclonic summer storms, and not by prolonged, low-intensity, winter rains. Chew Valley Lake, for example, built in 1953, provides large amounts of water for supply, so that a large storage volume is usually empty during the summer months (Fig. 12, C). Indeed, draw-down of the water-level within Chew Valley Lake during July through September is such that all high-magnitude discharges have been absorbed by the reservoir throughout the period of record. For example, although the storm of 10 July 1968 caused catastrophic flooding within neighbouring catchments, Chew Valley Lake absorbed the flood; the Lake was 0.9 m down prior to the storm, and rose by 0.5 m—an inflow of 2.14×10^3 m³, equivalent to 37 mm of runoff.

During winter, temporary storage above the spillweir plays an important

42

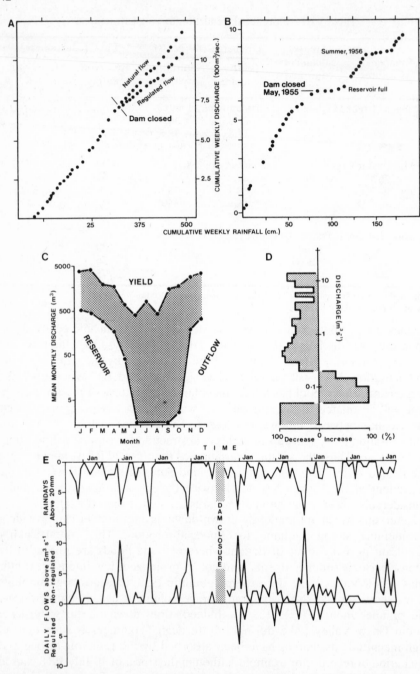

Fig. 12. Hydrological effects of selected British water-conservation structures: rainfall–runoff relationships for the River Avon (A) and River Yeo (B); inflow–outflow relationships for the River Chew (C) and, as a frequency of discharge-days, for the River Avon (D); and rainfall and runoff frequencies for the River Hodder (E).

role in reducing the maximum rate of outflow from the water-conservation reservoirs (Table IX), and peak flows passing through these reservoirs, with storage full to spillweir level, will be delayed and reduced in magnitude. Thus, the provision of compensation flows by most dams eliminates extreme low-flows (Fig. 12, D), and seasonal flood storage, together with the effects of reservoir lag, will reduce the frequency of high-flows (Fig. 12, E), so that downstream reaches will experience an increased number of days with intermediate discharges.

The comparison of discharges above a selected threshold under natural and impounded conditions suggests that it is the late-summer peak discharges which are most significantly reduced, and that flood regulation by conservation-structures is clearly seasonal. Separation of the inflow and outflow flood-frequency distribution, based upon the annual exceedance series'* (Chow, 1964), for the Stocks Reservoir, River Hodder, England, UK, demonstrates that the frequency distributions for the winter data are closely comparable, but that the magnitudes of summer-flow frequencies differ considerably (Fig. 13). The magnitude of the 5 years' return-period winter flood is reduced by only 8%, whilst the 5 years' summer flood is halved. Furthermore, in terms of the long-term flood-frequency distribution, small reservoirs may have an insignificant effect upon the magnitude of rare events, whilst regulating, or even eliminating, smaller floods. The reservoir outflow flood-frequency curve tends to be considerably below the inflow flood-frequency curve for more frequent events, and approaches it more closely for rare events.

MAJOR FLOW-REGULATION STRUCTURES

Multipurpose reservoirs are constructed with designated storage volumes, allocated for specific purposes, and operated to meet changing seasonal and daily demands. Operational procedures may be relatively simple for highly seasonal flow-regimes, the stored water resulting from flood-control in the wet season providing a reliable supply during the subsequent dry period. Thus, under predictably seasonal climatic conditions (e.g. in monsoon areas), flood control and supply needs (such as for power generation, navigation, or irrigation) can be satisfied with relative ease of management. Multipurpose reservoir developments often have hydroelectric power as the key role, but the imposed flow-regime may change over time with varying demand. Lake Kainji, for example, on the clearly seasonal River Niger in Nigeria, provides electricity, flood-control, navigation, and irrigation supply (Jansen et al., 1979).

Many reservoirs (e.g. Bighorn Lake, USA, cf. Soltero et al., 1973) are designed also to meet recreational needs, but the operational pattern is often flexible, allowing for procedural modifications in response to changing demands. For example, since the closure of Fort Randall Dam, Missouri River, the average monthly releases have followed a persistent pattern to satisfy navigation requirements (Livesey, 1963): the winter low-flows

* The 'annual exceedance series' uses the N highest flood peaks in a flow record of N years, not necessarily including one from each year, to determine flood frequency characteristics.

Table IX Flood routing through upland water-conservation reservoirs, UK (From Petts & Lewin, 1979. Reproduced by permission of G. E. Hollis)

River/Reservoir	Proportion of catchment inundated (%)	Spillweir length, for catchment of 1 km^2 (m)	Time-to-peak of inflow hydrograph (hrs)	Ratio of inflow–outflow hydrograph time-to-peak	Peak flow reduction (%)
W. Okement, Meldon Dam	1.30	0.64	11	1.36	9
Avon, Avon Dam	1.38	0.45	7	1.57	16
N. Teign, Fernworthy Dam	2.80	0.46	7	1.86	28
Campswater, Camps Dam	3.13	0.61	12	1.82	41
Rede, Catcleugh Dam	2.72	0.95	9	2.88	71
Hodder, Stocks Dam	3.70	0.91	11	2.09	70
Chew, Chew Valley Lake	8.33	0.44	27	2.89	73

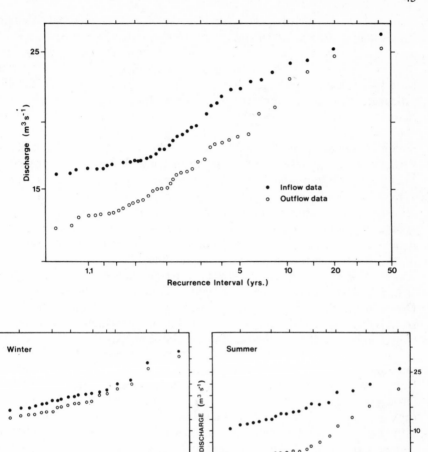

Fig. 13. Flood-frequency characteristics of the River Hodder above and below Stocks Reservoir, UK, demonstrating the variable seasonal efectiveness of flood regulation. (From Petts & Lewin, 1979. Reproduced by permission of G. E. Hollis.)

(150–300 m³ per s) and late summer high-flows (600–900 m³ per s) contrast with the late-spring flood regime of the natural river. Artificial seasonal flow-patterns may be imposed, but changes may also occur from year to year. The minimum flow released to the Zambezi from Lake Kariba has increased from 530 m³ per s in 1966 to 890 m³ per s in 1978—a 68% increase in response to an increased demand for electricity (Guy, 1981). Nevertheless, major flow-regulation reservoirs often provide total control of the flow-regime for a considerable distance below the dam, and for more than 150 km in the case of the Zambezi (Begg, 1973).

In Australia, the Murray–Darling system is regulated by nine principal storage reservoirs, sharing a storage capacity of 19 × 10⁹ m³ which serves irrigation, hydroelectric power, flood-control, and water-supply demands (Cadwallader, 1978). The uncontrolled river was high, cool, turbid, and fast-flowing in spring and early summer, and low, warm, and slow-flowing during the autumn. Dam construction effectively reversed this natural pattern. Reservoir releases for irrigation supply to downstream floodplain locations produce high-flows in summer and autumn (Fig. 14), and the dams store runoff in winter and spring—the seasons of natural high-flows. Below Hume Dam, the percentage of time that flows greater than 500×10^6 m^3 per month are equalled or exceeded has been reduced, and the duration of flows between 500×10^6 m^3 and 100×10^6 m^3 per month has been increased, for example by 22% for the 200×10^6 m^3 per month flow (Baker & Wright, 1978). The provision of regulated flows in summer and autumn months, releases reservoir storage space for flood absorption during winter and early spring, and flood stages have been reduced by over one metre. Summer floods—in June, July, and August—may also be completely eliminated but, in the process, the reservoir storage may be filled. Thus, in years when major floods occur in the September–October period, these floods pass through Lake Hume with little reduction in their peak flows (Baker & Wright, 1978). Downstream, the seasonal pattern progressively reverts to that of the natural river, as tributaries add to the winter and spring flows, and abstractions for irrigation reduce the summer and autumn discharges.

One of the most intensively-controlled river systems is the Colorado, USA. Impoundment of the main river began in 1935 with the completion of Hoover Dam, and today the river is controlled by a series of major dams located throughout its length. Major dams have also been constructed on the main tributaries—the Gunnison, San Juan, and Gila, Rivers—so that only the Little Colorado River remains uncontrolled. Its hydrological history is documented particularly by Dolan et al. (1974) and Turner & Karpiscak (1980). Before impoundment, annual runoff from the Colorado catchment was 16 000 hm^3†, but today less than 1% of its virgin flow reaches the river mouth. At Grand Canyon, annual runoff ranged between 5200 hm^3 and 24 500 hm^3 per yr; although this was reduced to 2000 hm^3 during the filling stage, under normal operating conditions annual runoff varies between 9900 hm^3 and 12 300 hm^3, and extremely high- and low-flow years have been eliminated. Also, monthly mean discharges prior to dam construction were clearly unimodal, with a marked peak in May and June in response to snow-melt floods, and the peak flow exceeded that for the driest month by a ratio of ten-to-one. Subsequent to the closure of Glen Canyon Dam, the maximum-minimum ratio was reduced to 1.8:1, and there is little monthly variation. Indeed, the whole range of flows have been altered for at least 200 km downstream of the dam (Table X): low-flows have been increased, and high-flows reduced. Prior to the closure of Glen Canyon Dam, maximum discharges at Lees Ferry were always greater than at Grand Canyon, due to storage along the 141 km channel-length, but maxima occurred during the same runoff event. Since dam closure, annual peak-flows have not been

† hm = hectometre

MONTHLY FLOW (Litres x 10⁶)

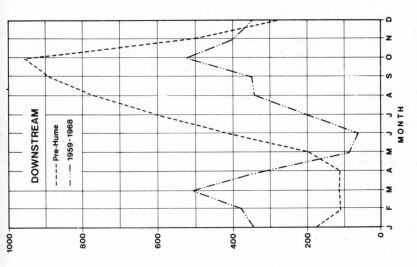

MONTHLY FLOW (Litres x 10⁶)

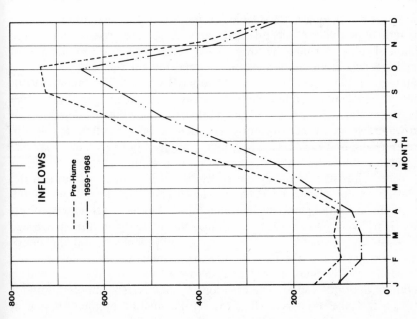

Fig. 14. Regulation of the River Murray by Hume Dam, Australia. (From Baker & Wright, 1978. Reproduced by permission of the Royal Society of Victoria.)

temporally correlated; the annual maxima at Grand Canyon are now greater than those at Lees Ferry; and the peaks at Grand Canyon exceed those at Lees Ferry by 10% at least once in every 4 years. These changes reflect the increased significance of the uncontrolled Little Colorado tributary contributions to discharges at Grand Canyon.

Table X   The hydrological characteristics of the Colorado River below Glen Canyon Dam. (Based on Dolan et al., 1974 and Turner & Karpiscak, 1980. Reproduced by permission of Sigma Xi, the Scientific Research Society and the US Geological Survey.)

	Lees Ferry (24 km downstream)		Grand Canyon (165 km downstream)	
	Pre-dam	Post-dam	Pre-dam	Post-dam
Daily average flow equalled or exceeded 95% of the time (m^3 per s)	102	156	113	167
Median discharge (m^3 per s)	209	345	232	362
Mean annual flood (m^3 per s)	2434	764	2434	792
10 years' flood (m^3 per s)	3481	849	3453	1132
Annual maximum stage (m)				
Mean	5.04	3.56	6.89	4.79
Standard deviation	0.96	0.17	0.35	0.15
Annual minimum stage (m)				
Mean	1.76	1.46	0.46	0.70
Standard deviation	1.40	0.23	0.85	0.45

The influence of major flow-regulation reservoirs may extend throughout the length of a river to the estuarine and near-shore, coastal zone. The incursion of salt water into the lower reaches of a river as a consequence of reduced freshwater discharges has been reported on the Zambezi River, Mozambique (Hall et al., 1977), and Dnieper River, USSR (Zalumi, 1970). In contrast, constant and relatively high releases from Gordon and Serpentine Dams, Tasmania, have prevented the penetration into the Gordon River of a salt-water wedge, which formerly intruded as an underflow for 48 km upstream (King & Tyler, 1982). Moreover, reduced but regulated river discharges can modify the water circulation pattern of the estuarine and near-shore zone. Thus the impoundment of the River Nile in 1964, through the completion of the Aswan High Dam, caused considerable changes of the hydrographic conditions over the continental shelf in the south-east Mediterranean (Din, 1977). Before 1964, the annual Nile flood discharged an average of 34×10^9 m^3 during the period August to November, and the floodwaters, passing through the Rosetta and Damietta estuaries, influenced the hydrographic and circulation pattern to about 80 km from the coast and to a maximum depth of 150 m. The creation of Lake Nasser imposed a major control on the Nile discharges (Fig. 15, A), augmented by the loss, due to evaporation and irrigation, of 60% of the Aswan High Dam releases. Before dam completion, the estuarine circulation pattern during the flood season was a two-layer flow, with Nile water over-passing a subsurface flow of Mediterranean water, and during the period of low river discharges, a one-layer sea-water-dominated flow. Subsequently, the one-layer flow-pattern has persisted during most of the year (Fig. 15, B). The natural flood-water formerly spread over the sea as a surface layer, and moved rapidly along the coast to the east; the 39°/oo isohaline was 80 km from the shore, with salinity values of less than 30°/oo to about 7–20 km offshore. Since 1964, the surface salinity of the near-shore zone during August and September has generally exceeded 39°/oo. (Fig. 15, C).

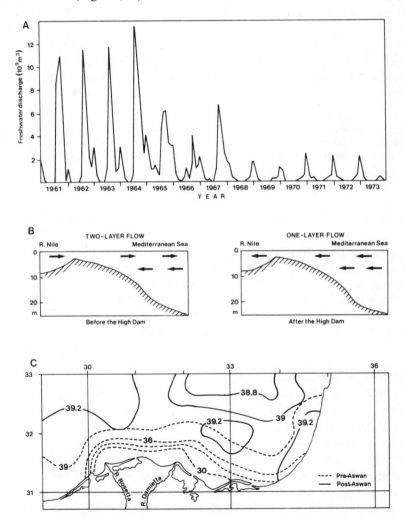

Fig. 15. Effects of Nile regulation by Lake Nasser (A) upon the flow circulation (B), and salinity distribution (°/oo) (C), in the south-east Mediterranean Sea. (From Din, 1977. Reproduced by permission of the American Society of Limnology and Oceanography, Inc.)

Impoundments within the St Lawrence River, Canada, have similarly changed the seasonal strength of the haline circulation, the salinity of the surface layer, and the seasonal temperature budget of the estuarine and near-shore zone; the nutrient supply from deeper ocean water to the surface layers is reduced, and the reproduction of many marine fish, and thus the structure of the biomass, has been affected (Neu, 1975).

Except in emergencies, discharge from major regulating reservoirs is controlled through the power penstocks and low-level outlet valves. The management of reservoir releases is related to downstream tributary discharges, both for the control of downstream floods and for the provision of minimum flows during the dry season. However, due consideration must also be given to the lotic, floodplain, and near-shore, systems—not least to the migratory fishes, which may depend upon correctly-timed high- or low-flows for the maintenance of viable populations. Unfortunately, such considerations may not be immediately compatible with economic or political demands.

Pulse Releases

A wide range of operational procedures can produce sudden fluctuations in discharge—fluctuations which often occur at abnormal rates (Table XI). Hydroelectric power demand, and irrigation demand, are the most common causes, but peak-discharge waves have been utilized for navigational purposes (Jansen et al., 1979), and to meet recreational needs. For many purposes, 'pulse releases' are made regularly: for example, daily releases through the power turbines of Fort Randall Dam produce stage fluctuations* of up to 2 m (Livesey, 1963). Similarly, the Colorado below Glen Canyon Dam now experiences daily stage changes of over 1.5 m (Fig. 16, A), in contrast to that of only several decimetres prior to impoundment. However, these are superimposed upon a 7-days periodicity, defined by a drop in stage on most Sundays, and with some striking drops on holidays—such as Christmas, New Year's Day, Memorial Day, and Labour Day—as a consequences of decreased power demand (Turner & Karpiscak, 1980). Other frequencies of pulse-release may reflect different purposes, such as recreation (Figure 16, B), but sudden changes of flow may also occur erratically. For example, after nearly 1 year of stable flow, at about 700 m³ per s, the flood-gates of Lake Kariba were opened and the discharge of the Zambezi rose suddenly to 4531 m³ per s (Hall et al., 1977).

Pulse releases have been used to remove sediment accumulations, usually deposited by uncontrolled tributary flows, from regulated channels. Below Granby Dam, flushing flows caused marked discharge pulses on the Colorado River (Eustis & Hillen, 1954): short-term flow fluctuations, from 0.57 m³ per s to between 5 and 9 m³ per s, over periods of about 15 minutes, successfully scoured sediment and Algae from below the dam. However, little is known about the effects of these unnatural pulses upon the lotic flora and fauna. In

* 'Stage' is a measure of water depth above an arbitrary datum.

A

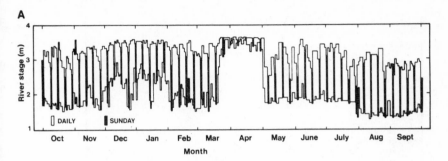

B

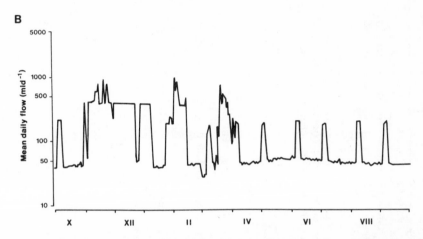

Fig. 16. Pulse releases from Glen Canyon Dam, Colorado River, USA (A) and from Lake Vyrnwy, UK (B). (From Turner & Karpiscak, 1980 (A). Reproduced by permission of the US Geological Survey. From data provided by the Severn–Trent Water Authority (B).)

many cases, the effects are localized because the wave may be attenuated by channel storage as it moves downstream. Stage increases of 2 m below Fort Randall Dam were reduced to 1.2 m 24 km downstream (Livesey, 1963), and a release from Granby Dam was reduced from 4.8 m^3 per s to 1.7 m^3 per s within 4 km (Eustis & Hillen, 1954). Nevertheless, the rates of stage change, together with the rates of associated change in water quality—particularly turbidity and temperature—may be more important than the magnitude of the release itself. Releases of 12.63 m^3 per s from Lake Celyn on the Upper Dee, UK, from base-flows of 0.368 m^3 per s, take 1 hour 30 minutes to travel 7.5 km, and cause a stage-rise here of nearly 0.4 m in less than half-an-hour. The flows appear to tap in-channel solute and sediment sources, causing an increase in the concentrations of some ions and also high turbidities, though the 'water-quality' pulses have a duration of only about 20 minutes before becoming diluted (Foulger & Petts, 1984).

Neel (1963) recognized that daily fluctuations in reservoir releases can discourage littoral stream-life, and may adversely affect a stream's carrying

Table XI   Examples of pulse-releases to regulated rivers.

Reservoir, River	Pulse and Cause	Source
Fort Randall Dam, Missouri River, USA	Daily fluctuations, from nearly zero to over 1000 m³ per s in summer, and 500 m³ per s in winter, due to navigation and power-peaking demand.	Livesey (1963)
Lake Kariba, Zambezi River, central Africa	Sudden changes of water-level in response to demand for electricity and flood-regulation: by opening one gated outlet the downstream flow can be increased from about 700 m³ per s to 2000 m³ per s in 30 minutes, producing a stage rise of 5 m below the dam.	Begg (1973)
Canyon Dam, Guadalupe River, USA	A constant flow of 4.5 m³ per s from early January through March increased suddenly to about 20 m³ per s in mid-April.	Hannan et al. (1979)
Glen Canyon Dam, Colorado River, USA	Vertical, daily stage fluctuations up to 4.5 m in response to power demand.	Dolan et al. (1974)
Tallowa Dam, Shoalhaven River, Australia	Normal constant flow of 45 m³ per s terminated: outlet valve closed over a period of 15 minutes and within 30 minutes the channel bed was dry, except for four deep pools, for 200 m below the dam.	K. A. Bishop & Bell (1978)
Jackson Lake, Snake River, USA	Releases for irrigation supply cause violent fluctuations of discharge: flow reductions from 2.8 to 0.3 m³ per s in less than 5 minutes, caused a drop in river stage by 0.3 m.	Kroger (1973)
Flaming Gorge Dam, Green River, USA	Discharge fluctuations from 10 to 70 m³ per s increased water depth of 65 cm 11.7 km below the dam, but this was attenuated to only 10 cm about 100 km further downstream.	Pearson et al. (1968)
Kakhovka Power Station, River Dnieper, USSR	Tailwater flow fluctuations by 1.5 m daily, due to changing power demand.	Zalumi (1970)

capacity for many life-forms. It has been suggested that judicious reservoir-level manipulation, to cause periodic spillweir flows, could be used to flush 'pest' insect larvae (especially *Simuliidae*) (Brown & Deom, 1973), and that artificial 'freshets' could be of ecological benefit through stimulating upstream migrations of salmonids (Banks, 1969). Although pulse-releases from dams could be used to maintain stream quality, little is as yet known of the consequences of pulses of different magnitude, duration, and timing, for the fauna and flora of the regulated river.

RESERVOIR LIFE-EXPECTANCY

As soon as a dam is operational, it will begin to trap sediment within its storage volume (*see* Chapter 4), and reservoir sedimentation will progressively alter the character of discharges downstream. The Huang Ho, China, has the highest sediment load in the world (1.6×10^9 tonnes per yr), and of its impoundments, the Heisonghi Reservoir lost nearly 20% of its storage capacity within 3 years of completion; even after operations to reduce the sedimentation rate, the expected life of the reservoir is less than 80 years (Walling, 1981). Other large multipurpose reservoirs in the same catchment have a similar fate: the Sanmanxia, for example, lost 40% of its storage capacity in the first 20 years after completion. Indeed, many reservoirs in areas of high sediment-yield have a life-expectancy of less than 100 years (Dorst, 1970). Consumption losses will decrease, and flood-peaks will increase, as the reservoir becomes infilled with sediments.

The small reservoirs of the Garza–Little Elm basin, USA, deplete the runoff yield by 10% during early years; but after the permanent pools of the flood-water retarding structures are filled with sediments (estimated at about 30 years), the depletion of the annual yield is expected to be less than 1% annually (Gilbert & Sauer, 1970). In southern California, Big Tujunga Dam lost 70% of its storage capacity in less than 4 years, as a result of high sediment yields, and major floods now pass the spillweir only slightly attenuated (Scott, 1973). Such progressive hydrological changes are typical of relatively small-capacity reservoirs which impound runoff from catchments with high rates of soil erosion. Large-capacity reservoirs, or reservoirs draining hard-rock or well-vegetated catchments, should be insignificantly affected by sedimentation; rates of storage loss of less than 0.01% per yr are common, so that a stable hydrological regime can be imposed for a long period of time.

CHAPTER III

The Quality of Reservoir Releases

The quality of natural river-water is controlled, predominantly, by the climatic and geological characteristics of the drainage basin. Water storage in open reservoirs induces physical, chemical, and biological, changes within the stored water. In consequence, water discharged from impoundments can be of different composition, and can show a different seasonal pattern, from that of the natural river. Whilst limnological studies have established common seasonal patterns of physico-chemical changes in natural lakes, for impoundments the differences in morphometry, stratification dynamics, and water movements, result in quite different physico-chemical patterns (Hannan *et al.*, 1979). Many factors influence the quality of reservoir discharges (Fig. 17), but those forces capable of generating reservoir stratification, and hence discharges of different quality from different release elevations, are particularly important.

Natural river-waters commonly contain four important cations (calcium, magnesium, sodium, and potassium), while chloride, sulphate, and bicarbonate, are the predominant anions. The relative importance of the different ions depends, in general terms, upon geographical location (Gibbs, 1970). Variations on a small scale have been related to effective precipitation and altitude (e.g. Douglas, 1972), lithology and land-use (e.g. Walling & Webb, 1975), and anthropogenic effluents which, from point- and diffuse-sources, are becoming increasingly important. Soil disturbance, drainage, and vegetation clearance, have increased concentrations of NO_3^-, SO_4^{2-} and Mg^{2+}, and urbanization can lead to the addition in stream-flows of phosphate, nitrogen compounds, and heavy-metals. Reservoirs act as thermal regulators and nutrient sinks, so that the seasonal and short-term fluctuations in water quality, which are characteristic of natural rivers, will be regulated. However, during certain periods of the year, reservoir discharges can have extremely low concentrations of dissolved oxygen, and unnaturally high concentrations of iron, manganese, and hydrogen sulphide. The dissolved-oxygen deficit created during summer by the Cherokee Dam, USA, downstream along the Holston River, has been equivalent to that produced by the effluent from a town of 3 500 000 people (Ingols, 1959). A dissolved-oxygen sag has been observed for 100 km below Hume Dam on the Murray River, Australia (Walker *et al.*, 1979). These undesirable impacts are primarily the result of lake stratification.

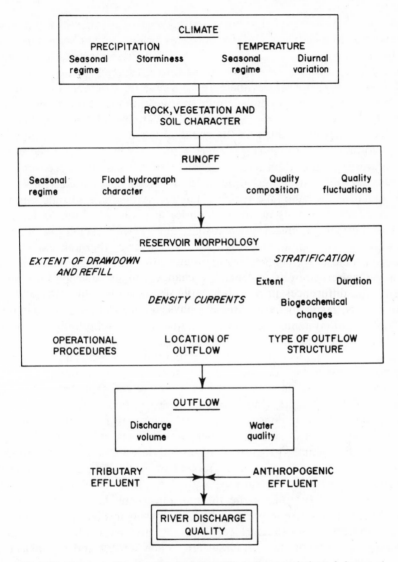

Fig. 17. Factors affecting the water quality characteristics of dammed rivers.

THERMAL STRATIFICATION

Water temperature is an important quality parameter for the assessment of reservoir impacts upon downstream aquatic habitats. Indeed, many important physical, chemical, and biological, processes are significantly influenced by the temperature of river-flows. The relatively small volume of water in any river section, together with turbulent mixing and the large surface-area in contact with the atmosphere, allows a rapid response of stream-water temperature to the prevailing meteorological conditions.

In reservoirs, however, the increased mass of relatively still water allows heat storage to take place, and this produces a characteristic seasonal pattern of thermal behaviour, due to temperature–density differences in the water (Zumberge & Ayers, 1964). From spring through summer, a lake gains heat, and as surface temperatures increase, a well-defined temperature gradient develops, and a highly stable summer stratification may become established (Fig. 18, A). An upper layer (epilimnion), of variable thickness, is characterized by temperatures which are higher than those of the underlying water, and wind-induced turbulence produces a surface zone of relatively uniform temperature. Evaporation-induced cooling of the surface layer, and local temperature differences caused by 'shadow' effects, associated with variable solar radiation over the lake surface, may further increase turbulent mixing.

Because water has its maximum density at about 4°C, and undergoes a normal thermal expansion—decreasing density with increasing temperature—from that point upwards, and an inverse thermal expansion—decreasing density with decreasing temperature—from that point downwards, the higher-temperature, low-density, water layer 'floats' on top of the lower, cooler, water. In natural lakes, the epilimnion is often observed to depths of 8–15 m below the surface; but it may be either shallower or deeper in artificial impoundments, depending, in particular, upon retention times, depths of water outlets, and withdrawal rates.

An intermediate layer of maximum temperature-gradient, the thermocline, may be established between the epilimnion and the cooler, denser, water of the hypolimnion, which may persist in a largely stagnant state. The rapid decrease of temperature with depth through the thermocline is associated with a rapid increase in density of the water. This major density-discontinuity is poorly developed in spring, but becomes thinner, and increasingly well-defined, as the summer progresses. At its full development, the thermocline commonly exhibits a temperature gradient of up to 2°C per m. Within the hypolimnion, temperatures are typically low, and decrease only a small amount from the bottom of the thermocline to the bed of the reservoir. During the autumn, the shortening days, decreasing insolation, and declining air temperatures, cool the surface waters. Progressive cooling produces an increase in density of the surface water, which settles, and is replaced by warmer water from below. This 'autumn overturn', characterized by convective circulation, produces isothermal conditions throughout the reservoir's depth. Because the density differences are also slight during this period, turbulence from wind action can provide mechanical mixing, which may reach the bottom. Indeed, the turnover usually begins shortly before surface temperatures of 4°C are reached, as a result of wind-induced turbulence.

Within man-made lakes, water movement can be induced by large outflows, and heat may be transported by eddy diffusion. The change in water-level is more variable in a reservoir than in a natural lake, so that advective heat-transfer, and vertical movement of the water-mass, become increasingly significant. Indeed, the importance of advective heat-transfer,

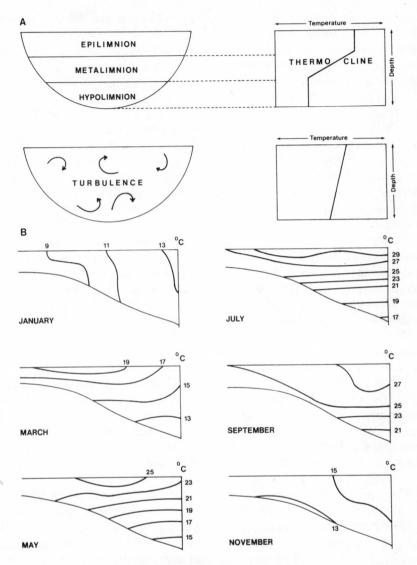

Fig. 18. Lake stratification: general characteristics (A), and seasonal variability of water temperature within Canyon Lake, Guadalupe River, USA (B). (From Hannan *et al.*, 1979. Reproduced by permission of Dr W. Junk Publishers.)

and vertical mixing, for the distribution of temperature in reservoirs, is such that, during summer, the temperature of hypolimnial water will be greater than that of a natural lake at the same depth.

The development and destruction, of a strong thermal stratification, will be influenced by the pattern of inflows, and by the relatively high outflows, in comparison with natural lake systems. Hannan *et al.* (1979), for Canyon

Lake on the Guadalupe River, USA, have demonstrated the internal variability of water temperature within a large, thermally-stratified reservoir (Fig. 18, B). A pronounced thermal stratification had been established throughout the reservoir by June, and a major temperature difference of 12°C occurred between the surface and the deepest layers in July. At the surface, temperatures exhibited an annual range of 18°C, whilst from April through September the temperatures of the hypolimnion increased 1–2°C each month at the deeper locations, which experienced total temperature increases of 8–9°C. However, at upstream locations, the water temperature was markedly influenced by the inflow water, and thermal stratification was periodically disrupted by particularly heavy inflow. The epilimnion was observed to increase in depth during the late summer, and this may have been caused by the withdrawal of cold water from the hypolimnion of the bottom-draining reservoir.

Wunderlich (1971) contends that vertical mixing will be restricted by density gradients, but horizontal movements are enhanced, so that during low-level hypolimnial draw-off, the stratified layers may resist mixing and induce a lowering of the epilimnion. Thus, Hannan et al. (1979) observed that the total volume of outflow from Canyon Lake, from June through September, was equivalent to the volumetric change of the hypolimnion. Epilimnial drawdown, as a result of low-level releases, also affected the timing of the autumnal overturn. In Canyon Lake, the overturn commenced in September and was complete by late October, but precise rates of change varied between locations. Complete mixing occurred first at the upstream locations, whilst the deeper areas were still stratified: the earlier overturn, at shallower locations, was induced by the epilimnion being drawn down to the reservoir bottom at those sites. The thermal characteristics of downstream river-flows will reflect, therefore, not only the seasonal pattern of stratification, but also the extent of epilimnial drawdown, which may be particularly significant for intermediate-depth withdrawal from dams.

This description of thermal behaviour provides an example of what has been considered to be the 'typical' reservoir. Many reservoirs require a period of several years before a stable thermal regime is developed within the impoundment, and, consequently, within the outflows. Some reservoirs may even experience thermal enrichment over time: for example, the mean temperature of releases from the shallow Ennis Reservoir to the Madison River, USA, in 1961 was elevated by 1.0°C (17.6°C cf. 16.6° for the influent stream), but by 1967 this had increased to 2.2°C (18.4° cf 16.2°) (Fraley, 1979). However, considerable differences exist in the thermal character of reservoirs in relation to geographical location, both at the large scale, in terms of latitude, and locally, associated with the topography and morphometry of the reservoir basin and the reservoir retention-time.

Latitudinal Variations

For temperate zones, thermal stratification during summer is common within reservoirs having a depth greater than 10 m. In spring, lakes will be homothermal, at a temperature of about 4°C, which is the temperature of maximum density. Subsequently, the air temperature, and incident solar radiation, increase, and warm the surface water-layer. The effectiveness of wind and wave action, for the promotion of vertical circulation, is dependent on the density difference between the upper and lower water-strata. Thus, the increasing density gradient during the summer provides progressively greater stability. During autumn and winter the epilimnion cools, so that the whole water-column overturns and becomes fairly well mixed—a condition which is maintained until the spring, when stratification is gradually re-established. However, in some areas, such as the UK, the magnitude of the temperature modification is suppressed by the prevailing climatic equability, compared with the more extreme conditions associated with 'continental' situations.

Arai (1973) suggests that reservoirs within temperate latitudes can be divided, on the basis of thermal characteristics, into two types, which are dependent upon the water retention-time in the impoundment. From two Japanese examples he demonstrated that, in reservoirs having a long retention-time, thermal stratification could be created in the summer season, but the thermal behaviour of short-retention impoundments is largely controlled by the inflowing water-mass. Within the Sakuma Reservoir, flows greater than 300 m³ per s are discharged in 10 days' travel-time; advective heat transfer is great, whereas thermal stratification is weak, and an isothermal layer extends over a depth of 80 m. Similarly, in Slapy Reservoir, Czechoslovakia, the effect of a short retention-time (38.5 days) is an early autumnal overturn, reduced vertical temperature-differences during summer, and highly variable winter temperatures (Straškraba, 1973): in dry years, an inversion results in the freezing of the reservoir surface, whereas in wet years, water is almost homothermal below 2°C when there is no ice-cover.

In northern latitudes, deep reservoirs exhibit an additional winter stratification resulting from the cooling of surface waters to temperatures below 4°C, and this produces an inverse temperature structure, with the colder, less dense, water at the surface, and water cooled to 4°C accumulated at the bottom. However, even at its greatest development during the coldest winters, the temperature difference cannot exceed the 4°C represented by freezing conditions at the surface and maximum density conditions at the bottom. Because the density differences are small, the winter stratification will be weak, and wind action may cause a high degree of mixing. Should an ice-cover develop, the wind disturbance will be stopped, the reservoir becomes stagnant, and a second overturn of water occurs during the spring, with the melt of the ice-cover, and a rise of surface temperatures. However, the formation and melting of ice has a major influence on the temperature of a reservoir. During the spring melt, heat is absorbed from the lake, and this exerts a regulatory control on the rate at which the water warms. Similar regulation occurs during ice formation, the release of latent heat into the underlying water preventing sudden reductions of temperature. As temperatures at the surface approach the 4°C threshold, the resulting convective sinking, aided by wind-induced stirring, re-establishes uniform vertical temperatures and a period of complete convective circulation.

A typical pattern has been described by Zhadin & Gerd (1963) within the Dzhezkazyan Reservoir, Kengir River, Kazakhstan, USSR. Thermal stratification occurs from May through August, and surface-water temperatures approach 28°C. In the autumn there is a sharp drop in temperature to 4.4°C, followed by homothermy close to 0°C. Under the ice-sheet, the temperature of the bottom layer rises to approximately 2.5°C, and a weak winter stratification is established until the April thaw, which initiates overturn, whereafter there is a period of spring homothermy, with temperatures of 8–9°C throughout the water-mass.

In the southern latitudes, average winter temperatures are above freezing, and reservoir waters remain mixed, being isothermal from top to bottom. Furthermore, the annual temperature regime may be insufficient to bring about summer stratification, in contrast to the situation in temperate lakes. Under tropical conditions, nearly all the solar energy absorbed is lost by evaporation, whereas only approximately 50% is dissipated in the evaporation of water in temperate latitudes. Tropical reservoirs are genrally sensitive to changing weather conditions. During the warm, rainy season the heavy rains cool the lake surface and, together with the high discharges, prohibit stratification; during the cool, dry season, stratification may be destroyed by the passage of cold fronts. Short-term destratification, in response to fluctuating weather conditions, has been observed in Subang Lake, Malaysia, particularly as a result of heavy rainfalls (Arumugam & Furtado, 1980). Such variable weather conditions may result in the lack of any definite seasonal pattern of thermal behaviour. Thus, in Brazil, Froehlich *et al.* (1978) observed that although the Americana Reservoir was stratified for most of the year, the stratification was unstable, and the water-mass was mixed throughout its depth at irregular intervals of time. Although a clear thermocline was seldom observed, and the temperature gradient was liable to rapid alteration—due to low retention-times (Wetzel, 1975)—the gradient was sufficient to maintain stratification, and to produce an anoxic hypolimnion. Seasonal flood-water can be important in destroying stratification, or in preventing its establishment (Sreenivasan, 1964; Imevbore, 1967); but, in some cases, stratification may nevertheless be initiated by the cooling of bottom waters by inflows during the rainy season, supplemented by descending cold currents induced in shallow embayments during nights by rapid fall in the land temperature.

Near the Equator, stratification can be weak, so that the wind and humidity regimes regulate the seasonal cooling had mixing process, whereas the overall controlling mechanism at higher latitudes is the variable input of solar energy (Talling, 1969). However, water density decreases exponentially with increasing temperature, so that the stability of a tropical lake will be greater, for a particular temperature difference, than is the case within a temperate reservoir (Golterman, 1975). Thus, lakes of comparable stability will exist with a 5°C temperature difference (e.g. between 25 and 30°C) in tropical latitudes, and with a 12°C difference (e.g. 20°C to 8°C) in temperate areas. Low temperature differences between surface and bottom waters associated with thermal stratification, have been reported for reservoirs in southern India (Ganapati, 1973). Because the change in water density for a temperature difference of 1°C between 29°C and 30°C is from two to three times as great as that for the same temperature difference between 14°C and 15°C (Ruttner, 1963), thermal stratification within the Stanley, Bhavani Sugar, Amaravati, and Hope, reservoirs becomes quite stable during the summer season of March to June, despite small vertical temperature differences of between 1.1°C and 6.6°C. Indeed, it is generally agreed that relatively small differences of temperature (e.g. 0.5°C for Lake Volta: Viner, 1969) may define a thermocline in tropical temperature ranges. Thus it has been concluded (Symons et al., 1965) that the density profile—not the temperature profile—is the dominant control of reservoir stratification.

Morphometric Variations

The climatic characteristics of a particular location will impart a major control upon the thermal behaviour of its reservoir waters. However, considerable variability may occur within a region, because of different topographies and different reservoir-basin morphometries. Shallow lakes respond most rapidly to fluctuations in atmospheric conditions. Such reservoirs warm rapidly in the spring and cool quickly in the autumn: and although the vertical variation of temperature is small, the mean temperature throughout each reservoir's depth fluctuates widely with a change of air temperature. Daily surface warming generally affects down to 7 m, but occasionally may influence a layer up to 15 m in thickness (Leentvaar, 1966; Lewis, 1974). Changes of 7°–9°C are not uncommon in the course of one week during the spring and fall in American prairie-region reservoirs (Hergenrader & Hammer, 1973), and variations of up to 5.7°C within a 24-hours period have been recorded within the shallow Lake Gorewada, India (Hussainy & Abdulappa, 1973).

Strong winds may produce thermocline oscillations. Volta Lake, Ghana, has a surface area of 8300 km² and is highly susceptible to strong winds (Viner, 1969). The effects of a sudden squall, with winds over 40 km per hr for short periods, can produce a general mixing of the epilimnial water, reducing surface temperatures by 1.5°C. In contrast, a strong, steady wind can lower the thermocline by 7 m in less than 24 hours, and produce turbulence which gradually increases in depth. The rapid sinking of the discontinuity layer suggests that bulk water movements may be initiated. Wind-induced thermocline oscillations with amplitudes of over 10 m may occur (e.g. Senuya et al., 1969), particularly where storms arise periodically.

The amount of work necessary for mixing depends on the length of the path through which water must be moved (i.e. lake depth), whilst the degree of exposure, lake surface-area, and orientation in relation to the dominant winds, will also be important. Thus, in large-surface-area, relatively shallow, reservoirs, stability is more easily destroyed than in deeper lakes. In USSR the 450 000 ha Rybinsk Reservoir on the upper Volga, has an average depth of only 5–6 m, the lake is well mixed by wind action, and vertical temperature differences are virtually eliminated. This contrasts with the 15-m-deep Alminke Reservoir, Crimea, having a surface area of only 75 ha, where a considerable vertical stratification of temperature exists, with a gradient from 23°C at the surface of 15.6° at the bottom, throughout the summer (Zhadin & Gerd, 1963). Similarly, within the Great Prairie region of North America, reservoirs characteristically have large surface-area: mean-depth ratios, and are exposed to constant winds. Permanent thermal stratification fails to develop and vertical temperature variations rarely exceed 2°C (Hergen-rader & Hammer, 1973). Drift currents generated by wind action can penetrate for 5 m, or more, and major compensating currents can develop at depth (Filatova & Kafejarv, 1973). Thus, whilst climate controls the characteristics of thermal stratification, wind and lake morphometry deter-mine whether or not stratification will develop.

CHEMICAL STRATIFICATION

The chemical composition of stream water that is released from a reservoir can be significantly different from that of the inflows—though, in some cases, releases can have a chemical composition which reflects that of the inflows and any precipitation which is received. In reservoirs of short retention-time, the dissolved oxygen and solute concentrations of the outflow discharges may approximate those of the influents. Under normal weather conditions, the Dubossary Reservoir on the middle Dniester River, USSR, for example, changes its water-mass 18–20 times annually, thus ensuring high oxygen contents of between 48% and 138% of normal saturation, following pract-ically unchanged solute loadings in the first 2 years of operation (Zhadin & Gerd, 1963). However, when once a thermal stratification has been estab-lished, chemical changes within the impounded water will follow.

Chemical changes within reservoirs have been attributed to a variety of factors that are typically associated with their flow-dynamics and biological activity. Major, biologically induced, water-quality changes occur within ther-mally stratified reservoirs. Phytoplankton often proliferate in the warm epil-imnion, releasing oxygen and maintaining concentrations at near-saturation levels for most of the year. Little mixing occurs below the thermocline, and sunlight, necessary for photosynthesis, does not penetrate to the hypolimnion which, deprived of re-aeration, cannot replenish the dissolved oxygen that is used in biochemical processes. Due to the settling of dead phytoplankton, and the presence of heterotrophic Bacteria, oxygen will be consumed in the hypolimnion, often to exhaustion. Thus, the process of organic-matter decay becomes anaerobic; hydrogen sulphide gas is produced; carbon dioxide is released; pH decreases; conductivity, alkalinity, and orthophosphate all increase; and the solution of iron and manganese occurs from the bottom sediments. The quality of the hypolimnial water becomes progressively worse, until the autumn overturn. However, in shallow lakes the existence of complete circulation for most of the year would maintain oxygen levels, and the process of decomposition would be characterized by aerobic breakdown.

Oxygen consumption is due to respiration, decomposition, and the oxidation of dissolved organic compounds, though the pattern of dissolved oxygen depletion within lakes is highly variable—horizontally, vertically, and seasonally (Straškraba, 1973). However, considerable periods of dissolved oxygen exhaustion often occur immediately after dam completion, and this is associated with a trophic upsurge as the reservoir fills, although a reduced, stable trophic state will subsequently establish itself (Rodhe, 1964; Baranov, 1966; Ostrofsky, 1978). Decomposition of dense, submerged vegetation will create a high oxygen demand during the early years, producing a depletion of oxygen at depth, whereupon the mass of stagnant water in the deepest part may become anoxic, reduced substances (such as sulphide, and ferrous and manganese ions) may accumulate and nutrients—particularly phosphorous—will be released biologically and leached from the flooded vegetation and soil (Maystrenko & Denisova, 1972; Wilson et al., 1975). Thus, during reservoir maturation, releases from the Kiev, Kremenchug, and Kakhovka, dams, in the USSR, underwent increases in their annual inorganic phosphorus loads by up to 75%, and inorganic nitrogen by 53% (Denisova, 1978), though oxygen demand and nutrient levels generally decrease over time as the organic matter and nutrients are slowly exhausted.

Some reservoirs may mature after as little as 3 or 4 years, but the trophic upsurge can take 6 years, and in some reservoirs a period of more than 20 years may be required for the development of a stable water-quality pattern. A model for total phosphorus concentration within La Grande–2 Reservoir, Canada (Grimard & Jones, 1982), predicted that peak concentrations of 22 mg m^{-3} (compared with 5 mg m^{-3} prior to dam completion) would be reached 1–6 years after impoundment, and that values would return to normal 4 years later. Measured local releases of phosphorus in the lake bottom waters revealed concentrations of over 100 mg m^{-3}. Reservoirs constructed on saline rocks may achieve particularly high solute levels during their early years of formation. The salinity of outflows from Lake Mead, Nevada, USA, for example, was greater than that of the inflows during its first 15 years of existence, due to the dissolution of large quantities of gypsum and rock salt (Love, 1961). However, solute loads subsequently declined, as the primary sources were depleted.

A unique example of a reservoir maturing is given by Zhadin & Gerd (1963). The Dnieper Reservoir, USSR, built in 1934, was destroyed in 1941. During the intervening years considerable organic deposits, up to 4.0 m deep, accumulated and were subsequently colonized rapidly by land vegetation. In 1947, the rebuilt hydroelectric power-dam was completed, the reservoir drowned the organic deposits, and the decomposition of these deposits had an appreciable effect on the chemical and biological conditions of the new reservoir. During summer, a distinct stratification occurred, with the surface beween 4.5 and 9.5°C warmer than the underlying layers, and for the early years oxygen conditions were unfavourable. Above the decaying, flooded vegetation, oxygen was totally absent, and the amount of free carbon dioxide approached 20 mg per litre. Although an oxygen deficiency, and a high concentration of carbon dioxide, became a permanent feature of the bottom, deeper-water layer in summer, oxygen conditions generally improved. This change was brought about by the complete mineralization of readily hydrolysed substances within a few years of dam closure, and the development of phytoplankton—which in turn resulted in an increase in oxygen, often to above saturation-point, in the surface layers.

Dissolved Gases

Physio-chemical changes during a reservoir's natal years are common to all geographical regions, and examples have been reported from Africa (Ewer, 1966), the USSR (Frey, 1967), and the USA (Neel, 1967). However, the rate of decay of organic matter tends to be greater in tropical reservoirs. The high bottom temperatures (22–30°C) increase the rate of biochemical reactions by up to nine times (Ruttner, 1963); anoxic conditions will develop, despite the low temperature-gradients; and deoxygenation may occur more rapidly, and may persist longer during the year, than within temperate environments (Mitchell & Marshall, 1974). Thus, Ganapati (1973) reported the establishment of severe anoxia within the hypolimnion of four major reservoirs in India, shortly after dam closure. On reaching maturity—that is, when once insufficient organic matter remains, or is supplied by the inflows and autochthonous primary production—an oxygen demand will not be created in winter, and the water-body will become completely oxygenated. Severe summer anoxia, however, may be maintained in tropical areas by the high rates of autochthonous organic decomposition (Ruttner, 1963).

A dissolved-oxygen profile may demonstrate stratification even if the reservoir is not thermally stratified. In the epilimnial layer water-mixing, by wind and wave action, combined with photosynthesis by Algae, maintains dissolved oxygen levels at near-saturation. The rate of photosynthesis in the upper layers of a lake will be limited by the supply of nutrients—particularly nitrogen and phosphate—and by light-penetration, influenced partly by the Algae themselves but also by suspended solids. Within Cherokee Reservoir, USA (Fig. 19), high dissolved-oxygen levels of 10 mg per litre are related to dense concentrations of phytoplankton, of between 30 and 60 thousand per ml, near the surface; below 10 m, however, conditions are unsuitable for phytoplankton, and dissolved oxygen levels are reduced to less than 1 mg per litre, whilst between 6 m and 10 m the rapid reduction in dissolved oxygen concentrations reflect the high number of zooplankton which graze on the phytoplankton (Churchill & Nicholas, 1967). A similar interrela-

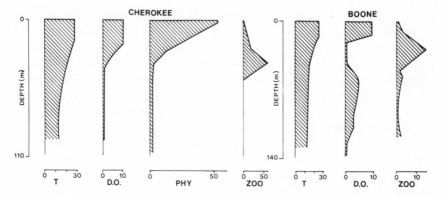

Fig. 19. Vertical water-quality profiles for summer in Cherokee and Boone Reservoirs, Holston River, USA: temperature (T), °C; dissolved oxygen (D.O.), mg per litre; phytoplankton (PHY), thousands per millilitre; and zooplankton (ZOO) number per litre. (From Churchill & Nicholas, 1967. Reproduced by permission of the American Society of Civil Engineers.)

tionship was observed within Boone Reservoir, Tennessee, although here the quantity of oxygen used in respiration by the zooplankton exceeds that produced by phytoplanktonic photosynthesis.

The daily variation in the photosynthetic activity of Algae, itself related to temperature and light intensity, can produce a diurnal pattern of oxygen production. Within a shallow reservior, Hussainy & Abdulappa (1973) identified a weak diurnal stratification, superimposed upon general year-round homothermy; it was caused by solar heating but was destroyed at night by nocturnal cooling or complete mixing. Daily dissolved-oxygen increases at the surface resulted from photosynthetic production by Algae. Viner (1969) reported a stronger dissolved-oxygen stratification, established from repeated diel cycles, within the Volta Lake, Ghana, where, during the morning, the surface water, at 29°C, was slightly warmer than the deeper layers and the oxygen content (78% saturation) above the hypolimnion was almost uniform. By early afternoon an upper layer of warm (31°C) and well-oxygenated (100% saturation) water had been established, spreading down to about 5 m. Below this, water temperature and dissolved oxygen had both risen slightly, suggesting the presence of limited mixing. Soon after midnight, both surface temperatures and oxygen contents had fallen to 28°C and 80% saturation, respectively, but the temperature and oxygen content of the deeper layers of the epilimnion had further increased, suggesting that extensive isothermal mixing within the epilimnion had taken place earlier during the night.

Biogeochemistry

Consequent upon the formation of a hypolimnion, concentrations of dissolved oxygen will decrease. As the dissolved oxygen levels approach zero, major changes in the chemical characteristics of the hypolimnial waters will occur, associated with the establishment of reducing conditions. Water from an anoxic hypolimnion will show a low redox potential, which indicates a strong reducing capacity. However, the redox potential provides only an indication of the potential magnitude of the reduction, and therefore a control upon the types of compounds that can be reduced. For a specific reaction, the quantity of electrons available governs the amount of matter which can be reduced. Nevertheless, the exhaustion of oxygen within hypolimnial water allows the existence of compounds which would otherwise be oxidized. Thus, a decrease in the redox potential is commonly associated with the release of manganese and iron from the bottom sediments, and most hypolimnia are also characterized by increasing concentrations of phosphate, ammonia, and silicate, as well as of Ca^{2+}, HCO_3^-, and H_2S. Increased iron and manganese concentrations have also been found just below, or even in, the metalimnion, where reducing conditions may prevail as a result of sinking organic debris—often dead algal cells (Senuya, 1972).

Concentrations of carbonates usually exceed those of the other anions present, and are commonly dominated by calcium bicarbonate, although sodium and magnesium bicarbonate may reach high concentrations, particularly in tropical waters. Bicarbonate ions play an important role within reservoirs, providing the primary buffer for regulating the hydrogen-ion concentration in water, and carbon dioxide for photosynthesis (Golterman, 1975). Thus, bicarbonate concentrations in reservoirs may demonstrate a highly dynamic longitudinal and vertical stratification.

Hannan et al. (1979) have described the variable nature of bicarbonate levels within Canyon Lake, Texas, USA, which receives calcium-rich water. During summer stratification, bicarbonate concentrations decreased in the epilimnion, due to photosynthetic uptake by phytoplankton. However, the removal of CO_2 from a solution of calcium bicarbonate in equilibrium induces the precipitation of calcium carbonate. The precipitate settles into the hypolimnion where the lower pH, and increased levels of CO_2 can induce the solution of the calcium carbonate. The high levels of carbon dioxide, from respiration and decomposition, enable large amounts of calcium bicarbonate and carbonic acid to be held in solution, and the dissolution of calcium carbonate will also occur from the bottom sediments. However, in the upstream portion of the reservoir, water chemistry will be strongly influenced by the inflows, as well as by photosynthesis and respiration. Thus, within Canyon Lake, alkalinity increased markedly during periods of high inflows, and decreased during June and August, accompanying low inflows and a high uptake of dissolved bicarbonates by the large algal population (Hannan et al., 1979). Because water containing dissolved material has a higher density than pure water of the same temperature, a 200 mg per litre increase in dissolved salt content with depth produces a considerably greater increase in density than that caused by a temperature difference of 4–6°C (Golterman, 1975). Therefore, if dissolved solids in the hypolimnion reach extreme concentrations during stratification, the stability of the water-body increases, and autumnal overturn may be delayed or even prevented.

Lakes act as nutrient sinks, and considerable attention has been directed to the eutrophication of natural lakes resulting from the accumulation—particularly of nitrogen and phosphorus, although potassium, magnesium, trace elements (such as iron, manganese, and copper), and organic growth-factors, also play a part. Vollenweider (1968) has suggested critical nutrient loadings to avoid eutrophication. Reservoirs can be efficient nutrient-traps: for example, mean inflow concentrations of orthophosphate in solution were reduced by 50%, and the total suspended sediment P decreased fourfold, in the outflows from Callahan Reservoir, Missouri, USA (Schreiber & Rausch, 1979).

In rare cases, eutrophication can lead to hypolimnial deoxygenation throughout the year (Marshall & Falconer, 1973). However, the relatively large amounts of outflow and short retention-times which characterize man-made lakes, considerably reduce the problem of eutrophication. Moreover, for reservoir discharges, particular nutrients will not always be in solution, and these will be unavailable to the river downstream of the dam. The turnover time for ionic phosphorus, for example, in the epilimnion of a lake during summer, is of the order of 1 minute (Rigler, 1964). Thus, there may be an extremely low concentration in the water at any one time, because the entire supply of a nutrient may be momentarily in storage in sediments, or in the bodies of organisms whether living or dead.

Temporal Patterns

The chemical composition of reservoir discharges will be related to the elevation of the outflow structure, and to the seasonal pattern of biogeochemical reactions, which may exhibit a spatial pattern of development within the lake. The nutrient concentrations within different layers of the reservoir will demonstrate marked seasonality, with oxidation reactions becoming dominant during periods of overturn and mixing, whilst reducing reactions occur within the hypolimnion during stratification. Thus, in the epilimnion, during the summer growing-season, high primary production rates by phytoplankton can produce oxygen supersaturation (Marshall & Falconer, 1973), and the water becomes depleted in bicarbonate, orthophosphate, and nitrate-nitrogen.

Consequent upon the autumnal overturn, PO_4^{3-} and NO_3^- concentrations increase (Soltero et al., 1974a). Such seasonal patterns are found, particularly, in reservoirs with long retention-times and large standing-crops of autotrophic organisms (Cushing, 1963; Hannan et al., 1973). The overall nutrient concentrations may only reach 'pollution' levels during the first few years of a reservoir's life, as the pattern of oxygen demand stabilizes. Thus, during the formative years of the Ucha Reservoir, USSR, the hypolimnion was characterized by high levels of iron, manganese, and hydrogen sulphide

(Zhadin & Gerd, 1963). Iron, for example, reached concentrations of 5 mg per litre, but after 3 years the mineralization of the flooded organic matter was completed, concentrations of dissolved oxygen improved, and iron levels were soon reduced to tenths of a milligram per litre.

A pattern of hypolimnial oxygen depletion, having a profound influence on the chemical composition of reservoir discharges (Hannan, 1979), has been reported within several reservoirs in the USA (e.g. Wiebe, 1938; Ebel & Koski, 1968; Gnilka, 1975), Africa (Bowmaker, 1976), and Europe (Fiala, 1966). The pattern is a function of hypolimnial volume, reservoir morphometry, oxygen demand, and drawdown. Furthermore, it demonstrates the importance, for the outflow chemistry, of lake hydrodynamics which influence the time-of-arrival of anoxic water at deep-release outflow points. One example is provided by Soltero et al. (1974b) for Bighorn Lake, Montana, USA. When once a hypolimnion has become established, anoxic conditions develop in the upstream end of the reservoir before the dam-end. Upstream parts of reservoir basins have relatively smaller volumes, but often receive greater amounts of oxidizable sediment, than down-reservoir areas, so that the upstream location first develops an anoxic hypolimnion before downstream parts (Hannan et al., 1979). Hypolimnial anoxia develops in a down-reservoir progression as summer continues, but overturn may occur before it reaches the dam or the outlet valves. Such a pattern is not smoothly progressive in its development, but will be influenced both by mixing, caused by inflow, and drawdown. By the time anoxic conditions reach the dam, upstream sections may have become oxygenated again, in response to inflows and deepening of the epilimnion produced by the release of hypolimnial water. Moreover, the data from Bighorn Lake suggest that anoxic water at the up-reservoir location could move suddenly down-reservoir with an internal density current. The occurrence of dissolved oxygen depletion and low pH should induce other chemical changes, such as an increase in dissolved Fe^{2+} and Mn^{2+}, elevated H_2S and PO_4^{3-}, and a change in the nitrogen 'species', in the reservoir discharges.

DENSITY CURRENTS

Water flowing into a reservoir is frequently different from that already present in terms of temperature, or in content of dissolved or suspended solids, or in some combination of these, and consequently in density. The inflow of water of different density to that stored in the reservoir will result in the formation of density currents within a lake. As a result, unusual and complex patterns of quality stratification may occur if the chemical composition of the inflowing water is different from that of the water already present (Fig. 20, A). Because of the density difference, the inflowing water does not immediately mix with the water of the reservoir, but moves through it, both downstream and laterally, as an overflow, underflow, or interflow. Simply, any inflow seeks its density level, moves along this level, and generates compensating upstream currents, whereupon a convergence zone will form

where the inflow and compensating currents meet. Moreover, if the level of
the density current coincides with that of the withdrawal zone, the inflow will
move directly through the reservoir and be released to the river downstream.

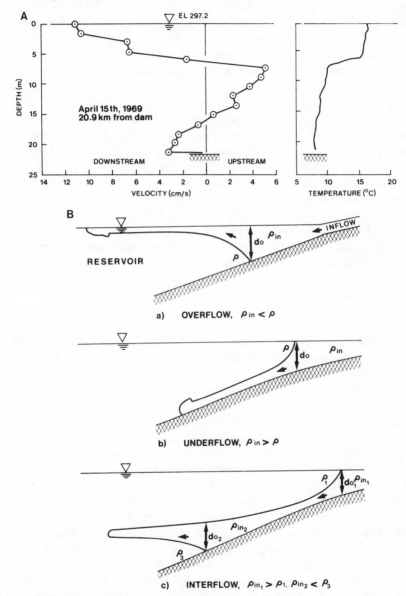

Fig. 20. Characteristics of density currents: overflow of warm inflow to
Douglas Reservoir, Tennessee Valley, USA (A); and conditions for
overflow, underflow, and interflow (B). (do = critical depth; ρ_x = water
density; ρ_{inx} = inflow water density.) (From Wunderlich & Elder, 1973.
Reproduced by permission of the American Geophysical Union.)

Wunderlich & Elder (1973) have described the flow characteristics of density currents which, independent of inflow type, show the deepening of the inflow until a critical depth is reached ('do' in Fig. 20, B). From this location underflow, or overflow, occurs downstream if the density of the inflow is, respectively, greater, or less, than that of the reservoir water. In deep reservoirs an underflow will probably encounter still colder water, so that underflow ceases and interflow results. The level at which the interflow occurs is dependent upon the density of the underflow, but at this level the flow depth increases (do_2), creating an energy head which drives the water along the density plane and into its final storage position.

In spring the inflow water is often warmer than that of the reservoir, and *vice versa* in the autumn. Consequently, the annual pattern of density currents may have a marked seasonality. During winter a reservoir may experience complete vertical mixing, and the relatively cold inflow passes through the reservoir as an underflow, inducing water circulation characterized by a reversed surface flow. In spring, inflows produced by snow-melt may be low in dissolved solids and will have a temperature similar to that of the reservoir; the inflow moves through the reservoir as an overflow and initiates a reverse circulation at depth. Reduced inflows during summer are often associated with increased salinities, and a weak interflow can be established. As temperatures decrease during the autumn, the inflow will sink, forming an underflow, until it reaches a level of equal density. At this point it will spread out as an interflow and induce two, often relatively strong, opposing currents. The seasonal variation may reflect changes of inflow density during wet and dry periods. Flood inflows are usually cold in relation to the reservoir water, but owe their greater density, primarily, to suspended solids, and produce underflow turbidity currents. Although density differences are normally only a few per cent, the formation of turbidity currents may significantly affect the flow-circulation pattern within reservoirs. In Lake Mead, Nevada, for example, turbidity currents have been observed to extend for the whole of the 160 km length of the Lake (Gould, 1960).

The complexity of the flow and stratification patterns in reservoirs of short retention-time have been described for the Slapy Reservoir, Czechoslovakia (Hrbáček, 1969). Considerable water-currents were observed below the thermocline during summer stratification, and for most of the year inflowing water plunged beneath the surface as an underflow, mixing to some degree with hypolimnial water, but not to any great extent with surface water. This situation is the reverse of the pattern observed in natural lakes of comparable depth, because hypolimnial anoxia may be prevented by such mixing. The tendency for the inflows to pass under the epilimnial water has been accentuated by the construction of another dam immediately upstream, which releases relatively cool water into Slapy Reservoir. Similar currents have been generated within the reservoir chain of the Tennessee Valley Authority (TVA), USA (Parker & Krenkel, 1969). All the reservoirs of the TVA system release hypolimnial water and are sufficiently closely spaced that the discharge from one reservoir is into the backwaters of the next. The underflows so produced were reported to have velocities of between 0.046 m per s and 0.107 m per s. Thus cold, solute-rich hypolimnial water could be rapidly transmitted through a series of deep-release reservoirs to the river downstream.

A large, deep reservoir will often dominate the characteristics of other reservoirs, and outflows, downstream. Dendy & Stroud (1949) cite several examples where a chain of reservoirs are dominated by one, large, deep-release dam. The Little Tennessee River, for example, is regulated by three line-of-the-river impoundments. Prior to the construction of the relatively young Fontana Reservoir upstream, water temperatures in summer below the two combined, low-capacity reservoirs were usually above 25°C, but after completion of Fontana Dam the water temperature fell to about 14°C. The data indicated that water released from Fontana Dam passes through the lower two reservoirs in less than 14 days, so that changes in the character of water leaving Fontana are felt quickly in the reservoirs downstream, and the influence of such changes are transmitted to the river below the dams.

WATER-QUALITY STABILIZATION

Despite the range of parameters which control water quality, the thermal and chemical regimes of rivers will be moderated as a result of upstream impoundment: annual variations will be reduced, short-term extremes will be virtually eliminated, and seasonal maxima, and minima, will be delayed (Table XII). Thus, the considerable fluctuations in conductivity of the influent streams to Bighorn Lake, Montana–Wyoming, USA, contrast with the stable conductivity of the effluent water, and the normal trend of conductivity was delayed in such a manner as to minimize the salinity hazards (Soltero et al., 1973). In general, the natural extremes of concentration are eliminated, but the effluent waters are lower in total dissolved solids below epilimnial release dams and higher in hypolimnial releases. Thermal modifications will also be clearly apparent, although mean annual temperatures have often not changed significantly. Indeed, mean water-quality data can mask important seasonal patterns due to stratification. Thus, below Cow Green Reservoir, River Tees, UK, mean annual potassium loadings have been reduced to 60% of normal, but maxima are only 30% of those of reservoir inflow levels (Crisp, 1977).

Long-retention reservoirs are often characterized by phytoplankton assimilation of nitrate-nitrogen, and by the anaerobic metabolism of nitrate by Bacteria in the hypolimnion. Nitrate loadings are reduced, whilst loads of ferrous iron can reach unusually high levels during stratification; the reduction of the insoluble ferric form of Fe to the soluble ferrous form is caused by high concentrations of CO_2, low pH, and the presence of organics within the anoxic hypolimnion (Hannan & Broz, 1976). In some cases, the effects of a reservoir may be limited to only 1 kilometre below the dam (Mackie et al., 1983), but often the effects have been transmitted for tens of kilometres downstream.

Table XII   Some observations of water quality changes below dams.

River, reservoir, location	Water quality change	Source
Missouri River, Fort Randall Dam, USA	Normal high-salinity run of river discharge was delayed.	Neel (1963)
Brazos River, Morris Shephard Dam, Texas, USA	Solute loads reduced to between 10% and 80% of influent-flow levels (despite hypolimnial releases).	Stanford & Ward (1979)
Bighorn River, Bighorn Lake, Montana–Wyoming, USA	Total nitrogen and total phosphorus reduced to 75% and 14%, respectively.	Soltero et al. (1973)
Niger River, Kainji Reservoir, Nigeria	Natural river low in chemical nutrients; bicarbonate, calcium, and sodium, increased in reservoir outflows.	Imevbore (1970)
River Tees, Cow Green Reservoir, UK	Mean calcium concentrations reduced by 50%, but annual maxima reduced by 75% from 37.2 mg per litre to 8.9 mg per litre; and minima increased by nearly 100%.	Crisp (1977)
Guadalupe River, Canyon Reservoir, USA	Mean specific conductance reduced by 34%, but concentrations of ferrous iron may be increased by up to 1800% during periods of stratification.	Hannan & Broz (1976)
South Saskatchewan River, Gardiner Dam, Canada	Increased winter temperatures prevented ice formation for 20 km downstream.	Lehmkuhl (1972)
Columbia River, Grand Coulee Dam, USA	Delayed season maximum temperatures.	Jaske & Goebel (1967)
Murray River, Hume Dam, Australia	Oxygen levels depressed for 100 km, and seasonal temperature changes delayed by 1 month for 200 km, below dam.	Walker et al. (1979)
Gordon River, Gordon Dam, Australia	Winter temperatures depressed by ca 6°C for more than 45 km downstream.	King & Tyler (1982)
River Lune, Grassholme Reservoir, UK	Daily temperature fluctuations in summer reduced from 12°C to 0.5°C.	Lavis & Smith (1971)
Japan	Annual range of mean monthly temperatures reduced below dam from 21.4°C to 11.8°C.	Nishizawa & Yamabe (1970)

The regulated water-quality pattern is well demonstrated by temperature variations: the seasonal variation will be reduced, and the annual maxima and minima will have a time-lag, sometimes of several months, in relation to the unregulated flows (Table XIII). The magnitude of the impact tends to increase as the time-period under consideration is compressed, particularly if the variability within the natural stream is high. At times of low-flows, relatively cold compensation water can form the dominant source of river-flow, whilst the small volume of water in the natural stream will permit the maximum response to long sunny days, and produce some of the highest temperatures of the year. Thus, during the filling of Flaming Gorge Reservoir, Wyoming-Utah, USA, between 1962 and 1966, the low discharges released from the reservoir could be warmed quickly by ambient conditions, so that summer temperatures increased from 5°C to 13°C shortly below the dam (Holden, 1979). However, after 1966, releases doubled and ambient warming was reduced to only a few degrees, with the result that the effects of impoundment were transmitted over longer distances downstream.

Lavis & Smith (1971) observed that the mean daily temperature-range below the deep-release Grassholme Reservoir, River Lune, England, UK, was only about 10% of that of the natural stream. Within the unimpounded River, daily temperature variations occurred with a range of up to 4°C in winter and 12°C in summer; when detectable, daily fluctuations below the Reservoir were of the order of only 0.5°C. The daily temperature-sequences for the reservoir outlet did not indicate a direct heating of the water, but merely reflected the general trend of water temperature for that month. On the rare occasion when a diurnal fluctuation did occur in summer, the maximum range observed had a magnitude of 1.3°C, and this variation probably reflected an internal circulation of water within the Reservoir.

Reservoirs will not only stabilize the daily temperature-fluctuation of the water immediately below the dam, but, with them, diurnal downstream temperature-gradients may also be established. For example, immediately below Cheesman Dam, South Platte River, USA, the diurnal changes of water temperature, dominated by the stable temperature of the hypolimnial releases, are negligible; 8.5 km downstream, the diurnal range is zero in December but increases to 6°C during July and August (Ward, 1974). Further downstream, atmospheric conditions and tributary water combine to return the River to its pre-impoundment state.

Where reservoir-discharge quality differs markedly from that of the natural discharges, thermal and chemical gradients may be established along the river, and the downstream extent of any such gradient will reflect the relative discharges from the reservoir and tributary sources. For thermal gradients to be established, reservoir releases must be of sufficient magnitude to overcome rapid ambient heating within the channel. Marked seasonal temperature-gradients, produced by hypolimnial releases (Table XIV), have been reported below Cheesman Dam, South Platte River, Colorado, USA (Ward, 1974). The reservoir exerts a regulating effect upon temperatures immediately below the dam, so that the primary natural peaks, expected during August and April, are not experienced. Hypolimnial releases during reservoir stratific-ation cause a thermal gradient to form between March and September, and temperatures increase downstream most rapidly in April and August. In October, uniform temperatures reflect the release of mixed water consequent upon the autumn overturn.

Table XIII Some examples of seasonal temperature-changes below dams.

	Winter		Summer	
	Change (±°C)	Lag (months)	Change (±°C)	Lag (months)
Hume Dam, Murray River, Australia: Walker et al. (1979)	+3	1	−3	1
Eildon Lake, Goulburn River, Australia: W. D. Williams (1967)	—	—	−15	—
Bighorn Lake, Bighorn River, USA: Soltero et al. (1973)	+7	4	−6	2
Cheesman Lake, South Platte River, USA: Ward (1974)	+1	1	−8	4
Gardiner Dam, S. Saskatchewan River, Canada: Lehmkuhl (1972)	+4	—	−10	1
Selset and Glassholme Res., River Lune, UK: Lavis & Smith (1971)	+4	2	−4	1
Cow Green Res., River Tees, UK: Crisp (1977)	+1	1	−2	1
Kassansai Lake, Kassansai River, USSR: Zhadin & Gerd (1963)	—	—	+8*	—

* Lake inflow is glacial meltwater.

Table XIV Water temperatures (°C) below Cheesman Lake, South Platte River, Colorado, USA. (Based on Ward, 1974. Reproduced by permission of E. Schweitzer-bart'sche Verlagsbuchhandlung.)

Month	Distance below Dam (km)			
	0.25	2.4	5	8.5
January	3	3	3	3
February	3	3	3	4
March	3	5	7	8
April	4	8	10	12
May	5	6	7	8
June	6	7	8	9
July	6	10	11	14
August	7	9	10	13
September	9	11	12	13
October	13	13	13	13
November	8	8	8	8
December	4	4	4	3

Epilimnial and Hypolimnial Releases

The location of the outflow facility will determine the quality of releases from stratified reservoirs, because withdrawal will occur from a relatively narrow layer. The effects of water outlets upon the reservoir storage–outflow relationship was first discussed by Bell (1942). Subsequently, the outflow hydraulics under stratified flow-conditions have been considered from a theoretical and experimental basis (Koh, 1964), and Churchill & Nicholas (1967) showed that outflows from Cherokee Reservoir, Holston River, Tennessee, USA, were withdrawn from a fairly narrow layer in the pool, located at the level of the intakes. Within a homogeneous fluid, point withdrawal from a selected depth will result in discharge from the full depth of the storage volume (Fig. 21). However, within density-stratified reservoirs, vertical movements are suppressed whilst horizontal movements are enhanced (Wunderlich, 1971). Water will be drawn from all layers in the first moments of a release (Elder & Wunderlich, 1968), but when once a steady rate of outflow has been achieved, a withdrawal layer of restricted vertical dimensions develops near the intake elevation, and 'continuity' is satisfied by the establishment of secondary currents. A relatively narrow layer of approximately constant density will be withdrawn, so that the water quality of the outflow will vary considerably if releases are abstracted from different depths.

During summer, a stratified surface-release reservoir will discharge well-oxygenated, warm, and nutrient-depleted, water whilst low-level outlets will produce relatively cold, oxygen-depleted, and nutrient-rich, releases which may contain high concentrations of iron, manganese, and/or hydrogen sulphide. Indeed, the release of iron and manganse may occur in sufficient concentrations to produce a precipitate on the stream-bed. Tyler & Buckney

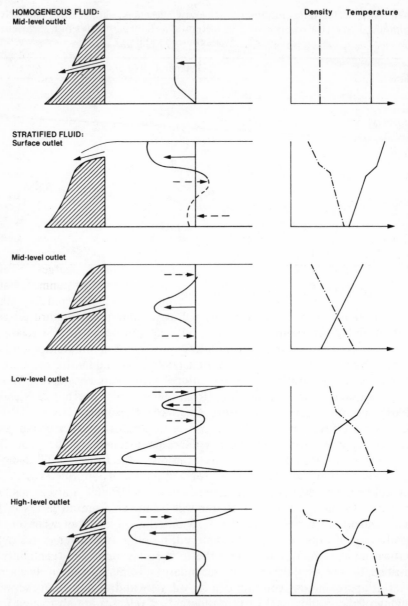

Fig. 21. Effects of selective withdrawal from various depths upon flow patterns, density, and temperature profiles, in homogenous and stratified lakes.

(1974) reported iron and manganese concentrations of 3.61 mg per litre and 1.10 mg per litre, respectively, below Lake Rowallan, Mersey River, Tasmania, Australia. The formation of organic deposits rich in iron and manganese on the bed of regulated rivers appears to be common below soft-water reservoirs. A black friable deposit was observed within the bed-materials of the lower River Elan, Wales, UK, below the Craig Goch reservoirs (Truesdale & Taylor, 1978). The deposit, between 1 and 10 mm thick, characteristically contained between 15% and 34% dry-weight of manganese and high levels of organic matter. A second deposit was observed during the summer low-flow period, and this contained low levels of manganese but large proportions of iron. Whilst the manganiferous deposits appear to be associated with microbiological activity (Wolfe, 1960; Tyler & Marshall, 1967), the iron accumulations may be formed by the flocculation of humic colloids (Ruttner, 1963).

Two detailed comparisons of the physico-chemical characteristics of surface- and deep-release reservoirs have been reported from North America. Martin & Arneson (1978) compared the outflows from a deep-discharge reservoir and a surface-discharge lake on the Madison River, Montana. Both lakes showed an epilimnial depletion, and hypolimnial accumulation, of NO_3^-, and salinity increased in the hypolimnion as summer progressed and thermal stratification intensified. Thus, the deep-outlet discharges contained high nutrient-levels and high salinities during the period of summer stratification, whilst the surface-outlet released nutrient-depleted and low-salinity water. Mixing of the reservoir storage during the autumn overturn increased the concentrations of nutrients and salts in the surface-water discharges, but reduced these in the deep-releases, so that opposing seasonal patterns, as well as different actual values, were produced by the different dams.

The second case exemplifies the particular effects of release-depth upon the outflows from a reservoir receiving flood inflows (Stroud & Martin, 1973). Heavy rain during June 1969 increased the level of the hypolimnial-release Barren Lake, Kentucky, by 5.8 m. The flood inflows transported large concentrations of suspended organic particles, and protracted storage led to a high oxygen demand, reduced dissolved-oxygen values, and high manganese and iron concentrations, in the hypolimnion, and to a marked deterioration in the quality of the outflows. Under an epilimnial discharge regime, the flood inflows would have caused the evacuation of high-quality surface water. However, oxygen reductions within the hypolimnion would probably have been much more severe than they were under the hypolimnion-release regime, and extended over a longer period of time, so that a pulse of low-quality water might occur consequent upon autumn overturn.

The fate of density currents, within the reservoir, and the vertical migration of lake strata during the different seasons, are also important for the quality of releases made from different depths. Inflowing river-water, which will have a variable quality, will either mix with the reservoir water to maintain a homogeneous body, or, in the presence of density gradients, will maintain its identity as a separate layer within the reservoir. Thus, the quality of reservoir releases will be determined by the location of the outlet at the dam and the level of any density current within the reservoir. It is possible that, as a result of such currents, waters may pass through the storage volume in relatively short periods of time without ever completely mixing with the epilimnial or hypolimnial waters, so that the downstream thermal and chemical characteristics may be similar to those of the inflows.

For deep-release dams, cold, dense, underflows may pass through the reservoir rapidly, but interflows, or overflows, may become part of the deepening epilimnion induced by the drawoff by hypolimnial water throughout the summer period of thermal stratification. If the rate of withdrawal is high, then the warm, well-oxygenated, surface water within the epilimnion could reach the elevation of lower-level outlets and be withdrawn. Thus Stroud & Martin (1973) found that, because of the release of oxygen-depleted water and induced epilimnial drawdown, the hypolimnial-release lake had favourable dissolved-oxygen concentrations during summer, with 36.5% of the maximum reservoir depth occupied by at least 4 mg per litre, in contrast to that of only 21.5% for the epilimnial-release reservoir. Moreover, Stroud & Martin reported a third case, Lake Russell, Kentucky, USA, which was operated with an epilimnial discharge initially and was then changed to a hypolimnial-release dam. Stratification was more pronounced during the first period than the second, owing to the influence of the epilimnial discharge during the early part of the year, and the hypolimnial summer temperatures were higher within the second phase of operation.

Intermediate-level outlets will provide releases of a quality that is dependent upon the occurrence of density flows, and upon the relative depth, and stability, of the thermocline. Fluctuations in the elevation of the thermocline over considerable depths, induced by wind, may have significance for short-term variations in outflow-water quality, but releases associated with density currents my have a strong seasonal variablity (Anderson & Pritchard, 1951). Nevertheless, the combination of inflow characteristics, outlet type, and reservoir hydrobiology, will interact to produce a variety of water-quality conditions downstream.

Low-quality Pulses

Early work (Purcell, 1939) indicated that artificial lakes and reservoirs normally reach a condition of stability, ensuring good-quality water downstream, a few years after their creation—a figure of between 5 and 10 years has often been quoted. After the initial effects produced by the submergence of vegetation and soils have disappeared, the water downstream from a non-polluted reservoir improves to the point where only the yearly increment of leaves, dead Algae, and other natural debris, causes a loss of dissolved oxygen in the hypolimnion. However, several situations may arise to produce the short-term release of low-quality water. These may be classified as repeated-, seasonal-, and infrequent-pulses. Repeated-pulses are produced by deliberate flow-fluctuation, most commonly associated with hydroelectric power operations. Reservoir releases can reduce the diurnal temperature-range: for example below Lake Vyrnwy, Wales, UK, from an average of

1.5°C to 0.4°C (Severn–Trent Water Authority, personal communication). Also, sudden changes of water temperature may occur in the receiving river as the release-wave moves downstream, while summer tailwater temperatures below deep-release dams may fluctuate 6–8°C as power releases peak and wane, which may occur two or three times each day (Pfitzer, 1967). Low-latitude reservoirs, in particular, can exert a pronounced cooling effect on streams *via* hypolimnial releases (Pearson & Franklin, 1968). However, low oxygen levels can also be produced by dam releases. Oxygen depletion within the regulated Catawbe River, South Carolina, USA, was caused by the first reservoir release after a period of stable low-flows (Ingols, 1959). Reduced dissolved-oxygen values occur because of the lower re-aeration of the greater depth of water. Sludges and slimes, which collect on stones or in quiet areas during the low-flow period, can be lifted from the bottom and suspended in the flow, causing a further drop in the dissolved-oxygen value.

Seasonal-pulses are climatically induced and may be caused by a major flood inflow which stirs the bottom sediments, releasing large quantities of orthophosphate into the water (Hannan & Broz, 1976), or by sudden destratification, which can produce extreme deoxygenation and the release of toxic water (Arumugam & Furtado, 1980). Particular water-quality changes can occur on overturn if an akinetic space develops within a hypolimnion (Fiala, 1966), and these have had profound effects upon the Zambezi River, Mozambique (Hall et al., 1977). Within the middle Zambezi, the major portion of the river-flow originates in Lake Kariba and local rains have a minimal influence upon water quality. Average nutrient concentrations have changed relatively little, but significant water-quality changes below the reservoir occur annually during July and August, associated with the autumnal overturn. Although these changes become less and less noticeable downstream, they have been observed only 200 km from the Indian Ocean—some 800 km below the dam. At average operating water-level, water for the turbines is drawn from around at least 20 m depth and during most of the year such a depth corresponds to the upper layers of the cooler, deoxygenated, hypolimnion. Because of the effects of selective withdrawal, water movement below the outlet level may be restricted, and an akinetic space may develop within the hypolimnion. Nutrients carried into Kariba by the main tributaries are temporarily trapped within such spaces, only becoming available during overturn, and the outflow from Lake Kariba experiences a sudden and extreme peak of phosphate, sulphate, and silica, during this period. Similar pulses of nutrient-rich water have been reported below the Ucha Reservoir, USSR, (Zhadin & Gerd, 1963), where releases during the autumn overturn were characterized by high concentrations of iron, manganese, and hydrogen sulphide.

Infrequent-pulses are more difficult than seasonal ones to predict, but can involve important changes of dissolved-oxygen concentration. Both supersaturated and anoxic water may be released. A periodic, explosive deterioration in water quality has been observed in an old reservoir—behind a 60-years-old dam—on the Catawba River, USA. (Ingols, 1959). During the summer of 1954 (a very dry season) the downstream water-quality was very good, but in 1955 the river downstream was depleted of dissolved oxygen and had a high manganese content. Low reservoir-levels during the summer of 1954 had exposed extensive mud-flats that were rapidly colonized by vegetation. This provided an excessive organic load when the mud-flats were resubmerged; an extreme oxygen demand was generated, and an anoxic hypolimnion rapidly developed. By 1957 the water quality had returned to normal. Therefore, a period of low-quality releases may be expected following drought years, as a result of the extreme lowering of reservoir water-level.

Bursts of phytoplankton activity have been identified as another cause of rapid variations between extremes of oxygen concentration. Within the shallow Tsimlyansk Reservoir, USSR, oxygen concentrations range between extremes of 206% saturation at the surface and zero at the bottom (Zhadin & Gerd, 1963). However, the extremes are often short-lived, being produced by algal blooms which are destroyed by wind-induced turbulence within the homothermal reservoir. At one location, a summer peak of 15.15 mg per litre oxygen concentration at the surface occurred during a water-bloom, but after only 4 days the phytoplankton had died and the concentration was reduced to 0.52 mg per litre. Nevertheless, discharge from the epilimnion could release supersaturated water to the river downstream. The supersaturation of air gases in reservoir releases could also be caused by the passage of water through turbines (Dominy, 1973) or by over-dam spillage (Beiningen & Ebel, 1970; Holden, 1979).

From tests conducted in Sweden, Lindroth (1957) found that no supersaturation occurred in water that had passed through turbines. MacDonald & Hyatt (1973), however, demonstrated that supersaturation of air gases could be caused by the passage of water through the turbine system, because at low generating levels air was vented into the system to reduce negative pressures. Thus, dissolved oxygen concentrations of generating water passing through the Mactaquac Dam, Saint John River, Canada, at low generating levels, increased from 8 mg per litre (66% saturation) to over 11 mg per litre (93–125% saturation). As the generating level increased, the degree of aeration declined, and the concentrations in the tailrace approached those in the reservoir.

Dilution Problems

One of the assumed benefits of impounded water is the increase in base-flow which, if aerated, provides dilution for various kinds of wastes. Within an irregular-bottomed stream, oxygen recovery is fast under low-flow conditions, but at high flows, or within smooth boundary channels, the opportunity for surface re-aeration is low, and oxygen depletion will persist for considerable distances downstream. Even if an anoxic hypolimnion develops in a reservoir, air drafts in the release ports, or turbulence in the tailwaters, can function to raise dissolved-oxygen levels in the receiving stream. Thus, despite the observed low dissolved oxygen, and high hydrogen sulphide, levels within the hypolimnion of Possum Kingdom Reservoir, Texas, USA, air drafts in the penstocks cause sufficient gas exchange to alleviate toxicity problems downstream (Stanford & Ward, 1979). Hannan & Broz (1976) observed a 6% increase of the annual mean percentage saturation due to turbulence within the tailrace of Canyon Reservoir, Guadalupe River, Texas, USA, despite the low-level outlet releasing hypolimnial water.

Within the Ohio River, USA, re-aeration is effected by numerous small navigation dams and weirs (Koryak, 1976). Montgomery Dam, on the upper Ohio, is a particularly efficient re-aeration structure: at flows of 550 m³ per s, the hydraulic structures of the dam add at least 90 000 kg of oxygen per day to the River. Moreover, during low-flows, the supersaturated releases are largely drawn from water of low dissolved-oxygen concentration within the pool. The effect of such re-aeration may be particularly significant below the last of a series of reservoirs linked by hypolimnial-release gates. The Monongahele River, Pennsylvania-West Virginia, USA, provides an example of such a series (Koryak, 1976). During low-flows, dissolved oxygen stratific-ation is extreme in the upper Opekiska Reservoir, with a vertical gradient ranging from 10 mg per litre at the surface of 0.1 mg per litre at the bottom, and this is accompanied by well-defined thermal stratification. Water dis-charged from the deep-level outlets, characterized by a marked dissolved-oxygen depression, flows directly into the lower, Hildebrand Reservoir. The outlets from this 'power' reservoir are considerably elevated above the down-stream reach, the sills are dentated, and the re-aeration of the outflowing water is sufficient to mitigate the dissolved-oxygen deficit created by the Opekiska Dam.

Despite these beneficial affects on re-aeration processes, dilution problems have been reported for three situations. Firstly, the release of water depleted in dissolved oxygen, has a reduced assimilation-capacity and a reduced flushing-capacity for domestic and industrial effluents, so that water-quality problems can result. Hypolimnial releases during summer will be character-ized also by low temperatures, and occasionally by high levels of iron and manganese. Ingols (1957) considered a dissolved-oxygen deficiency below dams as a major pollution problem, and a distance of 40 km, or more, may be required, in some cases, for the oxygen content to increase to an accept-able level (Churchill, 1957). Dissolved-oxygen concentrations in turbine releases may be as low as 0.5 mg per litre during summer and autumn, and 13 km of free-flowing stream below Fort Patrick Henry Dam is required for natural re-aeration up to 5 mg per litre (Ruane et al., 1977). Low dissolved-oxygen levels could also compound existing pollution problems. Indeed, the development of industries along rivers below dams—often encouraged by flood-control or power-provision—can contribute effluents of high biochemical demand (Dominy, 1973; Davies et al., 1975), and Hannan et al. (1973) observed that reduced flows downstream from impoundments on the

Guadalupe River, Texas, USA, in summer, decreased the dilution of effluent from sewage outflows.

Secondly, comparable situations arise from the loss of flushing ability with regard to tributary discharges into the regulated main-stream. The water quality of the middle Zambezi is mainly determined by the Kariba Dam, not only directly through reservoir releases, but also indirectly by increasing the influence of the tributary Shire River waters on the regulated river (Hall *et al.*, 1977). Prior to impoundment at Cabora Bassa in 1975, the Shire River discharges increased the average total alkalinity of the Zambezi by nearly 50%; as a result, average conductivity increased from 122 to 150 μS per cm* below the confluence. Major problems were expected subsequent to main-stream regulation. In northern California, flows from tributaries of the Sacramento River, contaminated with highly toxic chemicals leached from old mine-dumps, were diluted by the 'clean' flows of the unregulated main-stream. After the construction of Shasta Dam, in the upper reaches, the resultant flow-control allowed flushes of poisonous water from Spring Creek to pass undiluted down the main-stream, and these produced severe fish-kills for 24 km below the confluence (Armstrong, 1973).

Thirdly, river impoundment has significance for water salinity within the lower river. The reduction of runoff, resulting from abstraction from the reservoir etc., may allow the intrusion of salt water to estuarine and deltaic areas. However, for most impoundments the regulation of river discharges—and the provision of constant, relatively high summer flows—leads to reduced saline penetration. For example, prior to the impoundment of the Volta River, Ghana, the 100 ppm-salinity water penetrated 30 km upstream from the coast (Beadle, 1974). During reservoir filling, reduced river discharges allowed the penetration of this water to 50 km from the coast, but subsequent, continuous though low, discharges have restricted the penetration to 10 km from the coast of water giving 100 ppm salinity. More problematic is the severe salinification of naturally 'fresh' water that may occur below tropical dams because of discharge regulation.

Salinity and discharge are inversely related, and the Cl⁻ content of the lower Nile waters, has increased since the Aswan Dam changed the flow and silting regimes (Talling, 1976). Salinity problems are particularly severe in the River Murray Basin, Australia (Collet, 1978), where tertiary marine sediments contribute significant amounts of salt to the middle and lower reaches of the River. In the presence of upstream impoundments, flow salinities frequently exceed 500 ppm—the limit tolerated by irrigated crops. Salinification has also proved to be a major problem on floodplain wetlands in the absence of periodic flushing and dilution by flood-waters. Within the Pongola system of floodplain pans, in South Africa, salts released from the underlying marine deposits are no longer diluted by overbank discharges, because of upstream impoundments; sodium and chloride are the dominant cation and anion, respectively, and in one pan, conductivity reached 2750 μS per cm during 1970, after no flood-waters had been released from the

* Micro-siemens per centimetre.

dam for nearly 1 year (Heeg *et al.*, 1978). Problems of salinization may be severe below many new low-latitude reservoirs, and particular fears have been expressed for the River Lar downstream of a new dam in northern Iran (Coad, 1980).

Quality Control

The impoundment of a river causes two major changes in the character of the water, both of which have a marked effect upon the water quality of the releases: firstly, the creation of a reservoir greatly increases the travel-time of the water through the system; secondly, thermal or density stratification may occur. Density stratification can have severe consequences for downstream water quality, particularly if the reservoir releases are large in relation to the cumulative volume of tributary discharges. The regulation of the physico-chemical quality of river-flows may prove beneficial, but reduced dissolved-oxygen levels, in particular, can be detrimental. The control of release-water quality from impoundments, to maximize the inherent benefits and minimize the inherent detriments, is in all cases advisable, and essential in many. Numerous operational techniques and structures of different design have been used to control the quality of reservoir releases and to minimize the adverse effects of reservoir stratification. These may be classified as selective withdrawal techniques and destratification techniques, although for many rivers (e.g. Mitta Mitta River, Australia, cf. Burns, 1977) the use of both has been recommended in order to maintain minimum-quality standards in the outflows.

The provision of multiple-level drawoff points to facilitate selective withdrawal provides the simplest method of controlling water quality. Deoxygenated hypolimnial water, for example, can be drawn off slowly and blended with highly oxygenated water from the epilimnion. Furthermore, large discharges from outlets can generate mixing currents within the lake, which inhibit the development of anoxic conditions. The effectiveness of valve releases is well demonstrated by Gore (1977). Operational requirements of the Tongue River Reservoir Dam, Montana, USA, resulted in the control gates of the dam remaining open for most of the spring and summer of 1975. The continual high release of water prevented thermal stratification and the associated development of hypolimnial waters, so that the water quality of the releases returned to pre-impoundment conditions. Below the dam, discharges were observed to have temperatures only negligibly cooler than at the mouth of the river, approximately 150 km downstream, and the diurnal and monthly thermal fluctuations approached those expected for unregulated streams. The use of selected-depth releases may therefore prove to be beneficial.

Selected releases of water from an epilimnion, through surface outlets, during times of stratification will provide high-quality discharges until the lake overturns, when the benefit may be lost. Releases from a number of outlets would dilute the hypolimnial waters over a longer period, and provide less extreme water quality conditions downstream. However, selective withdrawal has been considered by some authors (e.g. Brooks & Koh, 1969; Fruh & Clay, 1973) to provide only a limited solution for highly stratified lakes within which mixing is not induced by the releases themselves. Indeed, water-quality problems during summer below Stanley Reservoir, River Cauvery, India, related to two outflow sources which continued to flow as separate and distinct streams for about 2 km below the dam (Ganapati, 1973). Water released from the high-level sluice was relatively warm, well oxygenated, and contained low concentrations of phosphates, silicates, and nitrates. Tailrace water from turbines abstracted at a depth of 30 m (up to 3°C cooler, containing 50% less dissolved oxygen, lower pH, and high phosphates, silicates, and nitrates) was released simultaneously; but mixing, and hence dilution, did not occur. After the autumn overturn, uniform releases were discharged from both sources.

Two approaches to artificially destratifying reservoirs have been commonly employed, namely air injection and mechanical pumping. The maintenance of the reservoir storage in a nearly isothermal condition would permit continued water circulation and, therefore, an available supply of dissolved oxygen throughout the entire water-mass. Anaerobic conditions would be eliminated, along with many of the problems which accompany them.

Artificial destratification and the mixing of epilimnial and hypolimnial waters during summer may often be desirable in order to obtain an improved and more uniform water quality within and, consequently, downstream from reservoirs. Indeed, the Quality Control in Reservoirs Committee in the USA recommend artificial destratification to water suppliers who are experiencing any water-quality deterioration in their reservoirs resulting from anaerobic conditions in the hypolimnion caused by thermal stratification (American Water Works Association, 1971). Several investigators have attempted artificial destratification by injecting diffused air into a reservoir's hypolimnion, so that the water at depth is raised to the reservoir surface where it mixes with the epilimnion. The mixing process, through interchange of heat from warmer to colder water, causes the water-mass to become nearly isothermal. Early attempts often failed to achieve a sufficient circulation (e.g. Derby, 1956; Schmitz & Hasler, 1958). In Sweden, Lake Langsjon experienced severe stratification, with a hypolimnion devoid of oxygen (Heath, 1961); but compressed-air injection effectively eliminated stratification. Similar benefits of forced circulation have been reported by Riddick (1957) for the Ossining Reservoir, New York, USA, and for Lake Wohlford, a subtropical reservoir in southern California (Ford, 1963). In the latter case, stratification was completely eliminated throughout the Lake after 7 days of operation, during which air injection occurred for 9 hours on each of the first 6 days and for 24 hours on the last.

More recently, an air diffusion system has been successfully applied to the destratification of a medium-sized lake (Rogers et al., 1973): Allatoona Lake,

Georgia, USA, which, under normal conditions, begins to stratify in mid-March and achieves complete stability in mid-July, with overturn in early October. In late summer, a maximum temperature difference of 17°C is established between the surface and bottom of the Lake, and dissolved oxygen may become exhausted from the upper 20% of the depth. During the summer of 1968, air was continuously supplied to the hypolimnion, and the reservoir was maintained in destratified condition, with adequate dissolved-oxygen concentrations throughout the Lake. Improvement in the quality of reservoir releases is particularly noticeable if comparisons are made between specific flow-conditions. Under high-flows, dissolved oxygen was slightly less during early summer in Allatoona Lake, but significantly higher during August and September under conditions of artificial hypolimnion aeration. Moreover, during low-flow conditions, levels of dissolved oxygen were generally maintained above 4 mg per litre, whereas prior to air injection, concentrations of less than this value were experienced for most of the summer, and temperatures were elevated by a maximum of 8°C. Although air injection can be applied to a considerable area, and a large amount of water can be set in motion, it is a relatively inefficient means of water transfer because it involves intermediate energy conversion for air compression.

Many of the detrimental aspects of water quality are related to the onset of thermal stratification, and artificial destratification, by pumping cold water from the bottom of a water-body and discharging it at the surface, has been suggested as an alternative method of control to air injection. The application of mechanical pumping techniques has been explored by Symons et al. (1965) and Irwin et al. (1966). The former employed mechanical pumping to destratify a small lake, increasing the temperature and dissolved-oxygen concentrations in the lower layers; manganese and sulphide concentrations were reduced to zero and the ammonia nitrogen concentration, resulting from anaerobic decomposition, was also reduced. Moreover, like the earlier results obtained from larger reservoirs, only about one-half of the original cold-water volume actually had to pass through the pump to mix the entire impoundment—thus reinforcing the opinion that currents generated by pumping are beneficial to the mixing operation.

Subsequent research into the efficiency of mechanical pumping procedures has demonstrated that pumping can be particularly useful if the operation is begun before a lake becomes stratified (Garton et al., 1976). An axial-flow pump was employed in midsummer to transfer water at $0.674m^3$ per s from the oxygen-rich epilimnion to the hypolimnion. Within 2 weeks the procedure had completely destratified water temperature in the lake, but a longer period of time was required to destratify dissolved-oxygen levels (Fig. 22). Significantly, the lake warmed uniformly at a particular depth, regardless of distance from the pump. Initially the dissolved-oxygen concentration of the surface water was markedly reduced, because of mixing with water from an anoxic hypolimnion; but more significant was the rapid rise in dissolved-oxygen levels at depth. After 7 weeks, well oxygenated water was observed throughout the profile; the time-lag in relation to temperature probably resulted from a large organic accumulation in the hypolimnion. The following year, pumping began before stratification had become established, although a 7°C temperature difference existed between the surface and bottom waters, and after only 2 days of pumping an isothermal profile was developed and the dissolved oxygen concentration was stabilized.

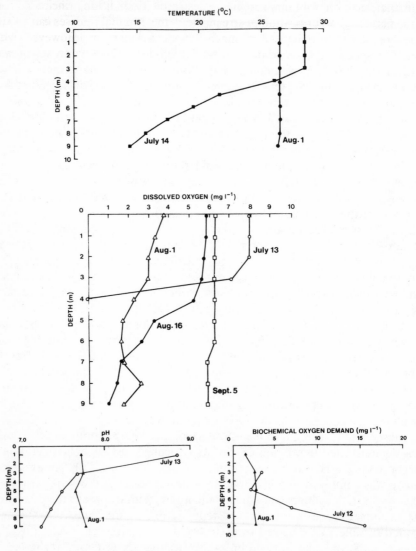

Fig. 22. Water-quality profiles for Ham Lake, Oklahoma, USA, before (July 12–14) and after artificial destratification. (From Garton *et al.*, 1976. Reproduced by permission of the Oklahoma Academy of Science.)

Changes of water quality induced by upstream impoundment can be adequately controlled, although rare pulse-effects may, on economic grounds, not be considered to warrant control. Moreover, the changes induced will be site-specific, depending upon geographical location, reservoir character, and operational procedures. Thermal stratification may be undesirable if reservoir releases derived from a cold, anoxic, hypolimnion are large in relation to the tributary supplies. Dissolved oxygen is the principal factor in the natural self-purification of streams, and hypolimnial releases can produce severe quality problems for long distances downstream. However, where oxygen levels are returned to natural levels, changes of the temperature regime can be the dominant water-quality parameter having responsibility for faunal changes (Ward, 1974).

CHAPTER IV

Seston Transport

Seston includes all of the mineral particles and non-living organic matter that is suspended in the flows, whether derived from allochthonous (tributory, etc., external) or autochthonous (living internal) sources. A general appreciation of the effects of upstream impoundment upon the seston of rivers may be gained by considering the turbidity of river water. Turbidity is an important ecological factor, influencing such factors as light-penetration, depth to which photosynthesis can proceed, and visibility of sight-feeding fishes. Although all suspended matter contributes to the turbidity of flowing water, in all but the largest, slowest-flowing rivers, inorganic particles are usually more important than organic debris or plankton. Under natural conditions, turbidity is strongly correlated with discharge, and in most cases it reflects both the seasonal and diurnal flow-variations, although glacier-fed streams, and streams draining heavily-cultivated watersheds, may remain cloudy for long periods, even under conditions of low-flow. Particularly high turbidities may be generated as a result of soil- and slope-disturbance during construction activities: the construction of Granby Dam, for example, released large quantities of sediment into the Colorado River, which accumulated along the channel margin downstream (Eustis & Hillen, 1954).

Dam construction works have been shown to increase suspended sediment yields by more than 50% within Swedish and Canadian rivers (Nilsson, 1976; Rasid, 1979). The sedimentation of the coarser particles may occur within the channel; but high turbidities, in relation to suspended sediment concentrations, can be maintained over long distances, and can increase downstream in relative terms. Turbidity is not directly reflected by suspended sediment concentration, because it is a function of the total exposed area of suspended materials—a greater amount of coarse material, and a lesser amount of fine debris, is needed to obtain a particular level of turbidity. Thus, below dam construction work on the River Lule, Sweden, Nilsson (1976) observed that turbidity increased with flow-time from the working area; 1 ppm of suspended material at the river's mouth gave double the turbidity of 1 ppm at a point 5 km below the construction area.

When once completed, however, reservoirs provide areas of relatively low current velocity, which allow the materials that are held in suspension to settle out, so that the turbidity of the river downstream is reduced. On the Bighorn River, Wyoming–Montana, USA, mean turbidity was reduced to one-sixtieth (Soltero *et al.*, 1973), while a 61% reduction of annual turbidity in the Tennessee River, USA, resulted from the closure of Cherokee and Douglas Dams (Kittrell & Quinn, 1949), and Hammerton (1972) noted an increase in the depth of light-penetration in the Blue Nile below Roseires Dam, Egypt. All except the finest inorganic particles are commonly deposited within the reservoir, and most of the allochthonous particulate organic matter will be deposited or processed. Indeed, discharges from hypolimnial-release dams are usually very low in organic particles (Ward, 1974, 1976*a*).

Reservoirs increase the overall organic processing of a river, defined as the difference between total particulate input and total particulate output, because of their ability to trap and decompose particulate organic matter. Webster *et al.* (1979) demonstrated that the output of particulate organic matters would be reduced from 6.60×10^3 t per yr to 3.07×10^3 t per yr in a simulation of the effects of river impoundment—an increase of processing efficiency by 2%. Thus, the construction of six headwater impoundments on the Missouri has produced clarified discharges in a river which, under natural conditions, had a high average suspended load of 1500 ppm. This has stimulated ecological changes within the channel downstream—particularly an increase of photosynthetic production (Whitley & Campbell, 1974). Indeed, if a reduction of river turbidity were the only change induced by reservoirs, Whitley & Campbell suggested that increased biological production, and a more effective harvest of some species of game fishes, should be expected. However, these advantageous effects of reservoir sedimentation may be masked by increased autochthonous (organic debris and plankton), allochthonous (tributary etc.), and anthropogenic (e.g. sewage) inputs.

Despite having an important clarifying effect upon the inflowing water, some reservoirs export large numbers of living, and moribund, limnoplankton (Cushing, 1963; Merkley, 1978; Simmons & Voshell, 1978). Indeed, Lind (1971) concluded that, although Lake Waco trapped 71% of the particulate organic matter transported by the Bosque River, Texas, USA, the reservoir had little effect on the total amount of organic matter transported; sedimented particulate organic matter was replaced in reservoir releases by living and dead limnoplankton. In the Missouri River system, Lewis and Clark Lakes alone supply 12 619 tonnes (wet-weight) of zooplankton to the River below the dams each year (Cowell, 1967).

In some cases, the suspended organic matter may increase to several times that carried by the river upstream of the reservoir (Spence & Hynes, 1971), and an increase in turbidity at times of low river discharges may result. Thus, Décamps *et al.* (1979) demonstrated the significance of reservoir-derived phytoplankton for increasing the turbidity of the River Lot, France, since dams were constructed. High turbidities (Table XV) were related to suspended mineral, and organic, particles ranging in size from 1 to 20 μm—a size-range composed mainly of phytoplankton, although daily discharge fluctuations, resulting from the hydroelectric powerworks, result in the alternate suspension and sedimentation of inorganic particles.

Table XV Suspended load of the River Lot, France. (From Décamps & Casanova-Batut, 1978. Reproduced by permission of Annales de Limnologie.)

Size class (μm)	Load (mg per litre)	Proportion of total load	
		Mineral (%)	Organic (%)
>32	1.35	8	6
20–32	1.83	11	7
1–20	6.76	26	42
1>	0	0	0
Totals	9.94	44	55

Changes of river turbidity downstream from impoundments will be related to the release, from the reservoir, of inorganic 'fines' and particulate organic matter (especially plankton); to the development of a lotic plankton; to the supply of particulate organic matter from riparian and aquatic vegetation; and to the interaction between the regulated discharges and turbid water, introduced by tributaries and effluent outflows. The integrated effects of these factors are well summarized by Hall et al. (1977) in a study of the middle Zambezi. Although turbid, flash-flooding tributaries have little influence upon the flow of the Zambezi, but markedly influence the transparency of the regulated river, keeping it down to 10 cm—as measured using a 30-cm Secchi disk—during the wet season (from late November to April). By May, however, the stream-flow becomes dominated by the clear-water releases from the dam, and the transparency increases to more than 60 cm. A marked reduction in water transparency occurs in the middle of the dry season (September–October); little sediment is derived from tributaries at this time, and the reduced transparency results from phytoplankton blooms, induced by high nutrient releases from the Kariba Reservoir, consequent upon reservoir overturn. When once the nutrients are exhausted, the bloom subsides, and transparency is increased. Turbidity changes will, therefore, be characterized by a dynamic, temporal pattern, which reflects the interaction of supplies of particulate matter from both allochthonous and autochthonous sources.

Pulse releases can effectively flush fine sediments, which accumulate in a channel under controlled low-flow conditions, producing a short-duration pulse of high turbidity. Because suspended sediment concentrations in reservoir releases are controlled by the availability of in-channel sediment sources, and not by discharge or flow-velocity, peak concentrations may occur some distance downstream, despite the progressive attenuation of the release wave. Thus, Beschta et al. (1981) found that a flushing release of 4.9 m³ per s (from a low-flow of 0.4 m³ per s) from Electric Lake Reservoir, Utah, USA, into

Huntington Creek, caused a peak suspended-sediment concentration of *ca* 800 mg per litre at a point 6.7 km below the dam, which contrasted with a peak of only *ca* 330 mg per litre at 1.4 km: the downstream station had 4.7 times as much channel length, as a potential source, as did the upstream station.

The magnitude, timing, and duration of the turbidity pulses will relate, primarily, to the distribution of available sediments along the channel. The first release after a season of low-flows may produce a high-magnitude turbidity pulse, which, downstream, becomes progressively lagged behind the peak stage, reflecting contributions from large areas of channel upstream. Subsequent releases, or sites only a short distance below a dam, may have available only local sediment sources producing lower turbidities, and peak values coincident with peak flows (Gilvear & Petts, in press). Characteristically, turbidity pulses are of short duration, and suspended sediment concentrations fall exponentially under conditions of sustained high-flow.

THE MINERAL COMPONENT

Reservoir Sedimentation

'Reservoir sedimentation has taken most waterworks men unawares. Twenty-five years ago the existence of the problem was barely recognised. Today, through the occurrence of destruction of reservoir storage, the matter is being brought forcibly to the industry's attention.'

Hudson *et al.* (1949 p. 913)

Reservoirs will permanently store almost the entire sediment-load supplied by the drainage basin. Although the problems of sediment storage differ according to the type of reservoir, in each case the progressive filling of the water-storage volume will modify the effectiveness with which the impoundment regulates flood discharges, and will alter the proportion of inflowing sediment that is released from the dam. Many small reservoirs have been completely filled with sediments, but reservoir sedimentation was not conceived as a problem until the mid-1920s. It is only during the last 20 years that reservoir designs have incorporated 'dead space' for sediment storage, and plans have often been made to raise a dam in order to increase the reservoir's storage capacity. Hydroelectric power-dams are built to provide a generating head; little storage is required to allow for diurnal fluctuations in demand; dead-storage is a large proportion of the total; and the slow filling of the reservoir with sediments is of little consequence. In contrast, water-storage, and flood-control, reservoirs often have a high proportion of live storage, so that sedimentation can have severe consequence—both ecological and economic.

The hydraulics of reservoir sedimentation have recently been reviewed by Graf (1983). The general pattern of sedimentation is essentially the same for all impoundments. As the relatively high-velocity, and turbulent, water of rivers feeding the lake is transferred into slow-flowing water within the lake basin, the sediment load is deposited. Part of the sediment is deposited in the storage volume proper, and part in the channel and valley-bottom upstream, as a result of backwater effects from the reservoir reducing velocities of river and floodplain flows. The coarser particles—including the bed-material load—settle out to form a delta, whilst the lighter particles, and especially the clays, are distributed further out into the lake, or may be retained in suspension. Some clays, such as of the montmorillonite group, may react with dissolved salts, producing early flocculation. In contrast, the kaolin clays may remain in suspension for long periods, and maintain turbidity throughout the entire reservoir (Bondurant & Livesey, 1973).

The reduction of reservoir storage capacity by over 1% per yr has been observed throughout Europe (Gvelesiani & Shmalkmzel, 1971), the USA (Frickel, 1972), Africa, and Asia (Buttling & Shaw, 1973). Average annual rates, however, are usually less: 0.51% in central Europe (Cyberski, 1973), 0.71% in central USSR (Yakovleva, 1965), 0.55% in Nigeria (Oyebande, 1981), and 0.2% in USA (Dendy et al., 1973). In detail, rates of reservoir sedimentation and concomitant storage loss are dependent upon three factors, namely: the size of the reservoir's drainage-basin, the characteristics of the basin which affect the sediment yield, and the ratio of the reservoir's storage-capacity to the river-flows. The scale factor—the absolute size of the drainage basin—is particularly important because sediment delivery decreases with increasing drainage area, most noticeably in small catchments with areas of up to 50 km^2.

Within a uniform drainage basin, the rate of sediment production per unit area of catchment decreases with increasing catchment size. Similarly, the peak rate of runoff per unit area of catchment decreases with increasing catchment size. For very large catchments, sediment deposition within the upstream channel network, and on the adjacent floodplain, will reduce sediment loads. Reservoirs with large catchment areas may, therefore, have relatively low rates of sedimentation when compared with a reservoir of smaller catchment area, but with a similar storage-capacity to catchment area ratio. Thus for the Sangamon River, Illinois, USA, Hudson et al. (1949) reported sedimentation rates of 333 m^3 per km^2 per yr in Lake Springfield, with a catchment of 686 km^2, and of 143 m^3 per km^2 per yr in Lake Decateur, having a drainage area of 2345.5 km^2. Similarly, but for a wider scale-range, on the Missouri River, F. E. Dendy et al. (1973) reported sedimentation rates of 3191 m^3 per km^2 per yr for reservoirs with small drainage-areas of less than 2.59 per km^2, which contrast with rates of 281 m^3 per km^2 per yr within reservoirs receiving drainage from areas of between 104 km^2 and 4000 km^2.

The catchment characteristics which influence sediment yield are obviously important controls of reservoir sedimentation rates. In general terms, resistant geologies and well-vegetated slopes produce low rates of reservoir storage loss, whilst high rates of sedimentation are associated with erodible

Table XVI Some observations on reservoir storage loss from four geographical regions

Environment	Location	Mean annual storage loss (%)	Source
Hard rock; humid climate			
	Cropston Reservoir, UK	0.005	Cummins & Potter (1972)
	Burrator Reservoir, UK	0.055	N.E.R.C. (1976)
	Sudety Mts Reservoir, Poland	0.04—0.12	Cyberski (1973)
Easily-erodible rock; humid climate			
	Carpathian reservoirs, Poland	1.0	Cyberski (1973)
	Southeastern reservoirs, USA	0.31–2.82	Brown (1944)
Receiving glacial meltwater streams			
	Austrian reservoirs	3.8–10.0	Cyberski (1973)
	Swiss reservoirs	7.0–10.0	Cyberski (1973)
Easily-erodible rock; semi-arid climate			
	Lake Austin, USA	10.375	Buttling & Shaw (1973)
	New Lake Austin, USA	7.354	Buttling & Shaw (1973)
	Lake Mead, USA	3.20	Thomas (1956)
	Grand Reservoir, USA	2.31	Buttling & Shaw (1973)
	Panchet Reservoir, India	5.429	Buttling & Shaw (1973)
	Habra Dam, Algeria	2.636	Buttling & Shaw (1973)
	Aswan High Dam, Egypt	0.002	Kinawy et al. (1973)

soils and semi-arid climates, where slopes lack a protective vegetation cover (Table XVI). Even within these major regions, local differences of climate and vegetation-cover will produce variable sedimentation-rates, so that large differences may occur between reservoirs within a single geographical region (Table XVII), depending also upon the initial storage-capacity of the reservoir. Thus, for their study of 1105 reservoirs in the USA, Dendy et al. (1973) found an inverse relationship between the rate of storage-loss and reservoir capacity: reservoirs with capacities of less than 12.35×10^3 m³ had an average annual storage-loss of 3.5%, whereas those greater than 12.35×10^3 m³ were associated with average annual losses of 0.16%

Reservoirs having a large storage-capacity will trap in excess of 95% of the sediment load transported by the river (Leopold et al., 1964). Early attempts to forecast a reservoir's trap efficiency (TE)—the percentage of incoming sediment deposited within the reservoir—were based upon ratios between reservoir storage capacity and drainage area. Brown (1944) represented the relationship as

$$TE\ (\%) = 100 \left(1 - \frac{1}{1 + 0.1C/DA} \right)$$

where C = reservoir capacity, DA = drainage area.

Table XVII Sedimentation within east African reservoirs: (A) from Jovanovic (1973), reproduced by permission of the Secrétaire Général, Commission Internationale des Grands Barrages: and (B) from Stromquist (1981), reproduced by permission of John Wiley and Sons, Ltd.

(A) Dam	Regional character	Inflow (Q) (10^6 m³ per km² per yr)	Sedimentation rate		Annual storage loss (%)
			(10^3 m³ per km² per yr)	(m³ per m³Q)	
Gajosa	Upland, flat, well vegetated	0.417	0.646	0.0016	0.52
Koka	Transition, well vegetated	0.136	0.455	0.0033	0.30
Ghinda	Transition, steep slopes, sparse vegetation	0.047	2.650	0.0563	37.50
Zula	Lowland, semi-desert, sparse vegetation	0.009	0.670	0.0750	8.03

(B) Dam	Drainage area (km²)	Average annual sediment yield (10^3 m³ per km² per yr)	Initial storage capacity (10^3 m³)	Annual storage loss (%)
Ikoma	640	0.191	3807	2.80
Kisongo	0.3	0.481	121	3.70
Matumkulu	15	0.581	333	2.60
Msalatu	8.7	0.556	421	1.15
Imagi	2.2	0.610	171.5	0.80

The relationship tends towards a curve which approaches a trap efficiency of 100% asymptotically as the C/DA ratio increases. In general terms, a low C/DA ratio suggests rapid sedimentation rates, and high ratios describe reservoirs having low rates of filling. In Illinois, USA, two reservoirs with C/DA ratios below 25 were losing 4.4% and 2.3% of their storage capacity annually, whilst impoundments with ratios exceeding 250 lost less than 0.8% of their capacity each year (Hudson et al., 1949). However, such relationships contain a large error due to the fact that reservoirs having identical C/DA ratios may have very different capacity–inflow ratios (Gottschalk, 1948). In semi-arid areas a low C/DA-ratio reservoir, receiving insufficient inflow to cause spillweir discharge, would trap 100% of the incoming sediment. A reservoir having the same C/DA ratio in a humid area may be full and overflowing for several months, so that a relatively small percentage of sediment is trapped. The C/DA ratio can be used within definite hydrologic regions, but for similar C/DA ratios, the trap efficiency will increase as the runoff per unit area decreases, so that it cannot be applied for wider-area comparisons.

The use of a C/inflow (C/I) ratio for determining reservoir storage requirements was introduced by Hazan (1914), and was adopted as an index of sediment-trap efficiency by Brune (1953), having been recognized as a more accurate index than the C/DA ratio by C. B. Brown (1950). For a survey of 44 reservoirs, Brune (1953) found trap efficiencies over the full range from 0 to 100%. Although the extreme values are theoretically impossible, they were found to relate to reservoirs with extreme C/I values. Reservoirs of very low C/I ratio may alternately fill and scour, depending upon streamflow conditions, and may have a trap efficiency of zero—or less—during some periods. Brune defined periods of erosion and deposition in terms of several years for Williams Reservoir, Indiana, USA. In contrast, reservoirs of very high C/I ratio experienced continuous sedimentation, giving rise to clear-water releases downstream.

Such reservoirs are generally found in semi-arid areas where evaporation and seepage are important, and where spillway flows occur only rarely. Velocities through the pool are very low, and even the finest particles may be retained in the reservoir. Thus, the reservoir retention-time has been identified as an important parameter in determining the trap efficiency of Imperial Dam Reservoir, Arizona, USA (Borland, 1951). Short retention-times (less than one day) commonly show trap efficiencies of between 40% and 90%, but for longer retention times, the trap efficiency usually stabilizes at 90% or above. 'Dry' flood-control reservoirs, with outlets at the base of the dam so that water is not held permanently, have been found to have trap efficiencies of between 65% and 94% (Roehl & Holeman, 1973), despite the relatively high flow-velocities within the pool. Nevertheless, Gottschalk (1964) has shown that measured trap-efficiencies agreed reasonably well with Brune's (1953) relationship, even though the estimated values were usually higher than the actual trap-efficiency.

For individual reservoirs, variable trap-efficiencies occur annually, seasonally, and from storm-to-storm, as a result of changing runoff and sediment yields—the latter being related to changing ground-cover and rainfall intensity. In general, bare soil, together with high-intensity rainfall, will lead to the transport of coarser particles, and to an increase of reservoir trap-efficiency. Thus Heinemann et al. (1973), and Rausch & Heinemann (1975), observed trap-efficiencies for individual storms ranging from 33% to 99%, and 60% to 90%, respectively. For similar retention-times, high-intensity storms—with high sediment loads—had high trap-efficiencies, whilst low-intensity storms—with low sediment concentrations—had a low trap-efficiency, due to the particle-sizes of the inflowing water, which ranged from greater than five microns for high-intensity storms, to less than two microns for low-intensity storms. Using data from forty-eight storms and three reservoirs in central Missouri, Rausch & Heinemann (1975) developed a multiple regression equation to forecast reservoir trap-efficiency:

$$TE = 100/e^{15.6e^{-0.0187TR - 0.32\ln Qp + 0.71\ln Qtot - 0.39\ln SY - 1.6\ln C - 0.43\ln DA}}$$

with $R^2 = 0.70$ and S.E. $= \pm 6.6\%$
where Q_p = peak inflow rate (m^3 per s)
Q_{tot} = total storm runoff (cm)
SY = storm sediment-yield (tonnes per ha)
C = reservoir storage-capacity (cm)
DA = drainage areas (ha)
T_R = retention time

The remaining variability in trap efficiency may be related to the flushing effect of successive storms. Highly turbid water in a reservoir, derived from storm runoff, may be displaced and carried to the spillweir by the inflow of further flood-water, thereby reducing the trap efficiency to only 32% for successive autumn storm-flows at 2-days intervals.

The transfer of large quantities of sediment towards the outlets of the dam may be effectively achieved by density currents. The dispersion of suspended solids in a reservoir is a three-dimensional process. Local currents are set up according to local fluid-density gradients, resulting from variations of temperature, salinity, and suspended solids concentration. When the inflowing stream-water has a similar density to that of the reservoir, then the suspended solids can quckly settle-out—within the upstream part of the lake. However, the inflowing stream, and its suspension, often remain essentially coherent, as a layer moving through the ambient reservoir-water. Turbid water has a greater density than the water within the reservoir, and tends to move as an intact layer rather than mixing with the water already present. The settling of suspended particles from a density current is a slow process, because fine silt particles are involved. Therefore, density currents, when once established and moving uniformly, have remarkable persistence, and sediments remain in suspension as long as they retain the intensity and scale of turbulence created by the fluid motion (Buttling & Shaw, 1973).

When reservoirs are not stratified, sediment-laden storm runoff will assume a level within the lake according to its density. Particularly during the early part of a flood, when the sediment concentration is the highest, dense flood-waters can form important density-currents. Later in the storm event, water of reduced sediment-concentration, and hence reduced density, will flow at a higher level than the denser flood-water. However, most reservoirs (even small ones) can become stratified for all, or part, of the summer months. Stratification is characterized by the separation of clearer, less dense, water near the surface, from denser, cooler, sediment- and nutrient-laden water beneath. Interflows, and underflows, may also be particularly common. For surface-spillweir release-dams, such flow-separation will result in the discharge of clear surface water, displaced upwards by the inflowing, dense, flood-water. Hypolimnial-release dams, in contrast, may release sediment-loaded water.

Density currents have been responsible for the distribution of silt throughout the basin of Lake Mead (Gould, 1951), and turbid water has been observed leaving the reservoir through flood-gates, whilst the surface water was relatively clear (Grover & Howard, 1938). Deep releases from a reservoir during storm inflows may extract the turbid flows, and leave clear water within the lake, so that after the passage of the flood, the bottom-withdrawal reservoir will contain mostly clear water, whilst the surface-discharge reservoir contains mostly sediment-loaded water (Rausch & Heinemann, 1975). Layers of turbid water, moving as density-currents, are rarely thick enough to avoid mixing with adjacent water if they are drawn towards an outlet. Nevertheless, a sequence of turbid density-currents can build-up a thickness of muddy water close to the dam, which may supply releases for long periods. Thus Neel (1963) recognized, for the USA, that muddy-water discharges are common from reservoirs that impound a catchment with fine, clayey soils affected by accelerated erosion—such as frequently result from agricultural development.

Inorganic Suspended Loads below Dams

Reservoir sedimentation trap-efficiency is affected by the retention time of storm runoff, and by factors governing sediment particle-size. The significance of reservoir sedimentation, however, is related to the volume of suspended sediment that is transported. Thus, Kellerhals & Gill (1973) concluded that the high sediment trap-efficiency of large Canadian reservoirs is inconsequential in the nearly sediment-free rivers of the Canadian Shield. For example, the Peace River's suspended load is small, so that the effect of the Bennett Dam is probably low—not least because many of the downstream tributaries are more heavily loaded than the main-stream. The impoundment of moderately- or heavily-loaded rivers, in contrast, can markedly influence the downstream suspended loads.

Although up to 50% of the sediment input may be transmitted through the lake (Table XVIII), the mean release from single impoundments is 26% of the sediment yielded by the catchment to the reservoir, and, because of the differential rates of settling within the lake, the sediment loads released from dams will have a relatively high concentration of fine clays. In the case of one small flood-control reservoir, the proportion of clay within the suspended load rose from 47% in the inflow, to 70% in the outflow (Schreiber & Rausch, 1979). The lower, unaffected part of a drainage basin produces, proportionately, only a small part of the total suspended-sediment yield, so that the effects of impoundment upon suspended-sediment loads may be observed for considerable distances downstream. Thus, suspended-sediment yields on the 'natural' River Ystwyth, UK, for two successive years, were 16 and 7 times those on the neighbouring Rheidol, despite the measurement station being 25 km downstream from the main Nant-y-Môch Dam, which collects sediment and runoff from 34% of the basin (Grimshaw & Lewin, 1980a).

Suspended sediment for the river below a dam will be derived from unimpounded tributaries, effluent outfalls, and the erosion of fine channel-bank materials. In many environments, erosion of the river-bank can form a major source of suspended sediment for downstream reaches, and particularly where the river actively meanders across a broad floodplain. The regulation of flows within impounded rivers can markedly decrease rates of bank erosion. Grimshaw & Lewin (1980b) reported that flow regulation by Nant-y-Môch Reservoir had induced increased bank stability, which contributes to the reduction in suspended-sediment yields of the Afon Rheidol subsequent to river impoundment. However, the imposition of an unnatural flow regime characterized by highly variable discharges, as a result of dam operations, can increase rates of bank erosion, and Guy (1981) reported the net erosion of some 230 ha of floodplain below Lake Kariba as a result of reservoir releases during a 7-years' period. Also, Ward (1976a) reported that the suspended-solids concentration of from 2.2 to 7.4 mg per litre at the outflow from Cheesman Reservoir, Colorado, USA, increased to from 29.5 to 36 mg per litre 40 km downstream. Further discussion of the effects of upstream impoundment upon channel erosion is given in Chapter 5.

Impounded rivers can experience considerable changes in suspended-sediment loads over time: changes in runoff, erosion, storage capacity, and operational procedures, can markedly alter trap-efficiency values. In 1941, approximately one-third of all reservoirs in the USA were losing storage-capacity at a rate of more than 1% annually (Hudson et al., 1949), and although the progressive construction of larger reservoirs since that time has reduced average annual rates of reservoir storage loss, sedimentation will progressively reduce reservoir storage-capacity, and decrease retention-times. Many recently-constructed dams have been designed so as to provide for a minimum life-expectancy in the face of sedimentation. This may be

Table XVIII Impoundment effects on suspended sediment loadings.

River/Reservoir	Country	Annual suspended sediments loads		Proportion of natural load (%)	Source
		Natural	Below dam		
Danube	Romania–Yugoslavia	23.8×10^6 t	3.5×10^6 t	15	Bruk et al. (1981)
S. Saskatchewan	Canada	1.81×10^6 t	0.7×10^6 t	37	Rasid (1979)
Ume Älv	Sweden	100×10^3 t	40–50×10^3 t	50	Nilsson (1976)
Ångermanälven	Sweden	92×10^3 t	34×10^3 t	37	Nilsson (1976)
Indalsälven	Sweden	76×10^3 t	26×10^3 t	34	Arnborg (1967)*
Rheidol	UK	26.7×10^3 t	2.7×10^3 t	10	Grimshaw & Lewin (1980b)
	WINTER	6.9×10^3 t	1.2×10^3 t	17	
	SUMMER	19.8×10^3 t	1.5×10^3 t	8	
	SUMMER†	4.7×10^3 t	1.0×10^3 t	21	
Vaal	South Africa	0.49×10^3 ppm	0.17×10^3 ppm	35	Schwartz (1969)
Missouri	USA	1.3<3.2×10^3 ppm	0.47<0.8×10^3 ppm	25	Neel (1963)
Nile	Egypt	0.6×10^3 ppm	0.05 ppm	8	Abul-Atta (1978)
	WET SEASON	1.15×10^3 ppm	0.05 ppm		
	DRY SEASON	0.06×10^3 ppm	0.05 ppm		
Callahan	USA	2.9×10^3 mg per litre	0.34×10^3 mg per litre	12	Schreiber & Rausch (1979)
Vltava	Czechoslovakia	0.64×10^3 mg per litre	0.3×10^3 mg per litre	46	Brádka (1966)
			0.06×10^3 mg per litre‡	9	

* Reported in Nilsson (1976).
† Excludes a severe overbank flood of August 1973, which produced total suspended-sediments loads for that month of 30 995 t on the natural Afon Ystwyth and 1035 t on the impounded Afon Rheidol.
‡ After construction of a second upstream impoundment.

achieved either by making provision to raise the dam, and thus increase storage, or by providing spare capacity at the time of construction. Lake Nasser, behind the Aswan High Dam, for example, was designed with nearly 25% spare capacity for silt storage—sufficient to last 500 years (Buttling & Shaw, 1973).

The depletion of storage space in reservoirs, due to the deposition of silt, is not only significant in economic terms, but also in ecological terms, because the progressive loss of storage-capacity will influence both the character of discharges and the suspended loads passing the dam. As a result of extensive sediment accumulation, water currents during floods can reactivate sediment transport, and carry material towards the outflow gates. Thus, within the Saalachsee Reservoir, Bavaria, mean values of sedimentation show a distinctly decreasing average annual loss of storage-capacity from 5.57% during the first 15 years to 2.51% over a 47-years' period, and the output of suspended solids nearly doubled (Bauer & Burz, 1968). In the short term, however, changing operational procedures may cause significant sediment-load changes. During the early years of operation of John Martin Reservoir, USA, large volumes of water and sediment were allowed to pass downstream with little restriction, and some 38% of the inflowing sediment was released through the dam (Brune, 1953). However, after 5 years the trap-efficiency was increased by reducing the outflow rate. Similarly, during the first years after closure of Lake Nasser, large volumes of suspended sediment—up to 100×10^6 m³ per yr—were flushed through the reservoir, and this continued until the water's depth was decreased to a level that was sufficient to replace the (previously dominant) turbulent flows with laminar flows, of lower velocity, which encourage rapid sedimentation (Entz, 1976).

The selective release of highly turbid waters from a reservoir is often used to combat sedimentation. Reservoir operations involve two procedures; namely, sluicing and venting. Sluicing involves the opening of large gates near the base of the dam to flush-out deposits. For the Conchas Reservoir, New Mexico, USA, sluicing operations decreased the trap efficiency from an estimated 97.3% to 95.8%, and increased the sediment loss from the reservoir by 56% (U.S.A.C.E., 1950). Similarly, Seavy (1948) reported that sluicing through bottom outlets, during periods of relatively low discharge, would reduce the trap efficiency of the Arrowrock Reservoir, Idaho, USA, from 93% to 90%, and to increase the sediment loads of releases by 39%. However, sluicing operations are ineffective for flushing sediment that has already settled—except in the immediate vicinity of the sluice intakes.

Venting refers to the operation of outflow gates, often synchronized to intercept density-currents, to 'waste' silt-laden flood-waters that are under-running the reservoir. Density-currents, for example, have been used to evacuate accumulated sediment, and maintain storage capacity, in Algerian reservoirs (Bondurant & Livesey, 1973). Rausch & Heinemann (1975) advocated the use of bottom withdrawals to decrease the retention time, by discharging storm-runoff density currents, whilst retaining the clean water which 'floats' above. For three reservoirs in central Mississippi, the latter authors found that a reduction in retention time from 30 to 2 days would decrease the trap-efficiency of the average storm from 90% to 82%. Bottom withdrawals increase sediment releases, both by reducing the retention time and by eliminating the 'dead' storage capacity, so that only the coarse sediment particles, deposited in the deltaic inflow area, will be trapped.

The operation of outlet gates during flood inflows could produce the venting of between 10% and 25% of the incoming sediment, without adversely affecting water storage in the reservoir (Brown, 1944). Venting of sediment-loaded water through a gate near the base of the Lake Issaquaeena Dam, Clemson, South Carolina, USA, just after a major flood in August 1940, resulted in a suspended sediment concentration of 552 ppm in the flow coming from the outlet pipe, while that passing over the spillweir was 183 ppm (Zwerner et al., 1942). However, the release of tonnes of sediment from behind the dam at Verbais, on the Rhone, France, had a disastrous effect on the river (Nisbet, 1961). The introduction of large quantities of fine silts and clays into permeable gravel substrates, for example, can have a catastrophic effect upon fish populations. Therefore, whilst the general clarifying effect of dams upon river water, associated with reduced loads of suspended sediment, is beneficial for the river flora and fauna, reservoir operations can result in extreme, and unnaturally high, concentrations of mineral particles, which may produce a major stress-effect upon the aquatic ecosystems concerned.

THE PLANKTON COMPONENT

Within lotic systems, phytoplankton production is often assumed to be insignificant, whereas in large rivers characterized by low current-speeds, phytoplankton production may become important. Some species can maintain populations throughout the year, provided that turbidity and light-penetration do not limit growth. Nevertheless, within most rivers, few truly 'potamoplanktonic' Algae exist. Phytoplankton can be derived from lakes, low-velocity backwaters—including beds of aquatic macrophytes, in the shallower areas of fast-flowing streams—and from benthic algal populations. In large, slow-flowing rivers, where nutrient availability does not limit phytoplankton development, a true lotic plankton can be formed, characteristically with the highest numbers always occurring during periods of low-flow, e.g. in the River Thames, England, UK (Lack, 1971). Within smaller, shallow streams, light will penetrate to the channel bed, benthic Algae will develop, and the total number of phytoplankton in suspension will be intimately connected with the bottom flora (Swale, 1964). Indeed, during high-flows, losses of periphytic Algae to the plankton can be extreme, and these can produce marked pulses of phytoplankton concentration. In contrast to lakes, the phytoplankton in rivers dominate over the zooplankton, and, of the latter, rotifers, as opposed to crustaceans, compose the largest proportion of the biomass (Hynes, 1970b). Furthermore, lentic plankton, originating in quiet backwaters, can be rapidly reduced in numbers upon entering the turbulent water of the main-stream. Natural rivers do contain free-floating microorganisms, but the plankton populations are inherently unstable and dependent upon the frequency of high discharges.

The introduction of a man-made lake into a river system, particularly as a result of impoundment in headwater areas, can markedly alter the plankton component of the lotic system below the dam. The significance of impoundments, as sources of planktonic organisms, has been recognized for many years: one of the first studies (Kofoid, 1903) identified impounding reservoirs as the main source of plankton for the Illinois River, USA. Consequent upon dam closure, the lentic system establishes itself rapidly as the reservoir fills, whereafter a microbial population explosion soon occurs, releasing nutrients from the newly-submerged organic matter, which stimulate an equally explosive development of the phytoplankton, that harnesses solar energy.

The enrichment of reservoir water by large amounts of nitrogen and phosphorus, as a result of the decay and mineralization of organic matter flooded by the lake, commonly leads to a multiplication of Cyanophyceae (blue-green Algae) during the first few years of a man-made lake's existence. Riverine plankton will be introduced into the filling lake by the inflowing streams. This input will include not only the true lotic 'potamoplankton', but also species originating in quiet backwaters. The rate of plankton development within the lake will reflect the character of the inflowing river from which the reservoir is inoculated. The plankton population of reservoirs constructed on a mountain stream or turbid lowland river, may require a long period of time to develop, whereas on a clean lowland stream, inoculation could be rapid, and followed by the development of an important plankton population. Some species will adapt to the lake conditions preferentially, and it is these lentic species which have significance for stimulating secondary production through invertebrates to fish etc.—not only within the lake, but also within the river channel downstream. The impoundment of the River Niger in Nigeria, for example, has more than doubled the annual peak phytoplankton density to about 2500 Algae per millilitre, and tripled the zooplankton peak within the river downstream (El-Zarka, 1973). In the USSR, the phytoplankton of the River Don, from the Tsimlyanskoe Dam to the river's delta, reflects the specific conditions of the reservoir, being dominated by Cyanophyceae (Zhadin & Gerd, 1963). Shiel (1978) contrasts the Rotifera-dominated zooplankton of the relatively natural River Darling, Australia—a 'typical' riverine assemblage—with the plankton of the impounded River Murray, having an important crustacean component that is characteristic of still waters. Therefore, river impoundment affects the plankton component of the lotic seston in two main ways: by changing the conditions affecting the development of autochthonous riverine plankton, and by altering, and commonly augmenting, the allochthonous supply from upstream. However, reservoirs experiencing eutrophication will be characterized by a progressively-changing phytoplankton composition, and increasing standing-crop, so that the lentic plankton contribution may increase over time.

Hergenrader (1980) has examined the eutrophication process as it occurred in a system of new reservoirs—The Salt Valley Reservoirs, Nebraska, USA. The intensively-cultivated prairie soils supply runoff to the reservoirs with high nutrient concentrations, and so promote eutrophication. All the reservoirs are shallow (maximum depth 10 m), and have large surface areas, which allow wind-induced mixing that prevents thermal stratification, and provide a relatively homogeneous heat and nutrient distribution—conditions highly suited for vigorous algal growth. One of the reservoirs, dominated by high levels of inorganic turbidity, developed only a poor phytoplankton population within which diatoms persisted, whereas three 'clear' reservoirs produced dense standing-crops of phytoplankton, and their surface-releases led to a continued increase in the nutrient pool, with a consequent increase in phytoplankton growth. The mean summer volume of phytoplankton (Y) for these 'clear' reservoirs was related to the reservoir age (X):

$$Y = 34.76 \ (X) - 75.91 \qquad R^2 = 0.83, \ p < 0.01.$$

The phytoplankton was dominated by three genera of Cyanophyceae (*Microcystis*, *Aphanizomenon*, and *Anabaena*).

The Lake Plankton Contribution

Man-made lakes tend to be morphometrically unique; the hydrological characteristics, and thermal and chemical regimes, are unique; and, as a consequence, the character of primary production within reservoirs is highly individualistic. Nevertheless, in all reservoirs, the primary production is mainly derived from the activity of phytoplankton. Blue-green Algae often predominate in numerical terms (*Microsystis* is particularly common), but they are often outweighed in biomass by diatoms (Bacillariophyceae), such as *Melosira* and *Asterionella*. Three factors govern the contribution of lentic plankton to lotic systems: the rate of water replacement within the reservoir (the retention time), the seasonal pattern of lentic plankton development, and the character of the outflows from the reservoir.

Retention-times for plankton development

The most important control upon the pattern of primary production, and of the significance of a man-made lake for the plankton population in the river downstream, is the retention-time of the inflow in relation to the sedimentation of inorganic and organic matter. The suspended materials directly affect light-penetration and the generation-time needed for full development of algal populations (Wetzel, 1973). Short retention-times are often associated with high turbulence, a mixed water-body, and a lack of thermal stratification—such that, if rates of water movement through a lake exceed a few millimetres per second, little plankton will develop (Hynes, 1970*b*). In the Pitlochry Reservoirs, UK, the retention-time is never more than 4 days, and only occasional planktonic organisms were found during a 4-years' period of observation (Brook & Woodward, 1956). The reservoirs were characterized by plankton having a rapid rate of replacement, and the composition of the plankton was determined by the hydrological characteristics of the lake basin. Such reservoirs will have little effect upon the planktonic component of the downstream lotic system, although sedimentation within the lake may actually reduce the concentration of organisms discharged to the river below the dam.

Impoundment of the Dnieper River, USSR, has impeded inoculation from upstream, so that the River below the dam receives only a very small amount of plankton. Such impounded rivers can, however, develop a plankton consisting largely of organisms washed-off the channel bottom (Zhadin & Gerd, 1963). Within the Marias River, Montana, USA, Stober (1964) reported that the construction of the Tiber Reservoir had little effect upon the planktonic component of the lotic system below the dam. Upstream of the reservoir, the River's phytoplankton was predominantly composed of Chrysophyceae (96%), with subordinate Chlorophyceae (green Algae), although in relatively wet years, when the total phytoplankton was reduced, the Chlorophyceae—particularly *Spirogyra* and *Closterium*—increased in proportion (13%). *Synedra, Asterionella,* and *Navicula,* were the dominant diatoms. Zooplankton were generally rare, and composed of Rotifera—particularly *Monostyla*—and nauplii of Copepoda. In the reservoir itself, both phytoplankton and zooplankton were extremely scarce. Downstream, the total average population of phytoplankton was slightly lower than in the upstream reach, with 955, as compared with 1 037, organisms per litre, but the Chrysophyceae maintained 90% of the population. The total average zooplankton population, although still small, was greater below the dam, and *Monostyla* again dominated. The only cladoceran found—*Daphnia*—was most numerous below the dam, and this pattern was followed also by the most abundant copepod, *Diaptomus*. However, on the Dniester River, USSR, the Dubossary Reservoir—despite having only poor phytoplankton, due to its sparsity in the upstream rivers, the turbid discharges, and the short retention-time—effectively creates a slow-moving, deep river, and the zooplankton is dominated by true planktonic rotifers and crustaceans, which reach a density of nearly 300 organisms per litre (Zhadin & Gerd, 1963).

Long-retention lakes favour the development of lentic plankton. Both zoo- and phytoplankton require a minimum retention-time to allow development, but Brook & Woodward (1956) suggested that this may be longer for the zooplankton than the phytoplankton. Nevertheless, the quantity of lake plankton tends to be inversely proportional to the speed of water-flow, so that a general transition may be observed from a small, shallow lake receiving runoff from a large catchment area, and having a low primary production, to a large and deep lake with a relatively small catchment, and supporting a high primary production (Rzóska *et al.,* 1955).

The full extent of the effect of upstream impoundment, as an interruption to the normal downstream pattern of change within the lotic environment, is demonstrated by Hammerton (1972), who described the significance of impoundments for the plankton of the Blue Nile River, Egypt. The seasonal development of a dense phytoplankton in the Blue Nile had been traced to the influence of the Roseires and Sennar dams upstream (Talling & Rzóska, 1967). Upstream of the reservoirs, the River contains only traces of rudimentary plankton, with usually fewer than 50–100 Algae per millilitre, and less than one zooplankter per litre. The Sennar reservoir caused a 100- to 200-fold increase in both the phytoplankton and zooplankton. Today, the River is dominated by the more recent Roseires Reservoir. Within 1 month of first filling, Cyanophyceae—dominated by *Microcystis flos-aquae*—became important within the reservoir, and after 3 years *M. flos-aquae* dominated in the River from the dam downstream to Khartoum. However, after 4 years, the Cyanophyceae disappeared, and the Chlorophyceae—most notably *Volvox aureus* and *Pediastrum* sp.—became prominent, although the ubiquitous diatom *Melosira granulata* was the dominant Alga.

The Blue Nile contrasts with the White Nile hydrologically: the former has a characteristic flood-season, whilst the latter is dominated by natural lakes and swamps, which produce a very regular flow-regime. Nevertheless, the creation of the Gebel Aulyia Reservoir has again increased the concentration of plankton 100-fold. Brook & Rzóska (1954) have described the influence of the Gebel Aulyia Dam on the plankton assemblage of the White Nile south of Khartoum (Table XIX). In the natural river above the reservoir, the plankton consisted of a sparse community of plants and animals, derived primarily from the natural lakes and swamps upstream. Of the plants, diatoms (principally *Melosira granulata*) constituted 75–90% of the population, with Chlorophyceae (mainly *Ankistrodesmus falcatus* and *Scenedesmus* spp.) next in importance, supplemented by a few Cyanophyceae and dislodged epiphytic Algae. The phytoplankton are associated with zooplankton dominated by Copepoda (up to 63%), with a few Cladocera. Adventitious zooplankton often comprised up to 24% of the whole population.

In the area of true lake conditions in the above sequence, the density of the plankton increased, and its composition changed. The adventitious zooplankters disappeared and crustacean plankton dominated, with the cladocerans forming up to 87% of the whole population at individual sites. A considerable increase in phytoplankton productivity occurred down-reservoir, towards the dam. As in the Blue Nile, the first phase of the annual phytoplankton development was associated with the upstream 'inoculum' *M. granulata* (Prowse & Talling, 1958). The relative proportion of diatoms fell from 93% to 31%, despite their increasing in actual numbers, while the Cyanophyceae became the dominant group, increasing from 5% to 86% of the population. Downstream of the dam, the composition of the plankton was dominated by these lake-forms: of the zooplankton, the cladoceran *Ceriodaphnia rigaudii* and the copepod *Cyclops* sp. remained important, and the phytoplankton was still dominated by the blue-green Algae *Anabaena*

Table XIX   Plankton abundance and composition along the White Nile (data from Brook & Rzóska, 1954. Reproduced by permission of Blackwell Scientific Publications).

Planktonic component	Organisms per litre			
	Upstream	Reservoir		Downstream
		Average	Surface maxima	
PHYTOPLANKTON ($\times 10^3$)	50	4490	6000	6590
Cyanophyceae	10	3530	5010	5800
Diatoms	40	870	940	970
Chlorophyceae	0	90	50	20
ZOOPLANKTON	1	84	282	72
Cladocera	*	36	110	29
Copepoda	1	33	125	40
Rotatoria (Rotifera)	0	15	47	13

* = less than one.

flos-aquae and Lyngbya limnetica, which together comprised 61% of the total phytoplankton population.

Seasonal Pattern of Lentic Plankton Development

The occurrence of lacustrine plankton assemblages below impoundments will vary seasonally, and with individual reservoirs, depending upon their geographical location. In cold climates the development of Algae, and the gradual increase in primary production, in reservoirs such as the Rybinsk on the upper Volga, USSR, begins after the breaking up of the ice at the end of April and the beginning of May (Butorin et al., 1973). Maximum phytoplankton development occurs between the end of June and the beginning of August, and is associated, primarily, with the intense development of Cyanophyceae. The zooplankton of the Rybinsk Reservoir vary markedly in both space and time, reaching two peaks (in June and August–September). The Cladocera (Daphnia and Bosmina spp.) dominate during the warm weather, but these are replaced by Copepoda during cooler periods.

In tropical lakes, there is no seasonal check to plankton growth comparable with the winter in temperate regions. Given the favourable thermal regime, the productivity of tropical lakes is mainly limited by the availability of nutrients, and the introduction of highly turbid waters during the 'wet season'. In Lake Kainji (Adeniji, 1973), the numbers of phytoplanktonic organisms start to rise in March–April, just before the start of the rainy season, and reach a peak in July. This is maintained through the autumn, with a minor reduction in September, due to the inflow of turbid flood-water. Similarly, the numbers of zooplanktonic organisms are low prior to the rainy season, but increase rapidly to a peak of over 13 organisms per litre in April, after which the numbers correlate with the change in lake-level.

Within tropical areas, rivers may be characterized by a marked seasonality of runoff, which produces particular patterns of plankton development within man-made lakes, such that significant pulses of plankton may be released to the river downstream. Imevbore (1967) provides an example of such a regime within the Eleiyele Reservoir on the Ona River, Nigeria. This ephemeral River, with a marked rainy season from March through November, experiences a clear seasonal pattern of plankton development, characterized by maximum plankton densities during the period of spillway discharge (Table XX)—a regime which Imevbore considered to be typical of the west coastlands of central Africa. The peaks of plankton density are strongly related to the two flood-periods of June–July and October. During the first flood-period, extreme turbidity largely prohibits plankton development, but the less intense floods during October cause the mixing of water within the lake, and resuspension of bottom sediments, thereby increasing the availability of nutrients within the surface layers. The annual growth of phytoplankton begins in July and, in October, reaches a major peak which is associated with maximum diatom development (particularly of *Melosira granulata* and *Synedra acus*). This precedes the zooplankton peak when maximum water-level favours the greatest growth for all the zooplankton groups, and when a food supply is provided by both the phytoplankton and suspended organic matter that is rich in Bacteria. Flood runoff during the wet season increases the water depth within the reservoir by up to 10 m, and in most years water passes the spillweir.

The duration of overflow varies each year, but inflow stops around the end of December. During the period of overflow, large quantities of plankton are transported into the river downstream—particularly diatoms, cladocerans, copepods, and rotifers—which achieve their maximum densities during this period. Indeed, Imevbore (1967) considered that the wash-out of plankton, during spillway flow, is likely to play a major role in determining the rate of phytoplankton decline during the late autumn. Subsidiary peaks of algal development during the dry season, e.g. the peak of Cyanophyceae (notably *Anabaena spiroides* and *Anabaenopsis raciborskii*) during January, are unlikely to influence the river downstream significantly, because the reservoir storage is below spillweir level, and the outflow is minimal.

Pulses of zooplankton output from reservoirs associated with specific groups, and even species, have also been reported by Peňáz et al. (1968) within the River Svratka, Czechoslovakia, where the seasonal development of plankton in the valley reservoirs has influenced the specific structure, and quantity, of plankton in the river. Although several species of zooplankton, common to the river above the reservoirs, were not found downstream, e.g. *Keratella serrulata* and *Acanthocyclops viridis*, potamoplankton (autogenic plankton of fluviatile locations) composed two-thirds of the zooplankton drift, and only one-third of the taxa were derived from the reservoir: these included *Daphnia cucullata*, *D. hyalina*, *Thermocyclops crassus*, and *Eudiaptomus gracilis*. Nevertheless, the total plankton drift below the dams reached five times that of the natural river. During the winter and spring the planktor drift included only Cladocera and Copepoda which, moreover, occurrec only below the dam (*D. hyalina, E. gracilis, Acanthocyclops vernalis,* anc *Mesocyclops leuckarti*). However, a major pulse of plankton was discharge from the reservoir in July. This was composed primarily of Cladocera, domi nated by *Chydorus sphaericus* and *Bosmina longirostris,* with some *D. hyalinc* and *D. cucullata,* and was preceded by a minor discharge of Cladocera anc Rotatoria (Rotifera) in June. Subordinate peaks of Rotatoria (October), anc Cladocera and Copepoda (November), occurred in the autumn. In this case the variable seasonal significance of plankton discharge from the lakes—associated with the pattern of lentic plankton development—wa$ related to temperature and turbidity changes, nutrient availability, and the movements of water within the lake.

Epilimnial and Hypolimnial Releases:

The characteristics of the plankton supplied by reservoirs to the river below the dam will certainly reflect the seasonal patterns described above, but this reflection will be modified, depending upon the depth within the lake from which water is released. This is exemplified, particularly, by Shiel's (1978) study of the River Goulburn, below Lake Eildon, Australia. Throughout the summer period of lake stratification, plankton were absent from the reservoir releases that were withdrawn from the bottom of the lakes. During the May–June overturn, large numbers of plankton survived passage through the turbines; and in winter, during spillway overflow, lacustrine plankton were again discharged into the downstream river, where they were recorded for 20 km below the dam. Within shallow, well mixed, reservoirs, the depth of withdrawal will not affect the contribution of lentic plankton to the river below the dam. For example, although the number and composition of Algae vary spatially within the Rybinsk Reservoir, the vertical distribution of Algae is nearly uniform, because of the almost constant mixing by wind (Butorin et al., 1973). However, reservoirs having long retention-times commonly experience thermal stratification for at least the summer months. The major component of the plankton is confined to the surface layers, so that water abstracted from different depths will supply different quantities of plankton to the lotic system downstream of the dam. Thus, the release of surface water from the Seine Dam, France, provides large quantities of phytoplankton—particularly *Asterionella formosa* and *Melosira granulata*—to the river downstream (Therenin *et al.*, 1973).

The Blue Nile River is dominated by the contribution of plankton from the Roseires Reservoir, and a 200-fold increase in the concentration of plankton sometimes occurs in the River. However, below the dam the River frequently carries only a fraction of the plankton density that is found within the Reservoir, and this contrasts markedly with the plankton contribution from the Sennar Reservoir, located downstream (Hammerton, 1972). The low density of plankton supplied to the River from the Roseires Reservoir was explained by the great depth of the Reservoir, which allows stratification to occur; this restricts plankton to the upper layers, whilst water is released from a depth of 30–40 m. The Roseires Reservoir was colonized by the previously unimportant cyanophycean *Microcystis flos-aquae*, which occupies the upper water layers, and is largely retained by the dam from which water is released at depth (Talling, 1976). The significance of the depth of withdrawal was confirmed by Stroud & Martin (1973), in a comparative study of the crustacean plankton (cladocerans and copepods) released from the Barren River Lake, with a hypolimnial drain, and the Nollin Lake, having epilimnetic discharge, in Kentucky, USA (Table XXI). Their data suggest that, during the late summer and autumn periods of stratification, the plankton contribution from the reservoir to the river downstream is maintained by the epilimnial discharge, but is severely reduced by hypolimnial releases. The location of the reservoir outlet will, therefore, substantially affect the rate of supply of an important food-source for both macroinvertebrates and fishes.

A detailed study of the plankton discharge from a thermally stratified reservoir with a hypolimnial drain has been made by Coutant (1963), on Perkiomen Creek, Pennsylvania, USA. Distinctly different plankton populations were observed within and below the reservoir (Table XXII). Within the reservoir, unicellular green Algae were abundant and five genera of diatoms occurred in numbers greater than 10 000 per litre: *Cyclotella*, *Cocconeis*, *Navicula*, *Nitzschia*, and *Achanthes*. Bacteria were the most numerous organisms, although amounting to somewhat less than 4% of the total volume of plankton, compared with diatoms (64%) and Chlorophyceae (33%). Downstream from the dam, the plankton was characterized by a vastly increased bacterial contribution, of more than 300×10^6 cells per litre, and apparently derived from the hypolimnion. The two dominant Bacteria—*Thiopedia* sp. and *Methanobacterium* sp.—are typical of anaerobic environments, and characteristic of lake hypolimnia. Furthermore, relatively few of the Bacillariophyceae and Cyanophyceae, which were abundant in the epilimnion of the reservoir, were found in the river downstream. Diatoms and green Algae were virtually eliminated from the plankton in the discharges that were released from the reservoir. The releases contributed, however, substantial numbers of moribund diatom cells—a condition indicated by extensive cytoplasmic contraction. These cells may have been derived either from the prolonged exposure of originally healthy cells to the adverse conditions of the hypolimnion, or from the settling of cells which were dying in the epilimnion. Given the stratified nature of the reservoir Coutant concluded that the diatom cells were most probably derived from dead or dying organisms, which settled into the hypolimnial draw-off zone.

Thus, considerable differences in the plankton contributed by reservoirs to the river downstream may be achieved by the selective withdrawal of lake water. Surface-(epilimnial)-release reservoirs generally yield an abundant

Table XX The seasonal pattern of runoff, reservoir condition, and plankton development, in the Eleiyele Reservoir, Nigeria (based on Imevbore, 1967. Reproduced by permission of Dr W. Junk publishers.)

Month	Runoff	Lake condition	Plankton status
June–July	Flood	Low dissolved-oxygen, intense mixing, and high turbidity	Low algal and zooplankton populations
August–September	Dry	Improving transparancy and stratification	Zooplankton and phyto-plankton develop
October	Flood	Resuspension of bottom sediments; mixing	Maximum algal density
November			Maximum zooplankton density
December	Dry	Strong stratification; bottom laters deoxygenated	Few Algae
January–February	Dry		Subsidiary algal peak
March–April	Dry		Subsidiary zooplankton peak
May	Dry		Low plankton

Table XXI Zooplankton abundance in releases from epilimnial and hypolimnial outlets (based on Stroud & Martin, 1973).

Ratio of zooplankton abundance (number per litre) to volume of effluent in:		Barren Lake (hypolimnial)	Nollin Lake (epilimnial)
April		0.46	0.50
May		1.40	1.40
June		2.30	2.10
July	Stratification begins	0.044	—
August		0.058	0.29
September		0.011	0.40

Table XXII Plankton contribution to Perkiomen Creek, Pennsylvania, USA, by a hypolimnial-release reservoir. (From Coutant, 1963. Reproduced by permission of the Pennsylvania Academy of Science.)

	Number of organisms ($\times 10^3$) per litre	
	Within the reservoir	Downstream
Protozda	5	38
Bacteria	8846	317 480
Cyanophyceae	0	18
Bacillariophyceae (Diatoms)	206	106
Green Algae	1281	0

plankton supply in contrast to hypolimnial-release dams, which provide only low concentrations of both phytoplankton (Hartman & Himes, 1961) and zooplankton (Ward, 1975) during the period of lake stratification. It has also been suggested that, even in surface releases, the concentration of zooplankton may be unexpectedly low, and the species diversity of the lake not fully represented, because of the ability of adult plankton crustaceans (copepods and cladocerans) to avoid lake outflows (Chandler, 1939; Brook & Rzóska, 1954; Brook & Woodward, 1956).

Although the concentration of living plankton is considerably limited by hypolimnial releases, a greater total number of individuals, and a greater volume of protoplasm per unit volume of water, will, in fact, be discharged to the river downstream, because of the release of large numbers of anaerobic Bacteria and moribund diatoms (Coutant, 1963). However, the lotic environment is highly unsuitable for anaerobic Bacteria, so that the plankton derived from hypolimnial-release reservoirs should be considered as largely consisting of organic debris, rather than functioning organisms. Nevertheless, the bacterial cells, and moribund diatoms, make an important contribution to the decomposable organic debris of the stream. In Perkiomen Creek, Pennsylvania, USA, organic enrichment from hypolimnial releases amounted to nearly one part per million (Coutant, 1963).

Plankton within Impounded Rivers

The flood-mitigating characteristic of river impoundment has had a significant effect in promoting the maintenance of relatively high plankton populations within regulated rivers. Flow-regulation imposed by Eildon Reservoir, Murray River, Australia, has allowed the increased development of phytoplankton within backwaters, billabongs, and fringing *Juncus* beds, creating an important source of phytoplankton (Shiel, 1978). The River's zooplankton is derived principally from the Reservoir, and the regulated flow and temperature regimes, together with the still water of the locks and weirs, contribute to the persistence of the plankton assemblage. Also, significant nutrient inputs from downstream towns and irrigation areas contribute to produce a slow-moving series of more-or-less discrete 'slugs' of water containing algal blooms of *Anabaena*, *Melosira*, *Microcystis*, and *Oscillatoria*. The plankton of an impounded river will be composed of both true lentic plankton, derived from the reservoir, and plankton supplied by the bed, backwaters, and tributaries, of the river below the dam.

For the River Tees, England, UK, the reservoir supplied *Daphnia hyalina*, *Cyclops* sp., *Canthocamptus staphylinus*, and *Bosmina coregoni*, in water from outlet valves at the top and bottom of the lake, but *Hydra vulgaris* reached a peak some 600 m below the dam, due to a contribution from the benthos (Armitage & Capper, 1976). Reservoirs can yield large quantities of plankton to the downstream river: small, turbid, alpine reservoirs yield between 1.4 and 25 tonnes per yr (wet-weight) of copepods and cladocerans

(Pechlander, 1964); 1–4 tonnes per yr are released from the low-nutrient-status, and highly turbid, Cow Green Reservoir (Armitage & Capper, 1976); but two relatively clear reservoirs on the Missouri River, USA, have yielded more than 12,500 tonnes per yr, and this supports an important tailwater fishery (Cowell, 1967). However, the morphological, hydrological, and anthropogenic, character of each impounded river will give rise to a different level of plankton persistence and production.

Downstream persistence of Lentic plankton

Within a reservoir, plankton development and distribution are directly related to water turbidity, i.e. light penetration, and to oxygen and nutrient availability. The fate of lake plankton entering lotic systems, producing a decrease in plankton population with distance downstream, has long been appreciated (Chandler, 1937, 1939; Reif, 1939). For example, Chandler (1937) observed a reduction in the phytoplankton component by 70% in 32 km, and by 60% in 24 km, along two impounded rivers. Both rivers were heavily vegetated with submerged aquatics, and the filtering action of these plants was identified as the primary cause of the observed phytoplankton decrease: species of *Najas* and *Chara* were found to be particularly effective filters, as was the periphyton. Stober (1964) similarly observed a decrease in the number of phytoplankton organisms per litre in the Marias River, Montana, USA—from 955 below the dam, to 471 some 67 km downstream from the Tiber Reservoir; but grazing, sedimentation, and mechanical destruction, as well as filtering, may have effected the decrease in numbers in this case. Mechanical destruction during passage over cataracts and rapids has been identified as the cause of plankton obliteration on the Nile (Rzóska, 1976). Similarly, during periods of peak abundance only 1–2% of the numbers of plankton individuals observed at the outfall of Cow Green Reservoir, England, UK, were found 6.5 km below the dam—apparently due, primarily, to the destructive effects of rapids (Armitage & Capper, 1976).

Lentic zooplankton are affected most severely by the change to lotic conditions, and rapidly decrease in numbers downstream (Stober, 1964). On one occasion in the River Svratka, Czechoslovakia, Peňáz et al. (1968) recorded that the number of individuals of *Daphnia hyalina* decreased by 62%, and of *Eudiaptomus gracilis* by 86%, within only 10 km of the dam. Furthermore, some 42% of the drifting *D. hyalina*, and 14.5% of the drifting *E. gracilis*, were damaged during the 10 km of transport, while the Rotatoria and Copepoda survived. This suggests that the persistence of lentic plankton downstream from dams is highly selective, so that a change of zooplankton composition, as well as of its density, may occur. Thus on the South Platte River below Cheesman Lake, Colorado, USA, Ward (1975) observed that the density of zooplankton decreased downstream, with annual means of 202, 111, 72, and 34 zooplankters per litre at four sites, respectively 0.25, 2.4, 5.0, and 8.5, km below the dam, and that whilst Copepoda dominated the zooplankton at the first site, within 5 km of the dam the rotifers were the most abundant organisms (Table XXIII). The reduction in total numbers by 80% within only 8.5 m is related, primarily, to the rapid loss of cladocerans and copepods, whilst the Rotatoria and Dinoflagellata decrease less quickly.

Table XXIII   Downstream persistence of zooplankton below Cheesman Lake, South Platte River, Colorado, USA. (Based on Ward, 1975. Reproduced by permission of E. Schweitzerbart'sche Verlagsbuchhandlung.)

Zooplankton	Size*	Distance below reservoir			
		0.25 km annual mean No. per litre	2.4 km	5.0 km	8.5 km
			Proportion remaining, with site 1 as 1.0		
CLADOCERA	ALL	3.9	0.28	0.28	0.20
Daphnia schodleri	L	1.2	0.17	0.00	0.00
Ceriodaphnia quadrangula	L	0.8	0.13	0.13	0.00
Bosmina coregoni	M	1.1	0.45	0.45	0.00
COPEPODA	ALL	93.3	0.38	0.22	0.09
Cyclops bicuspidatus thomasi	L	13.2	0.27	0.14	0.06
Diaptomus nudus	L	1.4	0.14	0.14	0.07
'Metanauplii'	M	28.9	0.33	0.12	0.03
'Nauplii'	S	49.8	0.44	0.30	0.14
ROTATORIA	ALL	34.2	0.81	0.62	0.40
Filinia longiseta	S	17.3	0.91	0.63	0.24
Keratella quadrata	S	11.8	0.64	0.58	0.46
Keratella cochlearis	S	3.2	0.97	0.69	0.78
Polyarthra sp.	S	1.8	0.72	0.67	0.44
DINOFLAGELLATA	ALL	37.2	0.72	0.56	0.24
SARCODINA	ALL	33.1	0.60	0.27	0.06

* L = large; M = medium; S = small.

Importantly, Ward (1975) demonstrated that the persistence of zooplankton, downstream from a lentic source, is primarily related to their size and form (in terms of their exoskeletal components). The larger organisms, having a smaller surface : volume ratio, will sink more rapidly, become entangled more easily, and be more liable to mechanical destruction and fragmentation, than the smaller ones. For three general size-categories, Ward observed that 64%, 34%, and 24%, of the small, medium, and large, organisms persisted for 2.4 km, and only 20%, 6%, and 5%, respectively, persisted to site 4, 8.5 km below the dam. The 'Copepoda nauplii' were five times as persistent as the larger 'Copeopoda metanauplii', and, even within the categories, size was important: Ward considered that the 32% greater persistence of Keratella cochlearis than K. quadrata is mainly a function of size. However, persistence downstream was also related to the exoskeletal structure and particularly to the form of the appendages, actively swimming crustaceans being more likely than rotifers to become caught up, because of their shape and structure. The larger size and additional appendages of the 'metanauplii', and of Daphnia and Ceriodaphnia, compared with 'nauplii' and *Bosmina*, respectively, would increase their chances of entanglement, and explain their lesser persistence downstream.

In general, the adult copepods in the above observations show greater persistence than the cladocerans of comparable size, because of their more rugged construction and streamlined shape (Ward, 1975). Ward also demonstrated that zooplankton persistence is closely, and positively, correlated to discharge, with a ten-fold variation in zooplankton persistence along the South Platte River between the 'wet' and 'dry' seasons. However, during the dry season the development of periphytic Algae may also play an important role (Chandler, 1937).

Enhancement of Downstream Phytoplankton Production

Reservoirs can provide important supplies of plankton to the river downstream, but river impoundment can also provide conditions to enhance the reproduction of plankton within the lotic habitats below the dam. Potamoplankton production is influenced by many variables, but discharge variability, low-flow velocity, water temperature, and turbidity, are probably the most significant. Turbidity, for instance, interferes with photosynthesis, and is also detrimental to some zooplankters. In large natural rivers, phytoplankton can form a major part of the primary production, although having a strongly seasonal pattern: in summer, the phytoplankton biomass can approach that of nutrient-rich lakes, but in winter, the plankton population is negligible (Hynes, 1970b). The seasonal cycle of phytoplankton development in rivers is in opposition to the peak of inorganic sediment, and non-algal organic debris—and, in summer, algal development—may be attenuated by the presence of suspended mineral particles associated with storm runoff. Therefore, upstream river impoundment, through flow-regulation, temperature moderation, and reducing turbidity, as well as reduction of effluent dilution, can enhance the development of potamoplankton, and stimulate its appearance in 'new' sections of rivers.

Reservoirs, by eliminating summer floods, allow the uninterrupted development of phytoplankton, although the products of excretion and decomposition can favour bacterial growth at this time, so decreasing the transparency of the water. Furthermore, as we have seen above, reservoirs can provide an important source of phytoplankton for downstream reaches, although the lentic species may be selectively eliminated by filtering, sedimentation, and destruction during transport by the river. Chlorophyceae including desmids, are particularly susceptible to selective elimination, whilst the Cyanophyceae and diatoms—especially species of *Cyclotella* and *Asterionella*—can persist (Hynes, 1970b).

Downstream from the Gebel Aulyia Dam on the River Nile, the relatively slow rate of water movement allows for the replenishment and reproduction—as well as selective elimination—of the plankton (Brook &

Rzóska, 1954). Indeed, the concentration of phytoplankton was found to continue increasing beyond the dam, reaching a peak of 6790×10^3 organisms per litre (compared with a maximum of 5200×10^3 in the reservoir), and this represents more than a 100-fold increase over the River upstream of the lake. The phytoplankton were again dominated by the lentic forms, especially the Cyanophyceae (83%) and diatoms (15%). Of the former group, *Anabaena flos-aquae* and *Lyngbya limnetica* together comprised 61% of the total phytoplankton population. The continued multiplication of phytoplankton downstream was suggested to relate to the mixing, and better aeration, of the water as it passed through the sluices. Downstream from the Sennar Dam, on the Blue Nile River, the dense plankton released by the reservoir maintained itself through several generations, and actually increased in density on occasion for 350 km—a travel time of 30 to 40 days (Talling & Rzóska, 1967). The principal components of the Blue Nile plankton are still recognizable for up to 200 km downstream from Khartoum (Hammerton, 1972).

Algae derived from upstream reservoirs have been found to proliferate in the slow-velocity, nutrient-rich, water along the downstream reaches of the regulated River Lot, France, where zooplankton are 'rare' (Décamps *et al.*, 1979). Almost 190 species and varieties were found, of which 60% were diatoms and 5% Euchlorophyceae. In August and September, blooms of Cyanophyceae were observed, composed of species of *Microcystis, Anabaena*, and *Aphanizomenon*, which had apparently been 'inoculated' from the reservoirs upstream. These algal developments have occurred despite an increase in turbidity (mainly organic particles), and short-term flow variability (lower reaches can experience instantaneous-discharge changes of from 6 m^3 per s to 100 m^3 per s to meet power demands) since the dams were built. Therefore, notwithstanding the significance of selective elimination within rivers, upstream reservoirs can constitute an important source of inoculation for downstream reaches, and plankton populations appear able to resume growth at locations of relatively slow current-speeds.

Uncontrolled tributaries entering the main-stream of a river below a dam, may provide a second important source of inoculation, although the deleterious effects of flood discharges, and the introduction of turbid water, may cause the potamoplankton population to be highly variable. Hartman & Himes (1961) have suggested that the increase of Chrysophyceae with distance below the Pymatuning Reservoir, along the Shenango River, Pennyslvania, USA, resulted, in part, from the addition and inoculation of plankton from the tributary streams. At sites of slow water-flow, reduced turbulence, and relatively wide channels,the phytoplankton appeared to be able to utilize available nutrients, allowing an increase in numbers. Downstream from these sites, a progressive reduction in population numbers may occur as a result of nutrient depletion, mechanical destruction in faster-flowing reaches, or increases of inorganic suspended solids below turbid tributaries. However, at low-flows, a stratification of the phytoplankton may develop within large rivers; the diatoms settle, whilst the Chrysophyceae and other mobile forms—Volvocales and Cryptophyceae—concentrate near the surface (Décamps et al., 1979).

Nitrogen, phosphate, and silicate, may be limiting factors for the phytoplankton, but different species require different amounts and ratios of nutrients, such that a nutrient supply-limited condition for one species may actually favour another species. It has been suggested that the construction of reservoirs in series would provide additional water storage, and hence lengthen the time available for plankton increase (Hammerton, 1972), but the development of a potamoplankton downstream from the dams may be limited by depleted nutrient supplies. The fertility of reservoir releases—based largely upon carbon, nitrogen, and phosphorus, loading—is again dependent upon the retention-time. Thus, for four impounded rivers in the Carpathians, Poland, having basins of comparable geology producing water dominated by Ca^{2+} and HCO_3^- ions, Bombówna et al. (1978) observed that the deep Solina Reservoir, having the longest retention-time, markedly reduced the fertility of the river downstream (by 82%), whereas the shallow Goczalkowice impoundment produced the least decrease (23%) and even an increase in fertility of 21% during winter. Below the Carpathian impoundments, the phytoplankton was dominated by diatoms, in contrast with natural eutrophic waters which are dominated by green Algae. Rivers play a positive role in the process of river self-purification, and Bombówna et al. suggested that the changed phytoplankton composition signified a favourable effect of the impoundments on the succession of microorganisms that are typical of mountain streams.

Many industrialized rivers, however, are characterized by several modifications to water-flow: from major dams to weirs and locks, built to control water-levels for supply, power generation, or navigation. Such structures, within impounded rivers in particular, provide suitable locations for the development of a plankton community. The lower reaches of the River Lot, in France, has a succession of weirs, constructed from the fourteenth century onwards to aid navigation, and the river appears as a series of small, slow-flowing water-bodies. These can stagnate when temperatures are high, sunlight is at a maximum, and water levels are low, so that Algae can proliferate in summer (Décamps et al., 1979). Dense phytoplankton communities develop, especially where a nutrient supply is readily available from agricultural, industrial, or domestic, sources. Such regulated, nutrient-rich rivers are highly suited to the growth of phytoplankton (Haslam, 1978), and dense developments of Cyanophyceae, particularly, may be found in regulated rivers below sewage outfalls (Bombówna et al., 1978).

CHAPTER V

Channel Morphology

'I find that the water, that falls at the foot of dams of rivers, places material towards the approach of the water, and carries away from the foot of the dam all the material on which it strikes as it falls.'

Leonardo da Vinci (In: MacCurdy, 1954)

Changes in sediment transport have often been identified as the most important impacts in any assessment of environmental problems within impounded rivers. These impacts arise because more than 90% of the sediment load—and effectively all of the coarser material—will be trapped behind the dam, at least during the early years of operation. The clear-water reservoir releases will be able to pick up available sediment particles from the stream's bed and banks, replenishing the sediment load that is trapped by the reservoir. Thus, severe erosion has been observed within the downstream channel (e.g. Einstein, 1961; Hales et al., 1970) and delta (e.g. Makkaveyev, 1970). Flow regulation, however, will reduce flood magnitudes and the competence of the discharges to transport sediment. Channel sedimentation (e.g. Peñáz et al., 1968), and channel (e.g. Graf, 1979), floodplain (Grimshaw & Lewin, 1980a), and canyon or valley slope (Turner & Karpiscak, 1980), stabilization, have also been reported.

Channel changes below dams are of significance, not least because they relate to the management of the river: aggradation and degradation relate to relocations, channel capacity, channel meandering, tailwater rating curves, and other downstream channel alterations and water-uses (Hathaway, 1948). Aggradation has reduced channel capacities to one-third of those anticipated during the design of reservoir operational procedures, and has increased damage from local flooding on the Republican River, Nebraska, USA (Northrop, 1965). Channel degradation and scour may have damaging effects on riverside structures (bridges, culverts, road embankments, etc.), and in West Pakistan, for example, channel degradation has decreased the water-head at diversion structures and reduced the efficiency of irrigation canals (Gill, 1968). Desirable effects may also arise. Thus channel incision was expected to produce maximum drainage and flood-control benefits in the north fork of the Broad River, and in Barber Creek, Georgia, USA (Miller, 1962). Also, degradation below the Danjiangkou Reservoir on the Chang Jiang, China, has proved to have a favourable effect for navigation

117

(Changming & Dakang, 1983), and the Arkansas River Project—25 dams along more than 700 km of river—has also realized considerable benefits in terms of navigation (Antonio, 1969). Thus, when once channel stability had been achieved, the clear-water releases were harnessed to maintain self-cleaning processes throughout the river's length; releases were used to flush tributary deposits and to maintain navigation depths, although bank stabilization and channel straightening were required along 364 km of river to prevent lateral migration and widening.

In coastal areas, however, erosional processes can produce only detrimental effects. Thus, fears have been expressed for the stability of the Mozambique coastline where fragile beaches may be vulnerable to the changed character of the delicate balance between erosion and deposition (Tinley, 1975). Indeed, the reduction of the sediment discharge to the delta of the Rioni River, USSR, subsequent to headwater impoundment, has resulted in shoreline erosion at a rate of 30 m per yr in this section of the Black Sea (Makkaveyev, 1970).

Rivers will establish an 'equilibrium' form by eroding sediment from some locations and depositing it at others, and the stable form will reflect the magnitude and frequency-distribution of discharges, the volume and particle-size distribution of the sediment load, and the constraints imposed by local conditions—particularly channel boundary materials and valley slope. The channel form is a dynamic features which changes continuously, but these changes will appear as fluctuations about average dimensions; this average form is defined as the 'quasi-equilibrium' channel (Langbein & Leopold, 1964). Within an individual channel-reach, the fluvial processes will relate to the interaction of main-stream and tributary inputs operating on an existing channel form. Under natural conditions, a major storm event can result in one of two sediment output states—depending upon the condition of the channel, which is itself a reflection of the relative recency, magnitude, and character, of preceeding storm discharges, and upon the nature of the sediment inputs to the reach. A single storm may produce abnormally high sediment-yields from tributary basins, and the rate of sediment supply could exceed the transport capability of the mainstream reach, so that deposition would occur within the channel. Alternatively, low sediment-yields may be generated, whereupon the supply to the main-stream could be less than its transport capability, and channel boundary materials could be eroded. In the long-term, erosional and depositional phases will form a balance, maintaining morphological quasi-equilibrium.

Channel change requires the readjustment of the channnel morphology from one quasi-equilibrium state to another, in response to a change in process regime which can be imposed by the impoundment. The readjustment process involves the migration of zones of erosion and/or deposition which restore morphological equilibrium. Thus, sediment transport discontinuities may be identified within impounded rivers, and these represent the transition between channel sections that are adjusted to the imposed flow and sediment-load conditions, and those that were adjusted to the previous regime. Channel adjustments below dams will be governed by the interaction of two discharge-frequencies in relation to the erodibility of the channel boundary and floodplain deposits. Both the rate and direction of channel change will be controlled by the interaction between (1) the frequency of sediment-loaded tributary events, affecting the construction of depositional forms within and along the main-stream, and (2) the frequency of competent reservoir releases, which are responsible for the erosion of existing and introduced deposits.

For the complete analysis of channel changes, all the variables that affect stream behaviour must be considered: discharge variability, water temperature, the composition of the bed and bank materials, bed roughness, the rate of sediment injection by tributaries, and the influence of riparian vegetation. Nevertheless, considerable progress in understanding channel changes below dams has been made in recent years. One important step has been to recognize that upstream river impoundment will initiate the complete readjustment—or metamorphosis (Schumm, 1969)—of channel morphology, not only immediately below the dam, but also throughout a significant length of the river. A second major step has been to appreciate the length of time required for the morphological response to be completed (Petts, 1980a). More than 5 years may be required before any channel response can be observed (Buma & Day, 1977); continuing channel-changes have been reported in some rivers more than 50 years after dam construction (Petts, 1978); and for the Colorado River, below Glen Canyon Dam, stability in terms of sediment transport and channel form may take 200 years (Laursen et al., 1975); certainly it requires more than 10 years but probably less than one thousand.

ACCELERATED EROSION

Reservoirs act as major sediment-traps within the drainage basins, and the downstream transport of bed-material load and coarse suspended load will be disrupted by them. For many rivers the upland, headwater, part of the drainage basin will provide more than 75% of the sediment load. River impoundment will isolate these sediment sources from the river downstream of the dam, and may considerably reduce the sediment yield from the catchment as a whole. This is illustrated by a comparison of two rivers in the UK (Grimshaw & Lewin, 1980a). The catchments are broadly similar in size and general characteristics; but a total of 98 km^2 (54%) of the Rheidol is impounded, while the neighbouring Ystwyth is considered to be natural. Measurements during 1973 to 1975 revealed that the sediment yield had been reduced by more than 90%: on the Rheidol, suspended sediment was only 4400 tonnes and bedload 216 tonnes, in contrast to loads of 55,500 tonnes and 6400 tonnes respectively, on the Ystwyth. Such changes of the sediment load have altered the patterns of erosion and alluvial sedimentation in many rivers.

Frequent and prolonged outflows of clear, sediment-free water from reservoirs into channels composed of transportable boundary materials can result in rapid erosion (degradation), which may extend for many kilometres downstream. Furthermore, channel-bed lowering, and the reduction of water levels in tributary streams during times of high-flow, may initiate tributary rejuvenation, so that accelerated erosion can occur throughout the stream network for a considerable distance below a dam. River-bed degradation at rates of more than 15 cm per year have been reported (Table XXIV), and large volumes of sediment have been removed.

Immediately below a dam, the impact and suction generated by water flowing over a spillweir into the channel can produce a considerable scour-force which may be augmented by the creation and separation of vertically-oriented vortices. Thus, the release of rapidly-pulsating flow from the hydro-electric power Tisalök barrage, Tisza River, Hungary, created scour-holes, from 5 to 10 metres below the former bed-level (Hamvas, 1963). Despite the construction of stilling pools, and structures designed to dissipate the energy of the spillweir flows and dam releases, erosion will occur as the discharges seek to replenish the sediment-load abstracted by the reservoir. Thus, on the Vistula below the Wloclawek dam, between 1969 and 1971, maximum erosion occurred about 600 m downstream (Szupryczyński, 1976): 2033 m^3 of sediment were eroded and the channel-bed was lowered by 2 metres.

Rates of channel-bed erosion below dams are typically greater than erosion rates within natural rivers, at least for several years after dam closure, and the erosional processes tend to dominate from year to year, whereas in natural rivers alternating erosional and depositional processes maintain the channel in quasi-equilibrium. Thus, below Gardiner Dam, South Saskatchewan River, Canada, accelerated erosion rates have been reported for 30 km downstream (Rasid, 1979). Average degradation, defined as the average lowering of the bed *minus* the average aggradation of the bed during the period of record, has accelerated within a 20-km reach: the mean average degradation-rate (33 mm per yr) during the 6 years after dam closure was nearly two-and-a-half times the pre-dam value (14 mm per yr based on the 1964–66 records). A similar, but more marked, change is revealed by the data for 'thalweg' degradation, that is, the amount of maximum deepening of the channel. During the pre-dam period, the mean rate was 30 mm per yr, but two sites were apparently stable; the maximum rate was 82 mm per yr. Subsequent to dam closure, all sites experienced thalweg degradation and the rate of degradation increased to a mean of 170 mm per year, with one extreme value of over 300 mm per yr.

Degradation can affect a considerable length of river—reservoir sediments have been sluiced for 150 km below Possum Kingdom Reservoir on the Brazos River, USA (Stanford & Ward, 1979). Over these long distances, the degradation process shows an asymptotic pattern downstream (Fig. 23, A): high rates of degradation may be observed for several kilometres below the

Table XXIV Observations on channel degradation below dams.

Source	River/Location	Degradation		
		Length of river affected (km)	Rate (mm per yr)	Volume (10^6m^3)
Hathaway (1948)	S. Canadian River, Conchas Dam, USA.	32.2	–	26.9
Komura & Simons (1967)	Colorado River, USA Hoover Dam Fort Peck Dam	145 39	217.7 110.8	– –
Pemberton (1975)	Colorado River, USA, small coffer-dam	9.7	–	2.17
Lawson (1925)	Rio Grande, Elephant Butte, USA	150		
Stanley (1951)	Colorado River, USA Parker Dam Imperial Dam	237	– <508	3.8 1.8
Livesey (1963)	Missouri River, USA, Fort Randall		304.8 for first 2 yrs then 30 for the next 8 yrs	
Patrick et al. (1982)	Missouri River, USA, Garrison Dam	64		
Leopold et al. (1964)	Red River, USA, Denison Dam	160	–	3.65
Leopold et al. (1964)	S. Canadian River, USA, Conchas Dam	–	10	–
Leopold et al. (1964)	Average rate during 10 to 15 years after dam closure in USA	–	30	–
Shulits (1934)	River Saalach, Reichenhall Reservoir, Bavaria	–	152.4	–
Kinawy et al. (1973)	River Nile, Nag Hammadi Barrage, Egypt	Bed lowered by 0.7 m after a few years of operation		
Warner (1981)	Hawkesbury–Nepean River, Australia	36	Area of long profile, increased by 16 000 m²	
Szupryczyński (1976)	River Vistula, Wloclawek Dam, Poland	9	–	4.195
Petts (1978)	UK River Hodder, Stocks Reservoir	0.2	Channel cross-sectional area doubled	
	Camps Water, Camps Reservoir	0.15	ditto	
	River Rede, Catcleugh Reservoir	0.15	ditto	

dam, but these accelerated rates will decrease downstream, initially rapidly and then less quickly. Thus, below Fort Peck Dam, Missouri River, USA, degradation rates of more than 100 mm per yr for the first 10 km are reduced to less than 10 mm per yr beyond 50 km (Hathaway, 1948).

Erosion is characteristically initiated in an upstream section close to the dam, and maximum degradation usually occurs between the tailwater and a

122

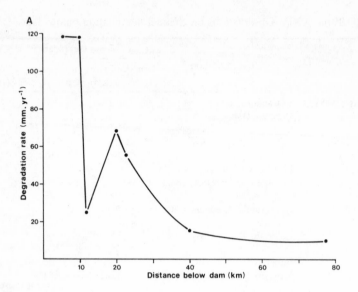

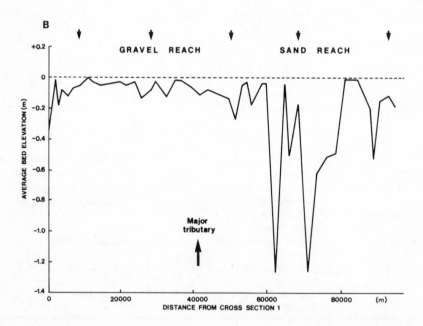

Fig. 23. Observations on channel degradation below dams: an asymptotic trend (A) below Fort Peck Dam, Missouri R., USA, and the predicted effects of variable substrate materials, (B) below Wabo Dam, Purari River, Papua New Guinea. (Based on Hathaway, 1948 (A): Reproduced by permission of the International Association of Hydraulic Research. From Pickup, 1980 (B): Reproduced by permission of John Wiley & Sons, Ltd.)

point sixty-nine channel-widths downstream (Wolman, 1967). The erosional front will move progressively downstream, at a rate of several kilometres per year in lowland streams and tens of kilometres per year in mountain streams (Fedorev, 1969), until the effects of impoundment are eliminated by the sediment loads derived from tributary sources. For sand-bed streams within the Indus system, the erosional front has moved at rates of 32–48 km per yr (Malhotra, 1951). However, the degradation process is variable in space and time, and this is illustrated by Pemberton (1975) for a study of the Colorado River below Glen Canyon Dam (Table XXV). Changes in tailwater levels were noted during the construction of the dam, which was completed in 1963. The small coffer-dam used during construction had little effect on the high-flows but trapped most of the sediment load: $2.17 \times 10^6 m^3$ of sediment were scoured from a 9.6 km reach below this dam, and maximum degradation lowered the channel bed by 1.83 m at a point 1 km downstream during the first 3 years of construction. By 1965, more than $7.5 \times 10^6 m^3$ of channel-bed sands had been removed, but between 1963 and 1965 degradation was greatest between 9 and 13.5 km below the dam. As a result of this accelerated erosion the channel became stabilized, and during the subsequent 10 years there was no significant degradation.

Table XXV Net degradation* of channel reaches along the Colorado River, below Glen Canyon Dam, USA. (Based on Pemberton, 1975.)

Period	Reach (km)				
	0–4.5	4.5–9	9–13.5	13.5–18	18–22.5
1956–59	0.99	0.55	0.25	0.18	0.12
1959–65	0.62	1.73	1.85	1.48	1.23
1965–75	0.005	0.005	0.005	0.05	0.05
1956–75	1.73	2.34	2.10	1.85	1.60

* Degradation values $\times 10^6$ m³.

Limitations to Channel Degradation

Channel degradation and scour below dams will persist until the reduction of channel slope reduces the flow velocity below the threshold for sediment transport. Thus, immediately below Gardiner Dam, South Saskatchewan River, Canada, Rasid (1979) observed that channel slope had been reduced from 1.3 m per km to 0.3 m per km within an 8-km reach 6 years after dam closure; downstream channel slope had altered little. However, field observations of channel degradation (e.g. Lawson, 1925; Lane, 1934; Borland & Miller, 1960) have demonstrated that the rate and magnitude of the degradation process varies considerably, both between impounded rivers and downstream within a single channel. Degradation is rarely able to progress freely, being limited by the existence or development of two conditions. Firstly, if the stream bed and banks are composed of material that is too large, or too cohesive, for the available stream-power to move, or is bound or protected by vegetation, the channel will be stable and degradation will not occur, although transportable sediments introduced by tributaries may be flushed downstream. Secondly, degradation will be limited by the local hydraulic conditions within the channel: the interaction of a low channel slope, large cross-section, and rough boundary, could reduce flow-velocity below the threshold for sediment transport. Thus, Hammad (1972) calculated that the River Nile below Aswan High Dam would attain a stable condition before an appreciable amount of degradation could occur. Indeed, 18 years after dam completion, field data revealed that total degradation below Aswan High Dam had lowered the bed-level by only 0.025 m (Shalash, 1983*a*, 1983*b*). The rate of degradation had not exceeded a few centimetres per year, and had decreased gradually with time.

Few rivers, if any, contain homogeneous bed-sediments, so that numerous interrelated hydraulic, sedimentological, and biotic, factors complicate the process of channel degradation (Tinney, 1962). In particular, where non-transportable particles are present in the bed sediments, the selective transport of the smaller sediment sizes results in the production of a coarse sediment layer at the surface which can protect the underlying material from erosion. This phenomenon of channel armouring, resulting from the erosion of heterogeneous bed-sediments, was identified from flume studies by Harrison (1950). Only a small percentage of non-transportable particle-sizes—gravel or even coarse sand—in the channel boundary materials is required for a natural 'armour' layer to be built up on the bed. A single-grain thickness of coarse materials may effectively prohibit degradation, although rare high-magnitude floods may disturb this surface layer in the long term.

Hydraulic sorting is affected by all the variables involved in the transport of sediment: discharge regime, fluid characteristics, energy slope, channel morphology, and sediment characteristics. Nevertheless, an incomplete gravel layer for several kilometres downstream from Fort Randell Dam, Missouri River, USA, has effected total armouring, and protected the under-lying, erodible sands (Livesey, 1963). Thus, the interaction between river discharges and the sediments which form the channel boundary, will finally result in a state of bed-equilibrium—not just as a result of slope adjustment, but also through changing channel roughness and bed armouring.

Because degradation occurs most rapidly in the upstream reach nearest to the dam, armouring and degradation will shift progressively downstream, until the channel hydraulics are adjusted, or until the bed is protected by an 'armoured' layer. For 170 km below Hoover Dam, Lower Colorado River, USA, $3.5 \times 10^6 \text{m}^3$ of sediment was scoured from the channel bed between 1935 and 1956, and the channel slope was noticeably reduced (Stanley, 1951). However, the rate of degradation varied during this period as the channel-bed progressively developed an 'armoured' layer. During the first 6 months of operation, channel-bed scour averaged over 1 m within a 2-km reach below the dam. The sediment load decreased annually, due to armouring, and this induced a rapid downstream migration of channel degradation. Stanley (1951) observed a similar response below Parker Dam on the Lower Colorado River: over 1 m of erosion within a 24-km reach during the first 6 months produced an 'armour' layer and stabilized the channel; but degradation moved downstream to a point 150 km below the dam, and maximum degradation was experienced 65 km downstream. Thus the extent, and rate, of channel degradation at any location will reflect the particle-size composition of the channel bed-materials in relation to the frequency of competent discharges, and the depth of scour is related to the percentage of non-moving particles within the bed materials, and to the number of particles required to form a single layer.

Hales *et al.* (1970) developed a mathematical–graphical procedure to predict river-bed degradation below three reservoirs in USA: namely, Gavins Point, Garrison, and Fort Randall, all on the Missouri River. The procedure demonstrated the asymptotic behaviour of the degradation process, but the degree of armouring predicted was found to vary considerably between sites. Maximum change was observed nearest the dam, and the degree of change decreased downstream. However, the degree of armouring was highly variable, and this was related to local conditions such as channel meandering, sediment influx from tributaries, and the differing proportions of non-transportable size-fractions within the bed sediments. Thus, a unique response of channel bed-slope would result for each impounded river. Komura & Simons (1967) made the first attempt to incorporate the size of the bed-material theoretically into an analytical solution for the problem of channel degradation. Channel width, the bed and water-surface slopes, and the particle-size distribution of the bed-materials, were employed to obtain a final equilibrium profile for river-bed degradation, and to estimate the effects of 'armouring'. However, their procedure made several assumptions: that the river banks are not erodible and do not constitute a source of sediment supply; that seasonal variations of discharge released from the dam do not occur, or do not produce a breakdown, and turnover, of the armour layer; that meandering does not occur; that sediment injections by tributaries do not occur; and that vegetation growth does not occur. At any location, one or all of these factors may operate to complicate, and commonly to reduce, the degradation of the channel bed. The importance of these factors has been emphasized by Tinney (1962) and Gill (1968).

The character of channel degradation is complicated further by the occurrence of isolated rock outcrops, or coarse 'lag' deposits, at a downstream location, because these can effectively control the slope of, and hence the degradation within, upstream reaches. Thus, Pemberton (1975) observed that degradation below Glen Canyon Dam, Colorado River, USA, had been controlled by a series of ten isolated stable (i.e. 'armoured') gravel-cobble bars within a 24-km reach. Also, the exposure of gravel and clay 'lenses' below Fort Peck Dam on the Missouri River prohibited channel-bed degradation (Einstein & Chien, 1958), while the exposure of rock outcrops and boulders can control degradation, which may be confined to short reaches of sand deposits (Hathaway, 1948; Stanley, 1951). Flow regulation may limit the competence of a river to move coarse deposits that have been produced by tributary processes, landslides, or historic floods. The reduction of peak flows from over 500 m³ per s to 170 m³ per s below Flaming Gorge Dam, Utah, USA, has crossed a significant sediment transport threshold, so that in Dinosaur National Monument—68 km downstream—93% of the rapids are now stable as geomorphic/hydraulic features, though only 62% were stable before dam closure (Graf, 1980).

Several models are currently available for the prediction of river-bed degradation downstream of a dam: for example, the Morphological River Model (Ackers & White, 1973; Bettess & White, 1981), and HEC-6 (HEC, 1977). Both these examples allow for the effects of bed-armouring. The former model requires as data a description of the river and sediment, and details of the discharges. It has been used to predict the degradation that took place downstream of the Milburn division dam in Nebraska, USA, between 1961 and 1964. The channel bed was observed to have been scoured down by nearly 1 m just below the dam, but this was reduced to only 25 cm about 2.5 km downstream. The model provided an accurate assessment of changes in bed-level, and an indication of the changes in the nature of the bed-sediments, occurring during the period of the investigation.

HEC-6 was designed originally for the prediction of erosion and deposition, and changes of the sediment load, above and below dams; but it simplifies river behaviour by treating the horizontal location of the channel banks, and the surface of the floodplain, as fixed. This model has been applied by Pickup (1980) to assess the likely extent and distribution of erosion in the River Purari, Papua New Guinea, subsequent to completion of the Wabo Dam. The results (Fig. 23B) provide estimates of the depth of scour required to develop an armour layer one grain thick on the bed. Over most of the River's length, the channel has a gravel bed and the model predicts that less than 0.2 m of degradation is required for an armour coat to develop, so that channel-bed erosion will be limited. However, a sand-bed reach downstream from cross-section 37—some 60 km below the dam—may be expected to scour up to 1.5 m. Thus, channels having bed-materials of variable size-composition may, as a result of armouring, experience maximum degradation at a location some considerable distance below the dam.

Bank Erosion and Floodplain Processes

Within 'armoured' channels having only minor bed-degradation, or channels receiving only limited quantities of sediment from tributaries, the flows may replenish the sediment load from bank sources. Thus for example Pickup (1980) expected extensive bank erosion to occur along the Purari River below Wabo Dam in Papua New Guinea. Other examples are given by Buma & Day

(1977), Hathaway (1948), and Kellerhals & Gill (1973). Indeed, Hathaway reported severe bank-caving, with little tendency for channel deepening, in actively meandering reaches of the Red River below Denison Dam, Oklahoma, USA. Channel-bank erosion is part of normal river-system dynamics, and is particularly significant within migrating river-meanders. Channel migration is associated with floodplain construction and the deposition of coarse channel sediments, and finer sediments on the floodplain surface. River impoundment can alter the character of migratory rivers in several ways: flood regulation might reduce rates of erosion; sediment abstraction, on the other hand, could accelerate erosion rates, while the depletion of fine suspended solids would reduce the rate of over-bank accretion, so that new floodplains would take longer and longer to mature, and the soils would remain infertile. Certainly, flow regulation by the Nant-y-Môch Reservoir has decreased rates of bank erosion and reduced rates of floodplain accretion along the Rheidol River, Wales, UK (Grimshaw & Lewin, 1980b).

The reduced rate of over-bank accretion consequent on impoundment has led to the development of lower floodplain elevations than would otherwise be expected, and to an absence of levées. Below Garrison Dam, Missouri River, USA, reduced rates of bank erosion within a 48-km reach have been related, in part, to flood regulation (Patrick et al., 1982). In contrast, beaches along the channel margin of the Colorado River were expected to erode for 480 km below Glen Canyon Dam, because the transport capacity of the regulated flows is in excess of the amount of beach material supplied (Laursen et al., 1975). Sediment transport is effectively stratified, with coarse sediments transported on or near the bed, while fine sediments are transported through the system; the intermediate sizes are important for the bars and beaches. Therefore the stability of the beaches depends upon the supply of the moderate-sized sediments, supplied under regulated conditions, primarily by tributary sources. Below Glen Canyon Dam, the capacity of the regulated flows for transporting the beach-building material is 12×10^6 t per yr compared with the tributary sediment supplies of only 2.7 x 10^6 t per yr. Bank erosion may also result in response to the lowering of in-channel water-levels in relation to the ground-water level within the floodplain (Miller, 1962), and in response to an increase in velocity and shear on the banks, as pools fill with sediments.

Thus, under some conditions, upstream impoundment can serve to hasten the processs of channel-bank erosion. From 1966 to 1973, some 230 ha of land were lost (after accounting for land gained by deposition) from 10% of the total bank-length of the mid-Zambezi, between Lake Kariba and the Zimbabwe–Mozambique border (Guy, 1981). Erosion has been particularly pronounced at alluvial sites with non-cohesive sandy bank materials. Guy speculated that the 'apparently excessive rate of erosion' (Guy 1981, p. 199) was related to four characteristics of impounded rivers: the release of silt-free water, the maintenance of unnatural flow-levels, sudden flow-fluctuations, and out-of-season flooding. The maintenance of relatively high, but in-channel, water levels for long periods has encouraged upper-bank erosion as a result of wave action, while out-of-season flooding has had an adverse effect upon the riparian vegetation—adapted to the natural flooding regime—and the effectiveness of the plants in reducing bank erosion has been lessened. The sudden closure of six flood-gates—and sudden cessation of reservoir releases—caused extensive bank-collapse as the ground-water level dropped, and at the same time many of the floodplain channels experienced bed-scour by ebb-flows as the water in them suddenly returned to the river. Repeated water-level fluctuations due to power-peaking releases may also cause accelerated bank-erosion. Such releases into the Connecticut River, producing water-level fluctuations of 1.5 m, are particularly destructive, and bank retreat by up to 8 m is forecast for some sites before stability will be achieved (Simons & Li, 1982).

Cohesive-boundary channels may be particularly prone to accelerated bank erosion if unnaturally high low-flow levels are maintained—a common consequence of river impoundment. The effectiveness of discharges for the erosion of cohesive materials has frequently been related to their degree of preconditioning and wetting (Hooke, 1980). Fifteen years after the construction of off-line storage reservoirs designed to augment summer low-flows on the River Ter, England, UK, channel dimensions have been increased for 12 km below the reservoirs, primarily because of channel-bank erosion (Petts & Pratts, 1983). At the three measurement sites immediately below the reservoir bank, erosion has occurred at a mean rate of 73 mm per yr, and this contrasts with rates of only 42.1 mm per yr for downstream locations. At the upstream sites, flow regulation has increased base-flows by over 1000%. However, base-flows are augmented by less than 300% at the lower end of the reach, as water is abstracted there for irrigation. Thus, the reduced sediment supply and artificial manipulation of water levels may disrupt the process of river migration and floodplain formation; many channels will widen, and local channel instability will result.

Bed-material Composition

An important characteristic of channel degradation below reservoirs is the significant change in the size-frequency distribution of the river-bed sediments. As a result of sediment deposition within the reservoir, the bed will degrade; the finer fractions will be removed from the surface by sorting, and these will be transported, leaving the bed armoured with coarse particles. Selective erosion results in an increase of the median grain-size (d_{50}) of the surface layer of the bed-material (Table XXVI), which becomes approximately equal to the range of the d_{80} through d_{95} size before closure of the dam, where d_{80} and d_{95} are the grain-sizes of the bed-material of which 80% and 95% by weight, respectively, are finer. In general, the change of bed-material size becomes less significant over time and with distance downstream (Table XXVII), and these trends follow the pattern of channel degradation.

Table XXVI   Grain size-changes of channel substrates within degraded reaches.

Source	River	Size parameter	Grain size (mm)	
			Before	After
Livesey (1963)	Missouri River, USA: Fort Randall Dam	d_{65}	0.20	1.0
Einstein & Chien (1958)	Colorado River, USA:			
	Hoover Dam	d_{50}	0.175	33
	Parker Dam	d_{50}	0.18	7
	Imperial Dam	d_{50}	0.125	0.390
Pemberton (1975)	Colorado River, USA: Glen Canyon Dam	d_{50}	0.32<20	100

Table XXVII   Composite data describing equivalent original bed-material average particle-size below Garrison Reservoir, Gavins Point Reservoir, and Fort Randall Dam, Missouri River, USA, after 1 year(*) and 10 years. (Based on Hales et al., 1970. Reproduced by permission of the American Geographical Union.)

Original bed-material (grain-size)	Immediately below dam		Distance downstream					
			5 km		12.5 km		20 km	
d_{10}*	(d_{25})	d_{60}	(d_{20})	d_{40}	(d_{15})	d_{25}	(d_{12})	d_{21}
d_{30}	(d_{57})	d_{92}	(d_{48})	d_{76}	(d_{40})	d_{64}	(d_{38})	d_{57}
d_{50}	(d_{75})	d_{95}	(d_{70})	d_{88}	(d_{65})	d_{82}	(d_{60})	d_{78}
d_{70}	(d_{90})	d_{97}	(d_{85})	d_{92}	(d_{82})	d_{90}	(d_{80})	d_{89}

* 'd' refers to the particle size of specific percentiles (e.g. 10, 50, 65 etc.) on a cumulative frequency distribution.

Marked coarsening of the bed-material occurs during the first few years after dam construction; maximum change is usually observed near the dam, and the degree of change decreases downstream in an asymptotic manner. Noticeable coarsening occurs even with relatively well-sorted sands. Within the South Saskatchewan River below Gardiner Dam, for example, the proportion of coarse material (medium and coarse sand, and fine gravel) has increased at the expense of finer particles (silt and fine sand): the pre-dam mean of 0.53 mm is significantly lower than the post-dam value (0.67 mm), and the proportion of the sediments that are greater than 2.0 mm has increased from less than 5% to 11%. Furthermore, these changes produce a significantly greater negative (coarse) skewness for the post-dam sediments.

SEDIMENTATION

In 1925, Salvador Arroyo wrote of fertile lowlands endangered by raising the bed of the Rio Grande (Arroyo, 1925), a process initiated by the completion of the International Elephant-Butte water storage and diversion project. Sedimentation from outwash of small tributaries occurred as a result of mainstream flow regulation. During the past 50 years, cases of sedimentation in regulated rivers have been increasingly reported, although few give data

Table XXVIII   Observations on channel sedimentation below dams.

River	Reservoir	Observation	Source
Bistula (Romania)	Izvoru Muntelini	120 mm per yr for 3 years along 15-km reach	Ichim & Radoane (1980)
Svratka (Czechoslovakia)	Vir Valley	1.0 m accumulation of fine, organic sediments: maximum accumulation 20 km below the dam	Peňáz et al. (1968)
Rheidol (UK)	Nant-y-Môch	100 mm per yr average rate for 10-years' period	Petts (1984)
Logan Draw (USA)	Wyoming	1.650 m³ yr deposition during first 2 years	King (1961)
Colorado (USA)	Parker	Aggradation in 65 km reach	Stanley (1951)
	Hoover	12 x 10⁶m³ per yr aggradation along a 30-km reach	Stanley (1951)
Colorado (USA)	Glen Canyon	2.6 m aggradation in the upper Grand Canyon	Howard & Dolan (1981)
Rio Grande (USA)	Elephant-Butte	Aggradation in 70 km reach	Lawson (1925)
Hawkesbury– Nepean (Australia)	Five reservoirs	5.6 m aggradation resulted from a single flash-flood May 1943	Warner (1981)
Peace (Canada)	Bennett	Tributary deltas observed after 4 years of flow regulation.	Kellerhals & Gill (1973)
		Aggradation below Pine River–Peace River confluence extended for 8 km	Kellerhals (1982)

to describe the magnitude or rate of sedimentation. Nevertheless, rates of aggradation of 1 m per year have been observed, and tens of kilometres of channel have been affected by sedimentation (Table XXVIII). Malhotra (1951) reported that, for sand-bed channels in the Indus system, the accretionary front moved downstream at a rate of up to 26 km per yr. The aggradation process, however, can be a much slower one than degradation, requiring as it does the introduction and/or redistribution of sediment. The deposit may be composed of finer material than the pre-reservoir deposits, due to the abstraction of the bed-material load and the reduction in competence of the main-stream flows. This would tend to produce a narrower channel of reduced capacity. This morphological response requires the deposition of berms along the channel margin, which may be stabilized by vegetation, to confine the flow to approximately the mean annual flood-level (Richards & Wood, 1977). The progressive realization that degradation is not as problematical as it was formerly considered, has led to a wider appreciation of these less dramatic, short-term changes which, nevertheless, may produce channels of markedly reduced dimensions.

Flow-regulation will reduce the capacity of a river to transport the coarser fractions of the sediment-load supplied by sources downstream from the impoundment. Five main sources of sediment supply have been identified (Table XXIX). Constructional practices may be responsible for an increase of sediment-supply initially, but this will cease when once the dam is complete. Nevertheless, the increased sediment-load may have a significant impact upon the channel downstream, and accumulations of sediment below the dam may provide sources to replenish the load and to satisfy channel degradation for a few years, at least, after dam closure. Prior to the closure of Grandby Dam in September 1949, a large quantity of sediment was introduced into the Colorado River (Eustis & Hillen, 1954). Flows during the construction period were insufficient to remove the material, deposition occurred, and some of the sediment remained as deposits along the channel sides, even after flooding, to form high terraces.

The importance of wind-blown sediment will be dependent upon the local environmental conditions, but may prove to be significant especially in semi-arid areas. Slope-wash, similarly, may provide important sources of fine silts, clays, and organics, which may fill pools in the absence of high-flows to flush the material. On the Svratka River, Czechoslovakia, Peňáz et al. (1968) reported the accumulation of 'boggy' sediments, of thicknesses of up to 1 m on the channel sides and 0.5 m on the channel bed, below the Vir Valley Reservoirs. Sedimentation affected particularly the calmer parts of the stream, but maximum sedimentation was observed nearly 20 km downstream of the dam. Sedimentation here was caused, not only by the reduction of high discharges, but also by a cessation of the erosive effects of drifting ice-floes which caused extensive scouring of the finer deposits under natural conditions.

Channel degradation and scour below dams can provide an important sediment-supply for the river downstream, and several reports have identified channel sedimentation at locations which receive sediment from this source. Degradation for 160 km below Parker Dam, Colorado, USA, has been responsible for sedimentation within the next 65-km reach (Stanley, 1951). Similarly, degradation below Elephant-Butte Reservoir on the Rio Grande released material which was deposited within a reach extending for over 70 km (Lawson, 1925). On a smaller scale, Hathaway (1948) observed a tendency for channels to braid below degraded sections of the Ohio River, due to the selective erosion of the finer sediments within the bed and bank materials, which leaves the coarse sediment forming a lag-deposit on the channel-bed. Hathaway (1948) also noted that, on the Arkansas River, sediment was often moved from narrow sections, which experienced degradation, to wide sections which provided foci for deposition. The redistribution of the channel boundary and floodplain sediments within actively migrating channels may prove particularly significant. Experimental releases from John Martin Dam, on the Arkansas River, USA, resulted in channel scour which was not characterized by the longitudinal transportation of sediment, but by

Table XXIX   Sources of material for channel sedimentation below dams.

	Dam construction	Upstream degradation	Redistribution of channel sediment and bank erosion	Wind-blown	Tributaries
Eustis & Hillen (1954) (USA)	Colorado River				
Stanley (1951) (USA)		Colorado River			
Lawson (1925) (USA)		Rio Grande			
Hathaway (1948) (USA)		Ohio River	Arkansas River	Arkansas River	S. Canadian River
Schumm (1969) (USA)			N. Platte River		
Kellerhals & Gill (1973) (Canada)			Peace River		Peace River
Northrop (1965) (USA)					Republican River
Makkaveyev (1970) (USSR)				Don River	
Petts (1978) (UK)	North Teign	Campswater	Campswater		River Derwent
		River Hodder	River Rede		River Okement
Petts (1984) (UK)					River Vyrnwy
Diacon et al. (1973) (Romania)		River Bistula	Daerwater		River Rheidol
Peňáz et al. (1968) (Czechoslovakia)					River Bistula
Warner (1981) (Australia)					River Svratka
					Hawkesbury–Nepean Rivers

local redistribution: the flows scoured the bed, in places by up to 600 mm, and the material moved laterally, to be deposited on the channel margins. Thus, the lateral migration of meanders, together with the associated differential erosion and redistribution of floodplain deposits, along the River Rede below Catcleugh Reservoir, UK, has produced a new floodplain at a lower elevation than that of the former river (Petts, 1978). However, the more common sediment source for channel sedimentation is the unregulated tributary.

Sedimentation at Tributary Confluences

Deposition within regulated rivers occurs because the regulation of high-magnitude floods artificially slows sediment transport, while the debris discharged from tributaries is, of course, unaffected or possibly increased. Tributaries having steep gradients can introduce coarse sediments into the main-stream, and on entering the relatively low-slope and wide channel, the flows lose competence; this results in aggradation. Prior to dam closure, floods on the main river would periodically have reworked the debris. Tributary confluences provide common loci for sediment accumulation within impounded rivers experiencing regulated flows. King (1961) reported channel-bed aggradation at a rate of 1665 m^3 per yr within a 400 m reach of Logan Draw, Wyoming, USA, during the first 7 years after headwater impoundment; Hathaway (1948) reported sedimentation on the South Canadian River at a tributary confluence 112 km below Conchas Dam, New Mexico, USA; and Warner (1981) identified deltas with dimensions of 180 m and 100 m in length and width, respectively, elevated 3 m above normal water-level, at uncontrolled tributary confluences within the regulated Hawkesbury–Nepean system, Australia.

The most extensively documented case of channel sedimentation, however, is the Colorado River below Glen Canyon Dam, USA (Dolan et al., 1974; Turner & Karpiscak, 1980; Howard & Dolan, 1981). Prior to dam closure in 1963, tributary sediment inputs were eroded by frequent main-stream flows exceeding 1400 m^3 per s. Subsequently, flows of this magnitude have been virtually eliminated: at Lees Ferry, 24 km below the dam, the 10-years' recurrence-interval flood has been reduced from 2437 m^3 per s to only 765 m^3 per s. Downstream from Lees Ferry, flash-flooding tributary sediment loads, supplemented by sediments derived from main-channel degradation upstream of the dam, have produced deposits up to 2.6 m thick within the upper Grand Canyon. Deltas located at every major tributary confluence below Lees Ferry have become significantly enlarged, and unstable canyon side-slopes have been stabilized by debris cones which are no longer undermined by main-stream floods.

Flood regulation will also lower the effective base-level for tributaries, and this can induce tributary rejuvenation. Consequently, the sediment yield from a tributary may be increased for several years after dam closure, as the slope of the side-stream adjusts to the reduced base-level. Under natural conditions the discharge regime of the Peace River, like most north Canadian rivers, is in phase with the regime of its tributaries. As a result of flow regulation, tributary floods are more likely to occur when the stage at the confluence is lower than under natural conditions, producing tributary degradation at their junctions (Kellerhals & Gill, 1973). A reduction of the stage by from 2 to 4 m for 1800 km below the W. A. C. Bennett Dam during spring floods, has resulted in extensive tributary rejuvenation along the Peace River, Canada. Thus for example, on Farrell Creek, degradation some 300 m upstream of the confluence has exposed the foundations of a road bridge (Bray & Kellerhals, 1979), and on Pine River, degradation has extended for 2 km above the confluence (Kellerhals, 1982).

An extreme example has been reported by Makkaveyev (1970): flow regulation by the Tsimlyansk Dam, USSR, lowered tributary base-levels, and induced tributary rejuvenation. In the first few years after dam completion, downcutting in the lower reaches of the Northern Donets increased tributary depths to three times their former extent, and the discharge of sediment caused widespread aggradation within the channels of the lower Don. The sediment loads delivered to the main-stream by tributaries may also be altered by secondary impacts resulting from river impoundment. Dam construction in remote areas may provide a stimulus for other human activities, which can markedly influence the channel of the regulated river. For example, increased sediment yields may be induced by road construction and logging practices (Beschta, 1978), and by drainage ditching (Newson, 1980a).

Initially, tributary-induced sedimentation will be in the form of a simple delta, although the river sediments may be deposited along the channel margins downstream or incorporated into the channel substrate. According to Schumm & Hadley (1967), in sand-bed, semi-arid streams, aggradation would continue until the bed of the channel became over-steepened; the deposit would then be trenched by headward erosion to form a channel that is once again in quasi-equilibrium with the discharge, while the tributary's sediment-load would be transported through the reach—to be deposited in the next reach downstream—and the cycle of erosion and deposition would be repeated. However, observations along British rivers (Petts, 1979) revealed that, in gravel-bed channels, changes of channel-shape, associated particularly with bar construction along the channel margin, may serve to confine the flow, producing an increase in velocity, and facilitating sediment transport through the reach. Indeed, the processes responsible for the selective distribution of sediment size-fractions within a channel have been shown to provide an important mechanism of channel change.

By adjusting channel shape, so as to increase the sediment transport capability of the section, the progressive downstream migration of the depositional front is achieved (Petts, 1982), and a clear pattern of bed-material change may occur. Within the River Rheidol, Wales, UK, channel change, sedimentation, and channel substrate character, are closely related (Fig. 24). The

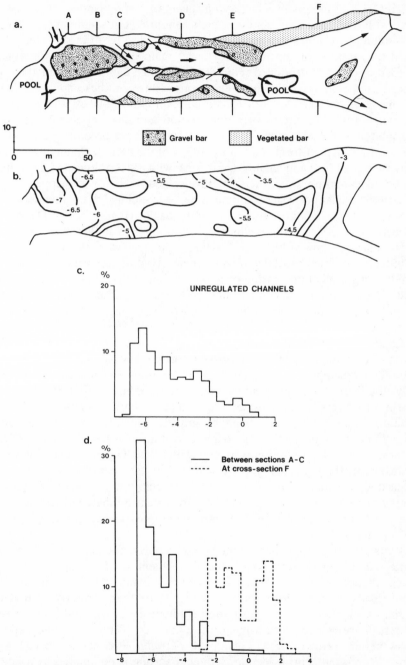

Fig. 24. Sedimentation within the River Rheidol below Nant-y-Môch Dam, UK: morphology (a), mean particle-size variations (phi values) of the surface layer (b), and (c and d) particle-size frequency distributions (phi values) for unregulated and regulated sections.

136

small (4.1 km) Peithnant tributary has been altered by land-drainage associ-ated with reafforestation, and now yields about 88 t of sediment per km^2 per yr—some seven times that expected from a natural basin in this area. Prior to dam construction upstream, periodic main-stream floods would have flushed down the introduced sediments; but since dam closure, in 1963, the dam has spilled on only three occasions. Between these spillweir flows, discharges have been limited to the compensation flow of 0.16 m^3 per s. Consequently, the pattern of sedimentation and channel change has been dominated by the way in which the tributary flows are accommodated within the existing channel-form. The morphology of the natural channel was relati-vely wide (ca 17 m) and shallow (ca 0.5 m), with a steep slope and a bed lined by boulders and pockets of coarse gravels. Today, the overall appearance is of a braided channel system, although a clearly-defined main channel mean-ders within the confines of the existing banks (Petts, 1984). Deposition occurred at the Peithnant confluences in response to sudden flow-expansion (particularly of flow-width) as the discharge emerged from the steep, confined tributary, into the regulated main-stream. Hydraulic sorting operated to produce deposits of decreasing size downstream of the confluence (Fig. 24b), and the generation of secondary flow-cells has caused selective, lateral sedi-ment dispersal, the construction of channel-side berms, and the confinement of the main flow-line.

Channel sedimentation within most rivers is localized, and the readjust-ment of channel morphology to provide for the downstream transmission of the sediment supplied will certainly require tens, and in some cases hundreds, of years. During this long readjustment period, sedimentation processes may operate to diversify the bed-material structure. Natural channels within the Rheidol basin typically are lined by gravels (Fig. 24a), which are poorly sorted and have an open-work coarse gravel layer at the surface, overlying a more compact and well-mixed gravel deposit. The frequency-by-weight distributions show a primary gravel mode, with the coarser end of the curve truncated at about -6.7ϕ† (ca 100 mm). This abrupt truncation relates to a limitation on the maximum particle-size by stream competence. The distribu-tions have a mean particle-size of -4.15ϕ (18 mm), and sand forms less than 16% of the sediments.

Within the regulated channel the bed-materials demonstrate a clear spatial pattern, and four substrate-types may be recognized: (1) Immediately below the confluence, the channel morphology has become adjusted to a narrower (and relatively deep) form with a steeper slope than formerly so that the sediment supplied is now transmitted through the reach. Indeed, the bed-materials are coarse, with a mean of -4.9ϕ (ca 30 mm) and the deposits show better sorting, with a clear peak between -7ϕ and -5.5ϕ (128–48 mm). Within the central section, sediment-sorting is dominated by a lateral rather than a longitudinal component, and this involves partial separation between coarser sediment (channels) and finer sediment (bars) (3). Neverthe-less, the bed-materials have a mean grain-size of -3.8ϕ (ca 14 mm), and

† The phi(ϕ) scale is defined as $-\log_2 d$ where d is particle-size in mm.

may contain up to 25% of grains smaller than 0ϕ (1 mm). It is within the third reach centred on Section F in Fig. 24, that the major bed-material changes have occurred. The mean particle-size is reduced to -2ϕ (4 mm) and up to 45% of the samples were smaller than 0ϕ. The frequency-by-weight distributions show a clear bimodality, with one peak within the range -0.5 to -2.5ϕ (1.4–5.6 mm) and a second between 0.5 and 2ϕ (0.71–0.25 mm).

Sites of active sedimentation are commonly associated with a reduction of bed-material size, but within impounded rivers this may be only a temporary change (although persisting perhaps for many years) because when once the channel has adjusted its cross-section and slope to the new flow-regime, the finer sediments will be transported through the reach, and the substrate will change from a depositional character to an erosional one. During this process the depositional front will migrate downstream. Beyond the downstream limit of aggradation (4), the existing channel bed-materials have been stabilized by the regulated flow, providing another discernible substrate-type.

The Role of Vegetation in Channel Change

Recent studies of the effectiveness of high-flows for erosion within river channels (Wolman & Gerson, 1978; Newson, 1980b) have directed attention towards the significance of riparian vegetation for channel change. Vegetation—particularly the composition and density of the plant cover—plays an important role in determining both the magnitude and frequency characteristics of effective events. The existence of a well-established vegetation cover will markedly influence the magnitude of a flood that may be required to induce any significant change in channel morphology. Thus, Graf (1979) demonstrated the influence of the biomass of vegetation on the valley floor as a threshold in the development of montane arroyos and gullies in the Front Range of Colorado. The rate of vegetation establishment in a given region is primarily dependent upon the availability of moisture and the existence of stable substrates. If the rate is rapid, then the channel boundary and floodplain sediments may become protected from erosion, and any new deposits may rapidly become stabilized and incorporated into the floodplain (Petts, 1982). Thus channel changes below dams will reflect, not only the imposed frequency of competent main-stream flows, and the frequency and magnitude of major sediment-producing tributary events, but also the rate of colonization and development of riparian and floodplain vegetation.

Reports of channel sedimentation at tributary confluences, of channel width-reduction in the absence of an immediate sediment supply, or of the stabilization of channel bank-materials when rapid erosion was expected, have emphasized the importance of vegetation both in stabilizing deposits and in trapping fine sediments (Table XXX). Prior to the construction of Harlan County Dam, for example, extensive growths of willows (*Salix* spp.) and cottonwoods (*Populus* spp.) along the Republican River, Nebraska, USA, were prohibited by frequent abrasive main-stream events. Ten years

138

Table XXX   Relationships between vegetation establishment, fluvial processes, and channel change, in the USA.

River	Dam or location	Observation	Source
Colorado	Glen Canyon	Invasion of Salt Cedar (*Tamarix* sp.) and willows, (*Salix* spp.) associated with expansion of debris-fans produced by tributary washes.	Dolan *et al.*, (1974)
		Stabilization of gullies, canyon slopes, and tributary deposits, by dense plant communities including Salt Cedar, Arrowhead (*Sagittaria* sp.) and Seep Willow (*Salix* sp).	R. M. Turner & Karpiscak (1980)
		Deposition of 'fines'—silts and clays—closely associated with encroaching vegetation.	Howard & Dolan (1981)
Logan Draw	Wyoming	Stabilization of sediments and further deposition induced by vegetation.	N. J. King (1961)
Republican	Trenton	Willows and cottonwoods stabilized bank materials and trapped up to 1 m depth of sand in 4 years.	Northrop (1965)
Peace	Bennett	Balsam poplars (*Populus* spp.) and willows stabilized gravel-bars and deltas.	Kellerhals & Gill (1973)
Sandstone Creek	Oklahoma	Establishment of permanent riparian vegetation stabilized channel-bank sediments.	Bergman & Sullivan (1963)
Green	Flaming Gorge	Channel narrowing from 1207 m to 46 m associated with establishment of trees on most of the floodplain.	Schumm (1969)
Tujunga	Tujunga	Failure of vegetation to become established on tributary deposits, contributed to high effectiveness of floods passing the dam and to channel instability.	Petts (1982)

after dam closure, the active channel had been occupied by woody growths, which trapped a depth of up to 1 m of fine sand and reduced the length of active channel erosion from 194 km to 32 km (Northrop, 1965). Moreover, the encroachment of willows, to occupy 60% of the bank-full channel, increased channel roughness, as described by Manning's '*n*',* from 0.036 for the natural sand-bed river to 0.20 for the vegetated channel.

The suspended sediment concentrations and grain-size distributions on the Colorado River, USA, at Grand Canyon are now similar to the pre-dam conditions, because of the flooding of below-dam tributaries (Howard & Dolan, 1981). The deposition of these sediments during 'high' stages along the channel margin has a close association with encroaching vegetation, providing moisture, nutrients, and a firm substrate. As the vegetation develops, further deposition of fine sediment will be encouraged, and the deposits will move towards an increasingly stable condition. Furthermore, the expected erosion of the sandy beaches along the channel subsequent to dam closure (Dolan *et al.*, 1974; Laursen *et al.*, 1975), due to the dramatic decrease of suspended-sediment load, has been observed on only a minor scale. Prior to dam completion, the large seasonal variations of water-stage, and the periodic scour and fill of the bank deposits, prohibited the development of riparian vegetation; subsequently the stabilization of water levels allowed the spread of woody species, especially Salt Cedar (*Tamarix* sp.), and these have protected the beaches from erosion.

The effectiveness of reservoir releases will thus depend not only on the particle-size composition of the deposits and the volume of sediment injected, but also upon the rate of vegetation establishment and growth. Under natural conditions the extensive growth of riparian and floodplain vegetation is often prohibited by relatively frequent flood-flows, which disturb the substrate and have an abrasive effect upon seedlings and, perhaps, larger plants, The reduction in frequency of flood discharges, and the provision of stable low-flows after improvement, may encourage vegetation encroachment which would stabilize deposits, trap further sediments, and reduce the length of active channel-erosion.

CHANNEL FORM

Although understanding of the relationships between channel form and process are as yet imprecise, it is generally agreed (Gregory, 1976*b*) that the frequency of flood discharges, and the magnitude and particle-size distribution of the sediment load, are the dominant controls of channel morphology. Reservoirs markedly alter the processes operating in the downstream river system by isolating headwater sediment sources, controlling floods, and regulating the annual flow-regime. Reduced icing in winter, and unnatural high-releases, may also have profound, but opposing, consequences. Nevertheless, in general terms, a reservoir may induce one of three potential adjustments within each channel reach below a dam (Fig. 25), depending upon the interaction of three factors: the degree of flow-regulation, the resistance of

---

* Manning's '*n*' is a measure of channel roughness related to mean flow velocity: $V = (R^{0.67} S^{0.5})/n$, where R = channel hydraulic radius (cross-sectional area÷wetted perimeter length) and S = channel slope.

140

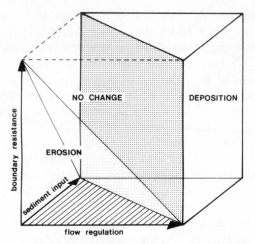

Fig. 25. Potential directions of channel change below impoundments. (From Petts, 1980c. Reproduced by permission of Elsevier Scientific Publishing Company.)

the channel-boundary materials to erosion, and the quantity and calibre of the sediment delivered by tributaries. The reduction in sediment load will tend to cause net scour, but this will be opposed by the effects of flow regulation which, by eliminating flood-peaks, reduces transport capacity. However, for a particular degree of flow regulation, the extent and rate of degradation will depend upon the resistance of the channel boundary and the contribution of sediment from tributary sources. Channels bounded by easily-erodible materials, and located below reservoirs having only a negligible effect upon flood magnitude and frequency, may undergo erosion in response to the release of clear-water discharges, which abruptly increase a river's capacity to transport sediment. Commonly, channel depth and capacity will be increased, and slope lowered, until the flow velocity is reduced below the threshold for sediment transport.

Channel degradation and scour can be limited by four factors. First, the channel bed and banks could be composed of material that is too large, or too cohesive, for the available stream-power to erode it. Second, channel slope could be so small that the velocity of the flows is insufficient to initiate sediment transport. Third, the channel cross-section could be too large, or boundary roughness too high, for the flows to achieve critical erosion-velocities. Fourth, the flows from the dam are so regulated that they may not be competent to transport the channel deposits. In the absence of a non-regulated sediment supply, and in situations in which the regulated discharges are incompetent to erode and redistribute the existing channel-boundary sediment, a simple 'accommodation' adjustment will occur (Petts, 1977). The channel-form is not altered, and the only adjustment occurs in the water cross-section, so that the only physical evidence of river impoundment will

Table XXXI   Observations on channel-capacity changes below dams.

River, reservoir, and country	Channel-capacity change	Source
Republican River, Harlan County, USA	Reduced by 66%	Northrop (1965)
Arkansas River, John Martin, USA	Reduced by 50%	Wolman (1967)
Rio Grande, Elephant-Butte, USA	Reduced by 64%	Lawson (1925)
River Tone, Clatworthy, UK	Reduced by 54%	Gregory & Park (1974)
Thirteen reservoirs, UK	Average reduction 52% range 35% to 84%)	Petts (1978)
River Nidd, Angram, UK	Reduced by 60%	Gregory & Park (1976)
River Bush, Burn, UK	Reduced by 34%	Gregory & Park (1976)
Rio Grande, Elephant-Butte, USA	Increased by 34%	Lawson (1925)
River Hodder, Stocks, UK	Increased by 42%	Petts (1978)
Campswater, Camps, UK	Increased by 31%	Petts (1978)

be a reduced frequency of over-bank and bank-full discharges, and of bed-material turnover, although at some locations a reduced rate of bank migration may also be anticipated (Richards & Wood, 1977). However, at sites of an accommodation adjustment, a high-magnitude, low-frequency, flood-discharge may act as a catalyst for channel change by inducing sediment transport within the regulated stream. Alternatively, sediment may be introduced from upstream or from tributary sources to initiate channel change. Thus, the cross-sectional area of the channel at bank-full stage (channel capacity) has been markedly altered on many rivers (Table XXXI).

The effect of each individual impoundment upon the fluvial processes downstream is produced by a unique combination of climate, geology, size of impoundment, and operational procedures, etc., so that a wide range of river-channel responses can be generated by river regulation. Wolman (1967) attempted to relate channel changes to flow regulation for eleven impounded rivers in the USA, and at a pre- to post-dam discharge ratio of 0.9, or higher, all channels experienced degradation, while for ratios below 0.75, channels tended to aggrade. However, at a ratio of 0.75, channel-capacity demonstrated changes ranging from increases of 40% to decreases of 50%. In a study of 14 impounded gravel-bed rivers in Britain, channel change was detectable along no more than 10 km of river, in some cases even after a period of 50 years (Petts, 1978). In contrast, the impoundment of suspended-load, predominantly sand-bed, channels can cause the most profound changes downstream. Leopold *et al.* (1964) reported channel narrowing and deepening for 550 km below Hoover Dam on the Colorado River, USA. Variable channel-changes also occur along individual impounded rivers. For example, Wolman (1967) examined 21 sites below Garrison Dam on the Missouri River, USA; ten had enlarged cross-sections—five by increasing width and five by increasing depth—and three had reduced cross-sections with smaller widths and depths. Moreover, in most cases the precise cause of channel change has not been quantifiable. The isolation of precise causes and effects in studies of channel change are hindered by the large number of variables involved, by the different rates of response of the different variables at different channel locations, by the time-lags involved in these responses, and, not least, by the dynamic behaviour of the channel-form.

Local and Short-term Variations of Channel Change

Natural river channels are seldom, if ever, in a true state of equilibrium. Not only does the discharge vary continuously, but the water temperature, the rate of sediment injection by tributaries, and even the particle-size distribution of the sediment complex, also vary. The result of this continuous change is a concomitant variation in the morphology of the channel; but over several years, scour and fill will balance to maintain stability. Upstream impoundment disrupts this natural process pattern. For example, the sand-bed of the Colorado River in Grand Canyon, under the natural flow regime, was subject to seasonal cycles of general scour and fill sometimes exceeding 1 m in amplitude (Howard & Dolan, 1981). But subsequent to the closure of Glen Canyon Dam, the reduced sediment-transport capacity has resulted in the domination of sedimentation processes, and up to 1.5 m of sand has covered the channel bed. However, Buma & Day (1977) found that a combination of short-term scour and fill and different channel changes, at cross-sections within a 2.22 km reach of Deer Creek below Deer Creek Reservoir, Canada, prevented the detection of any long-term trend. Of the eight cross-sections monitored over a 5-years' period, all experienced an increase of channel capacity, but this was achieved by variable increases of channel width and depth at different cross-sections.

Within changing channels, the sequence of scour and fill is locally dependent upon the shape of individual channel cross-sections (Andrews, 1982). Thus, downstream from a small reservoir in lowland Britain, 15 years after reservoir completion, Petts & Pratts (1983) observed that a two-fold increase in channel capacity has occurred immediately below the dam, but channel changes varied locally. The different channel dimensions prior to reservoir completion at each cross-section along the river explained most of this variability. As expected, in response to an erosional potential, the rate and magnitude of channel erosion was greatest at those sites which had the smallest dimensions before flow-regulation was introduced, whilst the initially 'over-sized' cross-sections did not experience change.

Short-term observations can, however, be misleading. The adjustment of channel morphology to a new stable 'equilibrium' is not commonly achieved by unidirectional change, but by a complex sequence of changes involving alternate periods of erosion and deposition. These reflect fluctuations about the trend imposed by the new process conditions, which are related to the stochastic nature of discharge events and sediment-load supplies. Observations of channel response in an experimental basin (Schumm, 1977) demonstrated that several alternating phases of erosion and deposition may occur as the channel system adjusts to the effects of change. Thus, Wolman (1967) recorded phases of aggradation and degradation of the channel-bed below Fort Peck Dam on the Missouri River, USA. Moreover, Malhotra (1951) reported that initially rapid channel degradation below dams on the Indus River, India, was followed, after from 20 to 30 years, by aggradation; while at Lees Ferry on the Colorado, below Glen Canyon Dam, USA, initial degradation (ca 3.4 m) has been followed by a slight recovery (ca 0.7 m) (Howard & Dolan, 1981). Thus Schumm (1969) concluded, from observations of the North Platte River, that channel degradation may only be the most immediate result of dam construction, and that, in time, a complete transformation, or metamorphosis, of channel morphology may be a consequence of river impoundment.

Long-term Variations of Channel Response

Variations of channel response along, and between, impounded rivers can be produced in response to slight differences of controls within individual reaches—such as bed-material size, bank-material cohesion, or valley slope. These local controls will modify the general trends produced by the downstream migration of depositional or erosional fronts. Different channel changes in response to the same stimulus may occur because of the large number of channel-form parameters that may change either individually or interactively: river channels are able to adjust their slope, bed-form, boundary roughness, channel capacity, cross-sectional shape, sinuosity, or meander dimensions. All may interact to produce a unique adjustment in response to upstream impoundment. Thus, even in the long term, the adjustment of channel morphology involves a variety of changes.

Channel degradation is often associated with increased channel-depths or channel straightening (Hathaway, 1948), whilst aggradation can produce reduced channel dimensions and increased channel sinuosity (Joglekar & Wadekar, 1951). In an attempt to examine the integrated response of channel-capacity and channel-shape in cross-section, Petts (1980b) employed a modification of the conveyance factor, used in the conveyance-slope method of discharge determination (Sokolov et al., 1976), to provide an improved estimate of channel change. The conveyance factor is a function of the channel-capacity and wetted-perimeter length, and incorporates the hydraulic radius which has an important influence on the efficiency of water-conveyance within a channel. For natural streams, the conveyance factor was found to increase the statistical similarity between cross-sections within a reach. In rivers below 14 reservoirs in the UK, the conveyance factor provided an improved description of the trend within channel response, and indicated a higher magnitude of change than was demonstrated by other morphological variables. The channel efficiency for water conveyance was markedly

144

reduced—on three rivers to less than 20%, and on all the rivers to less than 65%, of the natural capability.

Pratts (1983) has examined the interaction of the morphological variables, including bed-material size and boundary resistance, within a 6.5-km reach below Blackbrook Reservoir, UK. The data (Fig. 26) clearly demonstrate that degradational and aggradational processes can occur at each individual location to produce a channel of smaller cross-sectional area but with coarser bed-material. The degradational processes of selective erosion and armouring have influenced the bed-material composition and local slope, but accretion along the channel margins, enhanced by rapid vegetation establishment, and supported by the redistribution of the bed-material, have reduced channel dimensions. The response is manifested by a channel of reduced water-conveyance capacity, reflecting the virtual total removal of high-flows by the impoundment. Downstream, the addition of water, and sediment, discharges from tributary catchments, is reflected by the progressive return of the chan-nel-form variables to their natural values.

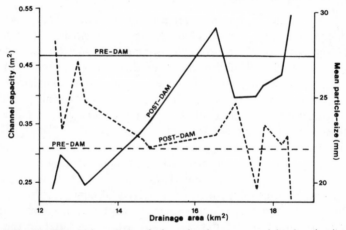

Fig. 26. Channel capacity (—) and substrate particle-size (----) variations below Blackbrook Reservoir, UK. (From Pratts, 1983. Reproduced by permission of J. D. Pratts.)

The diversity of the channel changes observed may be summarized within an empirical framework developed by Schumm (1969). Channel changes are related to the relative changes of discharge and sediment-load at any point on a regulated river, and four groups of changes may be recognized as indicated in Fig. 27. If the sediment load is greatly reduced but the discharges are affected relatively little, then channel degradation will dominate, the channel dimensions will be enlarged, and the slope reduced. However, all reservoirs have some effect upon stream-flow, and within meandering channels—that is, channels which have a local sediment-source from bed and bank erosion—slope may be reduced by degradation, but the construction of a new floodplain at a lower elevation than any former one, can produce

145

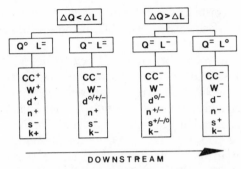

Fig. 27. Channel-form adjustments below dams. Different changes of discharge (Q) and sediment load (L) produce changes of channel capacity (CC), width (W), depth (d), roughness (n), slope (s), and conveyance (k).

a channel of reduced dimensions. At downstream locations, or within reaches below tributary confluences, regulated discharges may be associated with relatively high sediment-loads. Here, channel dimensions will be reduced by sedimentation processes, but channel slope will be steepened. Thus, downstream from small reservoirs, within sand-bed channels, or in environments where the rate of vegetation establishment is slow and growth minimal, clearwater dam releases should produce progressive degradation. In contrast, below large flood-control reservoirs, within gravel-bed rivers, or at locations of rapid vegetation establishment—and assuming a supply of sediment—a relatively rapid adjustment will be achieved, and this will be characterized by a reduction of channel width and of channel-capacity.

These different channel changes are found, not only within different impounded rivers, but also downstream along a single river, and this is a major problem in making a general assessment of the morphological consequences of impoundment—particularly on large rivers. The downstream pattern of channel changes relates both to the variable character of the drainage basin and its channel network, and to the progressive replenishment of the sediment load from channel sources. Furthermore, at downstream locations there will be a reduced tendency to degrade, mainly because the attenuation of surges in the releases reduces the capacity for transport (Laursen et al., 1975). Immediately below a dam, the sediment supply is zero and here the flow competence, rather than capacity, will determine the character of sediment movement and, hence, channel change. In rivers lined by coarse gravel, or boulders, the reduction in peak-flows by a dam will limit the competence of the river to move coarse sediments. Downstream, the sediment load will be progressively replenished from both tributary sources and channel erosion, and the character of sediment movement will be determined by flow-capacity. Although competence-controlled deposition may occur locally at tributary confluences, or at locations of active mass-movement within confined 'canyon' channels, channel changes will proceed until the flow characteristics (primarily depth and velocity) are so adjusted that the flow competence and capacity are capable of transporting all the sediments that are supplied.

Different responses have been observed below Glen Canyon Dam on the Colorado River, USA: channel-bed degradation and scour have character- ized a 24-km reach below the dam (Pemberton, 1975), but further down- stream channel aggradation has dominated (Turner & Karpiscak, 1980). A similar pattern of change has been observed below Elephant-Butte Reservoir on the Rio Grande, USA (Fig. 28, A). Channel degradation for 140 km, associated with an increase of channel capacity by 34%, is followed by channel sections of reduced size and increased slope: channel capacity has been reduced by up to 64%, and at many locations the channel is now higher than the adjacent floodplain, from which it is separated by levées (Lawson, 1925). Below the reservoir, the Rio Grande receives virtually no significant discharge inputs, although flash-flooding ephemeral tributaries contribute large volumes of sediment, and discharge is reduced downstream due to consumptive irrigation uses.

The effects of local conditions have been emphasized by Petts (1979) for two impounded rivers in Britain (Fig. 28, B and C). In both cases, channel degradation occurred immediately below the dam, and at downstream loca- tions channel dimensions were reduced, but the change of channel shape reflected three different processes: channel migration, bank accretion, and bed aggradation. The differential erosion of the channel-perimeter sediment, and the redistribution of the floodplain deposits within actively migrating meanders, characteristically result in a reduction of channel capacity— predominantly by a reduction of channel width. Below tributary confluences, channel-capacities are reduced, but the change can be produced by a primary reduction of channel width, by bank accretion (if suspended sediment, or sand bed-loads, are dominant), or by bed aggradation and depth reduction if coarse bed-load material is introduced. In many cases, channel change involves a combination of both width and depth.

The rate at which a new-equilibrium channel is formed will also be related to local conditions, and to the frequency, in real time, of competent reservoir releases and of sediment-loaded tributary discharges. Because the sediment load of many tributaries is governed by the rate of supply, rather than by channel hydraulics, it is virtually impossible to compute the transport rate. Only for simple situations can we calculate the bed-load transport capacity of the main river and its tributaries, and then the rate of channel change. Gravel-bed rivers require rare, high-magnitude events to induce sediment movement, so that a considerable time-lag may exist before a morphological change is induced. Within sand-bed channels, however, the response may be immediate; channel reaches below tributary confluences may aggrade rapidly, but reaches that are remote from such sources may not change for several years.

An erosional response related to the frequency of competent reservoir

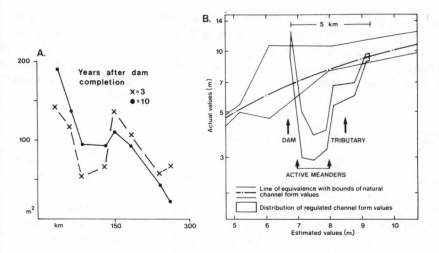

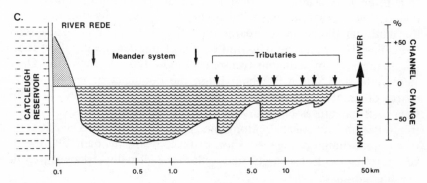

Fig. 28. Downstream variations of channel capacity: below Elephant-Butte Reservoir, Rio Grande, USA (A); Camps Dam, UK (B), and Catcleugh Reservoir, UK (C). (From Lawson, 1925 (A). From Petts, 1980*a* (B): Reproduced by permission of the Foundation for Environmental Conservation and Nicholas Polunin. From Petts, 1979 (C): Reproduced by permission of Edward Arnold (Publishers) Ltd.)

releases will be inversely related to the effectiveness of flow regulation. Thus, channel changes may occur more rapidly downstream from both reservoirs having a minimal effect upon the magnitude and frequency-distribution of flows, and from reservoirs that are operated to maintain high, or pulsating, discharges. For rivers having naturally low sediment-loads, even if dams are built in or immediately upstream from alluvial reaches, some degradation is to be expected, but this will be much slower than has been reported for heavily-loaded rivers in semi-arid areas.

For a depositional response below tributaries, the rate of change is related directly to the efficiency of flow-regulation, and is controlled by the frequency of sediment-loaded tributary 'events'. It is the interaction of these two opposing frequencies which produces the characteristic pattern of channel change below dams. For several years after dam closure, it may be that channel change will be observed at only a few well-defined locations; but the migration of erosional or depositional fronts will eventually establish a new, stable channel-form throughout the length of the regulated river.

Channel Form and Discharge Relationships

The morphology of river channels is related to a range of discharges (Pickup & Reiger, 1979), but a dominant discharge has often been identified, and, on average, the gross morphology of channels appears to be adjusted to the discharge having a frequency of 1.5 years on the annual series. Indeed, the frequency with which a channel is filled to bank-full is commonly within the range of 0.5 to 3 years. Adjustments of channel form may alter the frequency at which parts of the channel—and floodplain—are inundated. For six reservoirs in Britain (Table XXXII), the channel morphology downstream from three of the dams has adjusted to a discharge having a frequency greater than 1.5 years, whilst below the other three reservoirs the magnitude of the bank-full discharge is greater than the 1.5 year 'dominant-discharge' frequency. However, the bank-full discharge frequency will reflect the relative importance of sediment transport. Below Lake Vyrnwy, the Derwent Reservoirs, and Sutton Bingham Dam, the frequency of the bank-full discharge of less than 1.5 years may reflect the relatively high rates of sediment transport that are associated with channel sedimentation below important tributary sediment-sources. Along the Avon, Hodder, and Daer Rivers, the reach of maximum change is a short distance below the dam, and the relatively rare bank-full discharges may reflect the virtually sediment-free flows, and the 'armoured' channel-beds.

Table XXXII Frequency of bank-full discharges below dams in the UK. (From Petts, 1979. Reproduced by permission of Edward Arnold (Publishers) Ltd.)

Catchment	Recurrence interval of bank-full discharge (yrs)	Magnitude of the bank-full discharge (m^3 per s)	Gauged 1:5 yr flood (m^3 per s)
Avon	4.7 <5.5	10.8<12.26	7.84
Hodder	3.6<8.0	15.67<19.07	11.10
Vyrnwy	1.11<1.26	50.20<60.24	65.79
Daer	2.0<4.0	10.49<14.22	8.60
Derwent	< 1	17.29<29.37	50.00
Sutton Bingham	1.38<1.4	8.33<9.37	13.00

Adjustments of channel size, shape, and roughness, downstream from dams, have resulted in changes of the hydraulic-geometry (Leopold & Maddock, 1953), both at-a-station and downstream. The hydraulic-geometry approach considers the variation of water width, mean depth, and mean velocity, at a cross-section, or between cross-sections, as discharge changes. Channel roughness exerts an important control upon the hydraulic-geometry by influencing flow-velocity. Even a small change of bed-material size can produce a significant change of surface roughness, due also to changes of bed-form which relate to sediment size and local hydraulic conditions. Thus, below Imperial Dam on the Colorado River, USA, an increase in bed-material size (d_{50}) from 0.235 mm to 0.39 mm, produced by degradation—together with the associated larger bed-undulations—increased the Manning's roughness coefficient for a discharge of 340 m³ per s from 0.013 to 0.030 (Einstein & Chien, 1958). For this discharge the increased roughness produced a lower flow-velocity of 0.9 m per s and greater water-depth of 2.7 m compared with values of 1.53 m per s and 1.5 m, respectively, under the pre-impoundment regime.

Changes of channel shape will also exert an influence upon the hydraulic-geometry. On the West Okement, below Meldon Reservoir, UK, for example, changes of channel shape have resulted in a marked increase in the rate of variation of width as discharge increases, and minor reductions in both the depth and velocity change-rates (Petts, 1980b). At low-flow (1 m per s), the width of the channel inundated is reduced from 5.25 m to 4.4 m, whilst mean depth is maintained at about 0.23 m. During high-flow conditions (10 m per s), flow width is increased from 6.31 to 6.9 m and depth is reduced from 0.77 to 0.68 m. As discharge increases within braided channels—forms common, at least locally, within changing rivers—existing channels become wider, faster, and deeper, but additional channels become operative elsewhere on the river-bed and these can have similar physical characteristics to those of the original channels at lower discharges (Mosley, 1982).

The distribution of depths and velocities, and the change in water surface-area as discharge varies, are significant for the evaluation of stream habitats. For example, as water surface-area increases with rising discharge, the usable area for food production rises. Most fish-food is produced in shallow riffles with moderate flow-velocities, and in single-channel rivers water depths and velocities progressively increase with discharge throughout the cross-section. This produces a strong peak in usable area, so that an optimum discharge is definable. Large increases in total water surface-area as the flow rises may not provide a greatly enlarged area for food production, because additional area has excessive depths and/or velocities. Channel changes on the West Okement, UK, for example, would allow a higher optimum discharge and a larger usable area to be maintained before critical depths and velocities are exceeded. The significance of channel morphology for the assessment of in-stream flow-needs, and for the development of flow-regulation procedures, is discussed in Chapter 9.

CHAPTER VI

Vegetation Reaction and Structure

Within river systems, autochthonous organic matter is synthesized in primary production that is dominated by macrophytes and periphyton, significant phytoplankton production being found only in large, slow-moving rivers (*see* Chapter 4). Characteristically, however, river systems also receive considerable inputs of allochthonous organic matter from terrestrial sources within their catchments. In an extreme case, the supply of organic matter within streams draining woodland catchments is dominated by the terrestrial contribution—in this example, mainly of fallen tree-leaves. Riparian sources of organic matter are also important in large rivers where the river biota may be dependent, at least during part of their life-cycles, upon the seasonal or regular provision of organic detritus from the vegetation bordering the channel. This riparian habitat includes all of the land bordering the channel within which the biomass is dependent, in some way, upon regular or periodic inundation by floodwaters—for example to provide silt replenishment, or relief from grazing pressure.

AUTOCHTHONOUS PRIMARY PRODUCTION

The majority of aquatic plants are not adapted to lotic conditions; they lack adaptations to resist detachment or damage by even slight increases of flow-velocity, and significant losses of attached flora can result from substrate disturbance during high-flows. Flood frequency is important, as well as flood magnitude, because the frequency of floods which can be tolerated by any particular species depends, in part, upon the rate at which it can grow after damage. Regular, high discharges are important also because they retard the encroachment of terrestrial vegetation into the channel. Indeed, the removal of rooted aquatic plants from gravel substrates is a normal—and vital—factor in the river system, particularly as such gravels provide important spawning-beds for fish. Seasonal variations in water depth and flow commonly lead to seasonal changes in the aquatic plants, but light and water quality can also provide important controls on these patterns. Macrophytes can develop an enriched sediment around them, by trapping the organic and inorganic detritus transported by the river, and by collecting the organic debris from their own or their parts' death. They contribute significantly to primary production; they also act as a physical substrate for periphyton and invertebrates, facilitating overall food-chain production, and, therefore, providing a very positive benefit to the river ecosystem (Welch, 1980).

Attached Algae tend to dominate turbulent, clear water and fast-flowing, headwater streams, and they may develop profusely under conditions of stable summer flows. Angiosperms commonly increase in abundance in the middle reaches, where channel slope and flow-velocities decrease; but in downstream areas they become limited by the reduced light-penetration associated with higher loads of suspended solids, increased concentrations of dissolved organic matter, and increased water-depths. Westlake (1975) has suggested that because of the better nutrient supply, the growth of aquatic plants in fast-flowing streams might be more rapid than in slow-flowing, or still, waters. The lack of evidence for this expected relationship in natural rivers has been explained by the continual loss of plant fragments, broken off by the current, which obscures the benefits of higher velocities.

Within impounded rivers at least in temperate regions, the maintenance of higher summer discharges, the reduction of flood magnitude and frequency, reduced turbidities (increased light-penetration), and the regulation of the thermal regime, may allow the expected extensive and rapid growth in fast-flowing waters. In the naturally-regulated San Marcos River, Texas, USA—originating as a sub-tropical spring—Hannan & Dorris (1970) observed a year-round growth in response to the constant temperature regime and relatively regular discharge. Thus Bryophyta, which normally dominate the larger plants in upstream reaches having rock substrates and high velocities, have also been identified as dominating an increased phytobenthos below dams (Peňáz et al., 1968; Ward, 1976a). Armitage (1977) reported that a bryophyte cover had developed within the River Tees soon after the completion of Cow Green Reservoir and, subsequently, covered over 20% of the channel bed in the reach below the dam during autumn (Holmes & Whitton, 1981). Bryophytes require a firm, unsilted substrate and are favoured by turbulent water having an adequate supply of dissolved carbon dioxide, and, as a result, they are highly suited to locations immediately below dams.

Attached Algae

Algae may be attached to any submerged object, including larger plants, and appear as either thin films on substrate surfaces, or as strands or filaments. Some may be leaf-like, and other—especially Charophyceae—may resemble 'higher' plants, with which they may be classed ecologically. Such layers of attached plants in fresh waters are referred to as the periphyton or 'Aufwuchs'. Like the higher plants, attached Algae respond to particular combinations of temperature, light (i.e. turbidity), substrate stability, and current velocity, but they are also sensitive to water-quality changes. Many diatom species, for example, have definite ecological requirements and tolerances, and, consequently, can be used as indicators of environmental conditions (Mannion, 1982). This high degree of sensitivity is partly a reflection of their rapid reproduction-rate, which allows the preferential development of species favoured by a particular water-quality. However, few Algae have a well-defined habitat, and it is difficult to isolate the dominant cause of species variation in space and time.

Current velocity is certainly one important control variable, although the reason for this relates to several factors. Attached Algae often require a reasonable water-flow. High velocities can stimulate vegetative growth, because the rapid exchange of water around cell surfaces removes wastes and replenishes the nutrient flow—particularly during warm weather, when the dissolved gas contents (CO_2 and O_2) are low. Thus algal production, export, and biomass-turnover rate, are greater in faster currents than in slow ones (Welch, 1980). Different species become dominant at different current velocities, and McIntire (1966) reported the development of felt-like diatom growths at high velocities of 0.38 m per s, and of long filaments of green Algae under low-velocity conditions (0.09 m per s). However, at very high current velocities the substrate may become unstable and 'scour' may destroy the attached Algae.

The seasonal variation of water discharge, temperature, and light penetration, produce clear patterns of Algae (Hynes, 1970b): in temperate areas, the attached Algae are dominated in winter by diatoms and sometimes *Hydrurus*, changing to a mixture of other Chlorophyceae (green Algae), Cyanophyceae (blue-green Algae), and Rhodophyceae ('red' Algae), in summer; but in tropical areas the seasonal pattern is dominated by the annual flood-regime and associated turbidity changes, producing a complex pattern of algal development.

Within impounded rivers, an increase is apt to occur in the algal cover on rubble substrates of riffle habitats (Petts & Greenwood, 1981), and this probably results from the reduced flow-variability, the clarifying effects of the reservoir, the more stable substrate, increased nutrients, and higher winter temperatures (Stober, 1964; Ward, 1976c; Gore, 1977). However, the enhancement of low-flows will produce more habitat which remains productive for a longer period than formerly. Although a moderately swift current and stable flow appear to favour the luxuriant growth of periphyton, the effect of flow-regulation upon substrate stability may be the most important control. The periodic disruption of periphytic communities under natural, variable flow conditions may be eliminated, or decreased in frequency, as a result of flow-regulation. This allows the full development of a periphyton assemblage, at least in channels of relatively steep slope where moderate current-speeds can be maintained. Epilithic Algae within the Gordon River, Tasmania, Australia, below Lake Gordon, have been practically eliminated as a result of increased flows, with high levels of organic colour, severely impairing light-penetration (King & Tyler, 1982). However, extensive developments of attached Algae have been observed below dams, and appear to be more typical of impounded rivers. Ward (1976a), for example, observed that the perilithic standing-crops in the regulated portion of the South Platte River below Cheesman Reservoir, Colorado, USA, were from three to twenty times as great as in an unregulated channel. The attached Algae of lotic systems are normally dominated by diatoms, but the Chlorophyceae usually dominate below dams. Impoundment of the River Glama, Norway, has reduced summer flows and increased water temperature; turbidity has also been reduced and flows have been stabilized, while nutrients have increased slightly. Dense carpets of filamentous green Algae (mainly *Zygnema* spp. and *Ulothrix* spp.) have developed in areas of stony substrate and the faster flows, with single-celled benthic species of Chlorophyceae (*Scenedesmus* and *Chlamydomonas* spp.) and diatoms (*Tabellaria* spp. and *Pinnularia* spp.) in areas of slow-flow with sand or mud substrate (Skulberg & Kotai, 1978). Two species of filamentous green Algae have been widely observed below dams: *Cladophora glomerata* and *Ulothrix zonata*. These two species also dominate naturally stable rivers, such as the Metolius River, Oregon, USA, within which temperature, flow-velocity, turbidity, and solutes, fluctuate within only narrow limits (Sherman & Phinney, 1971). The development of *Cladophora* spp. suggests some form of nutrient enrichment (Whitton, 1975), and *U. zonata* is a rheobiont of cool, well-oxygenated streams (Blum, 1960). Reservoirs act as 'sinks' for dissolved solids, and regulate water temperatures, so that expansive development of these species may be expected below deep-release reservoirs.

The reservoir plankton will provide an important source of inocula for the channel immediately below the dam (Hynes, 1970b). For example, an increase in the population of *Cyclotella meneghiniana*, a common planktonic diatom, within Huntingdon Creek below Electric Lake, Utah, USA, was most probably introduced as a result of the drift of *Cyclotella* frustules that had been washed out of the reservoir (Ross & Rushforth, 1980), so that they do not represent true members of the local periphyton. Periphyton development below dams may be stimulated by the high nutrient content of hypolimnial-release reservoirs. Impoundments providing considerable amounts of organic matter, alone or together with inorganic nutrients, would be expected to stimulate species of phytobenths that are normally associated with nutrient-rich streams (Spence & Hynes, 1971; Lawson & Rushforth, 1975). The tailwaters of the naturally saline Brazos River, Texas, USA, receive hypolimnial discharges, and are inhabited by an extraordinarily diverse flora dominated by *Cladophora* spp. (Stanford & Ward, 1979). Moreover, the stabilized river regime may allow the year-round establishment of attached Algae. Thus, the Grand River, Ontario, Canada, is always covered with dense layers of diatoms below Shand and Belwood reservoirs (Spence & Hynes, 1971).

Downstream from hypolimnial-release reservoirs, the composition of the attached Algae, and the proportion of substrate covered, will change as temperature, turbidity, and substrate stability, vary in response to tributary

154

and anthropogenic inputs. Also, algal growths typically occur in the channel immediately downstream from dams, because of the nutrient-loading of the reservoir releases, and diminish downstream due to processes of self-purification. Thus Gore (1977) observed increased periphyton, including dense mats of *Cladophora* spp., in the cold water and low turbidity area below the Tongue River Reservoir, Montana, USA. *Cladophora* spp. dominated the Algae for 30 km below the dam, but became progressively less important downstream, due it was thought to increasing turbidity, and were further reduced through increased grazing, and chemical changes, to be replaced by the cyanophycean *Nostoc* sp. within 120 km of the dam. Similarly, Ross & Rushforth (1980) found no statistically significant difference between 'before and after' surveys of periphytic diatom communities 8 km below Electric Lake, Huntingdon Creek, Utah, at a confluence with a major tributary.

The characteristic changes in the attached algal community composition below dams are exemplified by Ward (1976a) below Cheesman Lake (Table XXXIII). The Cyanophyceae are generally unimportant, comprising 1% or less of the epilithic samples, and diatoms exhibit a stable relative abundance (15.5%), although different genera assume dominance: *Diatomella*, *Synedra*, *Frustulia*, and *Cocconeis*, dominate below the dam; and *Gomphonema*, *Tabellaria*, *Frustulia*, and *Cocconeis*, are prominent downstream from the North Fall tributary. *Hydrurus*, a mainly cold-water stenothermic genus, composes 11% of the samples below the dam, but only 3% at the downstream sites. The major changes, however, are related to the changing relative importance of the Chlorophyceae and detrital material, which display an inverse relationship. The Chlorophyceae decrease rapidly in abundance downstream, though maintaining the same species-composition, while the detrital fraction increases in importance, though both are relatively stable below the tributary confluence. However, considerable differences may be expected below hypolimnial- and epilimnial- release reservoirs. In the case of the latter, lentic phytoplankton productivity can rapidly deplete 'free'

Table XXXIII Epilithic standing-crop, and percentage composition values, below Cheesman Lake, South Platte River, Colorado, USA. (From Ward, 1976a. Reproduced by permission of E. Schweizerbart'sche Verlagsbuchhandlung.)

Distance below dam (km)	Dry-weight per 5-minute sample	Percentage composition		
		Chlorophyceae	Diatoms	Detritus
0.25	9.73	72	11	11
2.4	7.51	50	18	15
5	12.02	38	20	27
North Fork tributary				
32.5	0.64	14	21	60
36.4	0.70	17	16	64
38	2.25	23	13	61

nutrients, so that the release-water is nutrient-poor during periods of stratification, and this could limit periphyton productivity downstream (Lowe, 1979). Despite the different methods employed in the description of attached Algae, which inhibit direct comparison, two studies justify further discussion: that by Peňáz et al. (1968) on the River Svratka, Czechoslovakia, below a deep-release dam, and that by Holmes & Whitton (1981) on the River Tees, UK, downstream from a surface-release reservoir. In contrast to the Vir Valley Reservoir, on the River Svratka, Cow Green, on the River Tees, is generally well-mixed and, in most years, a well-formed hypolimnion fails to develop, while the outflows from the two reservoirs differ in terms of the concentration of some ions. Calcium concentrations, for example, are relatively high in the releases from the Vir Reservoir, as are nitrates and free CO_2, yet Cow Green Reservoir serves to regulate the supply of important nutrients—particularly PO_4-P—so that a moderate level of supply is maintained throughout the year. The regulated thermal regimes are comparable, and both rivers receive important effluent inputs at the downstream sites (Table XXXIV). Intensive assimilation and aeration in the rivers, together with the influence of tributaries, causes the water quality to approximate the original values within a relatively short distance of the dams, but further downstream the introduction of domestic and industrial effluents create eutrophic conditions in the lower reaches. Moreover, within the River Svratka, Peňáz et al., have observed the accumulation of fine sediment, supplied by tributaries, which has produced muddy substrates for 20 km below the dam.

As a result of successful flood-control below the two reservoirs, the channel substrates have become stabilized. This has allowed the extensive development of attached Algae (Table XXXV), encouraged by the buffering effect of the reservoirs upon the water quality of the rivers. Immediately below the Vir Dam, Algae cover some 88% of the channel bed—more than double that of the upstream reach. In contrast, below Cow Green Reservoir, the variation in plant cover is just as great as within natural sites, but although the mean monthly cover is only 41% of the substrate (cf. 57% upstream), the maximm monthly cover of ca 97% is higher than that for the upstream locations (ca 90%). Upstream from both reservoirs, the attached Algae are dominated by closely-fitted communities.

Above Cow Green Reservoir, the attached Algae were dominated by Calothrix spp. and Cymbella spp. The latter were also common within the Svratka, where Homeothrix varians and Melosira varians were comparably important. Downstream from both reservoirs, the diatom-dominated natural populations were replaced by extensive growths of filamentous Algae —particularly Chlorophyceae, of which Oedogonium sp. and Spirogyra spp. are important in both cases. Cyanophyceae, important within both natural Rivers (Homeothrix sp. in the Svratka, and Tolypothrix sp. in the Tees), are also reduced significantly below the dams. Holmes & Whitton (1981) recognized the importance of inocula from upstream sources in controlling

Table XXXIV   Comparison of the water quality along the River Svratka, Czechoslovakia, and River Tees, UK. (Based on Peñáz et al., 1968 (Reproduced by permission of Academia (Czechoslovakia)), and Holmes & Whitton, 1981 (Reproduced by permission of Blackwell Scientific Publications.))

	River Svratka				River Tees				
	Upstream	Distance below dam (km)			Upstream	Distance below dam (km)			
		0.1	10.5	39.9*		0.1	22	68	78*
Temperature	10.1	9.4	9.6	13.5	8.4	6.9	8.6	9.4	9.6
pH	6.8	7.2	7.4	7.5	7.75	7.2	7.5	7.7	7.7
Conductivity at 25°C (µS per cm)					237	85	152	304	607
Na (mg per litre)	6.4	11.4	8.7	8.7	3.6	2.6	4.2	7.8	30.4
K (mg per litre)	3.7	6.2	4.3	6.4	1.01	0.52	0.96	2.16	3.71
Mg (mg per litre)	1.8	4.2	5.5	5.4	2.3	0.86	2.20	5.27	12.78
Ca (mg per litre)	12.3	19.9	14.3	17.4	36.3	8.47	19.4	32.4	51.3
SO$_4$-S (mg per litre)	19.4	31.1	18.5	21.8	5.2	4.05	5.4	9.52	25.36
NO$_3$-N (mg per litre)	5.0	7.0	8.0	11.3	0.13	0.20	0.35	0.89	1.05
NH$_4$-N (mg per litre)	0.3	0.3	0.6	0.4	0.02	0.04	0.02	0.03	0.30
PO$_4$-P (mg per litre)	0.16	0.13	—	—	0.025	0.028	0.045	0.126	1.272

* Site influenced by sewage effluent.

156

Table XXXV Major changes in the composition of the phytobenthos* along the River Svratka, Czechoslovakia, and River Tees, UK. (Based on Peñáz et al., 1968 (Reproduced by permission of Academia (Czechoslovakia)), and Holmes & Whitton, 1981 (Reproduced by permission of Blackwell Scientific Publications.))

| | River Svratka | | | | River Tees | | | | |
| | | Distance below dam (km) | | | | Distance below dam (km) | | | |
	Upstream	0.1	10.5	39.9	Upstream	0.1	22	68	78
DIATOMEA									
Diatoma elongatum	1	3	3	2	2	3	2	2	1
Calothrix papietrina	NR	NR	NR	NR	3	—	1	1	1
Navicula avenacea	3	3	3	3	1	1	2	4	4
Cymbella spp.	2	2	3	3	3	2	3	2	2
Melosira varians	3	4	3	3	—	—	—	—	—
CHLOROPHYCEAE									
Ulothrix zonata	—	1	1	1	2	4	2	1	1
Microspora amoena	—	3	1	—	NR	NR	NR	NR	NR
Oedogenium sp.	1	3	3	—	1	3	1	1	1
Stigeoclonium tenue	2	1	3	1	—	—	—	3	2
Cladophora glomerata	—	1	2	3	—	—	2	4	4
XANTHOPHYCEAE									
Vaucheria sp.	—	2	1	—	—	1	—	1	2
RHODOPHYCEAE									
Lemanea sp.	3	1	4	1	1	1	4	2	—
CONJUGATOPHYCEAE	1	3	2	1	2	3	1	1	1
CYANOPHYCEAE									
Homeothrix varians	3	—	1	3	—	—	1	—	—
Lyngbya sp.	—	2	1	—	NR	NR	NR	NR	NR
Phormidium sp.	2	1	1	1	1	3	2	2	1
Tolypothrix penicillata	NR	NR	NR	NR	3	1	—	2	1

* relative abundance increases on scale 0–4.
NR = not recorded.

the composition of attached Algae below dams. Several of the diatom species that were common within the natural streams are still present below the dams, but the diatom flora is generally reduced, probably because the reservoir forms a barrier that prevents direct inoculation from upstream sites. In contrast, species that are abundant within the reservoir plankton also abound below the dams (e.g. *Diatoma elongatum*), due to constant replenishment. Holmes & Whitton also recognized that the absence of some species of Algae may be related to the persistence of higher plants subsequent to flow-regulation: on the Tees, the growth of *Hildebrandia rivularis*, *Heribaudiella fluviatilis*, and *Hydrurus foetidus*, may have been inhibited by the development of an extensive and permanent bryophyte cover. Despite flow-regulation, the widest range in species total on the River Tees is found immediately below the dam. This is largely due to the development of an extensive periphyton community within which species of *Ulothrix*, *Spirogyra*, *Oedogonium*, *Zygnema*, *Mougeotia* (filamentous green Algae), *Closterium*, and *Micrasterias* (desmids), are important. The extensive development of *Phormidium* sp. was related directly to the effect of flow-regulation, and the elimination or delay of destructive floods may also explain the occurrence of *Lyngbya* sp. on the Svratka.

Downstream from the dams, the periphyton changes progressively as the influence of the reservoir upon discharges, water quality, and sediment loads, is reduced by tributary inputs. These changes are particularly reflected by a reduction in the area of substrate covered: this decreases to a 'low' of about 15%, 22 km below Cow Green Reservoir, and to less than 50% some 30 km below Vir Valley Reservoir. Moreover, the changing species composition testifies to the change of habitat condition. In both cases, *Oedogonium* sp., *Vaucheria* sp., and *Spirogyra* sp., are uncommon more than 20 km from the dams; *Diatoma elongatum*, *Ulothrix zonata*, *Microspora amoena*, and *Phormidium* sp., generally decrease downstream; and *Cymbella* spp., *Homeothrix varians*, and *Lemanea* sp., recover at least to the abundance found in the natural channel—again within about 20 km of the reservoirs. In both cases the development of eutrophic water downstream was associated with the filamentous green Alga *Cladophora glomerata*. Thus, the general pattern of changes within the attached Algae of lotic systems below dams appears to be similar below reservoirs of different character.

A relatively steady current is the major contributing factor that leads to increased densities of periphytic Algae, and indeed short-term flow manipulation can have a severely destructive effect upon the growth of Algae. Different patterns may be found, for example, below hydroelectric power-dams, or irrigation supply-dams, where rapid fluctuations of discharge can have an adverse effect on attached Algae and even inhibit their development. Luxuriant growths of filamentous green Algae (predominantly *Cladophora* spp.), which developed for 8 km below Glen Canyon Dam, Colorado, USA, within 6 years of dam closure, have been destroyed by the scouring action of daily discharge fluctuations of up to 140 m^3 per s (Mullan *et al.*, 1976).

Unnatural discharge fluctuations, in response to the changing demand for irrigation-water released from Jackson Lake, were also reported to have destroyed the Algae within the Snake River, Wyoming, USA (Kroger, 1973). Attached Algae provide an important microhabitat for invertebrate fauna, and the genus *Cladophora* in particular has been widely found to be an important food-source for trout fisheries. However, extensive beds of attached Algae can be unacceptable to water managers. Several problems can arise: an undesirable taste and odour can be created; the decay of the beds can lead to dissolved-oxygen depletion; water intakes can be clogged; intra-gravel flow can be retarded; and fishing and boating can be restricted. Under natural conditions, seasonal variation in light, temperature, and flow, together with the regular destruction of attached Algae by floods, inhibit excessive 'nuisance' growths, although short-term problems may arise. Within stable impounded rivers, and particularly those enriched with organic nutrients, the dominance of certain prolific periphyton species, and the increase in algal production, may produce extensive and dense beds of attached Algae. Thus, Skulberg & Kotai (1978) reported that dense algal carpets and drifting filaments of *Zygnema*, *Ulothrix*, and single-celled benthic species, had made the River Glama, Norway, less suitable for supply purposes, bathing, and fishing, than hitherto.

Higher Plants

The classification 'higher plants' (or vascular plants) relates generally to the Pteridophyta and Spermatophyta, and in rivers the higher plants are dominated by the Angiospermae—the flowering plants, though Charophyceae may also be important. The spatial distribution of higher plants is related to the interaction of several physical and chemical attributes: discharge variability, flow-depth and velocity; turbidity and light-penetration; substrate size and stability; and dissolved chemicals. It is physiologically advantageous for plants to grow in flowing water (cf. still water), because the effects of even low dissolved-oxygen levels are to some extent offset by water movement, but only a limited number of species occur in truly running-water habitats (Hynes, 1970b). Although water depth and light-penetration are important controls upon the composition and spatial patterns of angiosperms, current velocity and the susceptibility of the substrate to scouring are the dominant controls upon the distribution of higher plants. In natural rivers, angiosperms dominate in middle, gravel-bed reaches, e.g. with species of *Ranunculus* and *Myriophyllum*, which have adventitious roots and tough stems, while more fragile plants such as *Callitriche* spp. and some *Potamogeton* spp. are found in low-velocity, silt-substrate, areas (Haslam, 1978).

In oligotrophic waters, which characterize most upland streams, phosphorus, nitrogen, and potassium may limit plant growth, but in lowland rivers an excess of these nutrients is often found. Most species can grow at least slowly in any water, although some species may dominate in waters of particular type because of successful competition in a specific habitat. It is unlikely that changes in temperature, dissolved gases, and nutrients—resulting from upstream impoundment—will be sufficient to affect the higher plants, so that the influence of a dam upon the hydrology and related factors will usually be more important. There are, however, two exceptional cases: (1) where the effluent discharges from industrial, domestic, or agricultural, sources comprise a high proportion of the flow; or (2) where river discharges are supplied by a eutrophic hypolimnial-release reservoir, in either of which extensive growths of aquatic macrophytes can develop. Thus, Churchill & Nicholas (1967) reported heavy growths of aquatic macrophytes for 65 km below highly eutrophic reservoirs on the Holston River, Tennessee, USA. Large daily fluctuations of discharge served only to modify the plant communities, and to produce a heavy and continuous organic load. The second exception is the effect of a regulated thermal regime upon high-latitude, or high-altitude, rivers which develop a winter ice-cover under natural conditions. The maintenance of an ice-free river can allow the extensive development of aquatic macrophytes by eliminating the scouring effect of river ice. Thus the persistence, through the winter, of extensive beds of rooted aquatics (*Potamogeton filiformis, P. crispus*, and *Ranunculus aquatilus*), has been observed below the hypolimnial-release Cheesman Lake, Colorado, USA (Ward, 1974).

River turbidity controls the depth of light-penetration and, therefore, photosynthesis. In most cases turbidity is reduced by upstream impoundment, so that an expansion of aquatic macrophytes would normally be expected. Thus for the Sutlej River, India, high turbidity prevented the development of aquatic macrophytes under natural conditions, but subsequent to the completion of the Bhakra Reservoir, clear-water releases have allowed an explosive macrophyte development (Ras & Palta, 1973). Under natural conditions, if turbid storm-discharges occur predominantly during the winter, and are least frequent during the plants' growing-season, the decrease in photosynthesis that they produce will have little effect on the aquatic macrophytes. The phytoplankton and zooplankton in epilimnial reservoir releases during summer, and the injection of turbid water from tributaries, can effectively increase the turbidity of regulated rivers. The deposition of silt on plant leaves can reduce light-acceptance and concomitantly decrease photosynthesis, so that, for some rivers, upstream impoundment could have a detrimental effect upon aquatic macrophyte communities.

Of particular significance is the general increase in bed-stability downstream from dams: compared with the situation in the 'free' river, the root systems of macrophytes experience reduced effects of scour, the plants themselves suffer less stress from high discharges, and the rate of channel migration is reduced, so that an area of channel-bed available for the development of aquatic macrophytes can be stabilized. Thus, the growth of *Oenanthe crocata* below Burrater Reservoir, UK, has been related to the reduction of substrate mobility consequent upon flow-regulation (Petts & Greenwood, 1981). Similarly, in the years since the creation of Lake Kariba, flow-regulation has allowed the rapid development of rooted macrophytes (*Panicum repens* and *Phragmites mauritianus* within the Zambezi (Jackson & Davies, 1976).

Flow-regulation not only decreases the competence of discharges and inhibits bed-material movement, but also induces the deposition of the finer sediments where supplies are available from tributary or effluent sources. Indeed, channel sedimentation, particularly involving nutrient-rich silt, can markedly alter plant distributions. For example, sedimentation is often associated with the invasion and spread of *Zannichellia palustris*, which traps further sediments as it develops. The significance of improved substrate stability, and sedimentation, for the aquatic macroflora is exemplified by the changed character of the Volta River, Ghana, subsequent to upstream impoundment (Hall & Pople, 1968). Prior to the closure of the Volta Dam at Akosombo, on the Lower Volta River, there was only a sparse, specialized flora which existed on the higher sandbanks that were exposed during the dry season. Only three truly aquatic macrophytes were known from the Volta River: *Tristicha trifaria* in fast-flowing rocky streams, *Vallisneria aethiopica* in fast-flowing sections with stable substrate, and *Ceratophyllum demersum* in slow-flowing, quiet-water areas. Under natural conditions, peak flows up to 14 000 m^3 per s scoured the channel-bed annually, preventing the development of a stable flora. After dam closure in 1964, river flows were reduced to between 20 m^3 per s and 100 m^3 per s, and were much less turbid than formerly, so that the submerged aquatic macrophytes could develop uninterruptedly. However, within the 20 km sampling reach, starting 65 km below the dam, no angiosperms appeared until the autumn of 1966. *Vallisneria aethiopica* was observed first, followed by *Potamogeton octandrus*. By April 1967 both species had undergone an explosive development to form extensive beds in sandy shallows and in water up to 2 m deep. Sandbanks largely devoid of vegetation prior to dam closure became colonized by *Cyperus articulatus* and *Typha australis*; but annual plants were suppressed by the dense weed-growth, and *Chloris robusta* (which was formerly found in damp areas during the dry season) has disappeared.

The time-lag between dam closure and plant establishment will reflect both the time required for a favourable habitat to develop, and the rate of plant dispersal. Within the Volta, *V. aethiopica* and *P. octandrus* both favour sandy substrates, and the time-lag here may reflect a slow rate of sedimentation. However, this was the first discovery of *P. octandrus* in Ghana, even though it is widely distributed throughout the tropical areas of Africa and Asia (Hall & Pople, 1968). The dispersal of seeds, vegetatively reproducing propagules, or fragments, within rivers can occur rapidly in the downstream direction, but upstream dispersal and inter-river disperal can only be effected by a transporting agent—birds can play a particularly important role. River impoundment can isolate headwater areas—a major source of inoculation for downstream reaches—so that a time-lag of several years may occur before aquatic macrophytes become established.

The time-lag required for a stable macrophyte community to develop may be dependent upon the frequency of tributary sources for inoculation. A few small tributaries may have an insignificant effect, whilst larger tributaries may provide for a rapid response. In the former case, macrophytes may 'spread upstream' from the lower river. Thus, below Cow Green Reservoir, UK, Holmes & Whitton (1977) considered that, even after 4 years, a steady state had not been reached. Just prior to dam completion a major flood removed most of the aquatic macrophytes for at lest 75 km below the dam-site. Subsequent to dam closure, two species did not return: namely, *Potamogeton perfoliatus* and *P. pusillus*. These species did not have a source in tributary streams which could inoculate the main river. A third species, *Elodea canadensis*, also disappeared briefly but re-established itself in the river, as in this case a tributary source was available. However, the floristic changes in the Tees have been most markedly affected by a major tributary 72.5 km below the dam. Four angiosperms had become established above the confluence 4 years after dam closure; one, *Ranunculus penicillatus*, had invaded, but three others are spreading upstream from below the River Skerne confluence, namely *Potamogeton crispus, Zannichellia palustrius*, and *Myriophyllum spicatum*. The upstream spread of the submerged angiosperms was continuing at the time of the study, in response to the removal of the primary factors responsible for limiting macrophytic growth—the extreme current velocities and the scouring effects of floods. However, even the most spatially advanced species (*P. crispus*) was still apparently absent for 50 km below the dam.

Changes within the macrophyte community, and especially in the percentage of substrate covered, may occur during the early years after dam closure. During filling, reservoirs can absorb all flood discharges; but when once they are full, they may overflow and allow high discharges to pass into the river downstream, so that the high degree of substrate stability consequent upon dam closure may be only a temporary state. Thus, Hilsenhoff (1971) reported that an almost complete cover of *Potamogeton crispus, P. foliosus, Ranunculus trichophyllus*, and *Elodea canadensis*, developed below a small reservoir on Mill Creek, Wisconsin, USA, during the first summer of impoundment. Three years after dam closure the cover was reduced to between 30% and 50%, and this was only slightly more prevalent than before impoundment. A change of the reservoir release operation may, also, affect the aquatic macrophytes and Algae adversely. If a major release is made from the reservoir, for example to supply irrigation or power-demand, or for scouring the sediment within the reservoir, the effect upon the aquatic vegetation will be parallel to that of a damaging flood-discharge, and recovery may take several years. Alternatively, frequent discharge variations may expose aquatic species, and land plants may invade. Rapid flow-variations from 2.8 to 0.3 m^3 per s, due to irrigation demand, on the Snake River, Wyoming, USA, exposed Algae and higher plants, and led to the destruction of the primary producers (Kroger, 1973).

Despite short-term flow-variations, Haslam (1978) has suggested that, because both drought and storm-flows are usually decreased below dams, and the channel substrate is stabilized, vegetation that is characteristic of a slower flow-type will develop. She suggested that, in hill streams, plants should increase—including *Callitriche* spp., *Ranunculus* spp., *Myriophyllum* spp., and *Sparganium emersum*—whilst an impounded lowland stream should change from a *Ranunculus–Zannichellia palustris*-dominated community to an *Enteromorpha–Sparganium emersum* community. However, within large impounded rivers, the macrophyte community which becomes established subsequent to dam closure will reflect the interaction of several components—especially the degree of flow-regulation, the accumulation of nutrient-rich sediments and fine inorganic substrates, and the enrichment in nutritive materials by effluent discharges. Décamps *et al.* (1979) considered that these factors allowed the development of extensive beds of *Ranunculus* on impounded rivers in France: 30 km of the Dordogne are infested with *Ranunculus fluitans, Potamogeton fluitans, P. crispus, Myriophyllum spicatum, Callitriche hamulata, Fontinalis antipyretica, Hygrohypnum luridum,* and *Leptodyctium riparium.*

Beds of aquatic macrophytes play a significant role within river ecosystems. The development of *Typha latifolia* and *T. domingensis* in submerged deposits subsequent to flow-regulation on the Colorado River, USA, has provided an important food-source for the Beaver (*Castor canadensis*) and Muskrat (*Ondatra zibethicus*), and shelter and nesting for birds (Turner & Karpiscak, 1980). Below deep-release dams, discharges may be oxygenated by plants during the day, and weed-beds provide a diverse microhabitat. Macroinvertebrates can have density values up to fifty times as great as in areas lacking macrophytes (Décamps *et al.*, 1979). However, the development of aquatic plants is not always favourable. The lack of regular flushing by floods on the Zambezi, for example, has encouraged the development of dense aquatic macrophytes to the detriment of living-space for *Hippopotamus amphibius*, crocodiles, and wildfowl (Attwell, 1970). Similarly, stabilized flows on the Tuolumne River, California, USA, have contributed to favourable conditions for luxurious growths of *Eichhornia crassipes* (Water-hyacinth) in some years, which can impair, and totally block in extreme cases, the salmonid migration (Fraser, 1972).

Floating plants are of little importance in temperate latitudes, but in the tropics, where turbid water inhibits the growth of submerged species, floating plants are often important (Hynes, 1970*b*). The elimination of high discharges to flush the problem species has allowed the extensive development of *Eichhornia crassipes* and *Salvinia molesta* in both Africa (Jackson & Rogers, 1976) and Australia (Walker, 1979). *E. crassipes* infested the lower reaches of the Fitzroy River, Australia, after upstream dam construction (Mitchell, 1978). The effectiveness of floods was reduced, and the stabilized flows prevented salt water incursions to the upper tidal areas. The infestations caused deoxygenation of the water, clogged irrigation and water-supply outtakes, provided breeding ground for mosquitoes, and interfered with recreation. These growths may be supplemented by the discharge of floating weeds from infested reservoirs. Thus, Davies (1979) estimated that 150 000 *S. molesta* mats per hour, supplied by Lake Kariba, passed the Luanga confluence on the middle Zambezi in January 1974.

RIPARIAN COMMUNITIES

Under natural conditions, rivers experience a wide range of discharges, and the low-lying land adjacent to the river, often produced by the migration of the river itself, will be regularly inundated. The width of this 'floodplain' which is periodically submerged by high-flows, can vary from little more than 1 m to several or many km. A continuum of morphological river-types exists, ranging from confined rivers, flowing within deep canyons and having only a narrow riparian zone, to true floodplain rivers. The Zambezi, for example, before impoundment regularly flooded up to 16 km on either side of the channel, and the Amazon has vast alluvial plains up to 100 km in width. For these major floodplain rivers, the annual floods can inundate vast areas (Table XXXVI). Sabol (1974) estimated that the area inundated by the 100-years' flood in the United States is 343 000 km²—6% of the total land area. In many cases the 'flood-season' may last for many months: 13 000 km² of the Magdelana River floodplain, Colombia, are inundated periodically for up to 6 months, and in West Africa the main Niger River flood-season may persist for up to 5 months.

Table XXXVI Some floodplain areas that are regularly inundated.

River	Location	Floodplain area (km²)
Danube	Eastern Europe	264 500
Orinoco	Central basin, Venezuela	70 000
Paraguay	Gran Pantaral, Argentina	100 000
Magdalena	Colombia	20 000
Niger	Timbuktu, Mali	30 000
Senegal	Senegal	5000
Zambezi	Barotse plains, Zambia	10 000
Ganges and Brahmaputra	Bangladesh	93 000
Mekong	Kampuchea/S. Vietnam	53 000
Sepik	Papua New Guinea	7500
Irrawaddy	Burma	31 000

The characteristics of the riparian communities are controlled by the dynamic interaction of flooding and sedimentation. Floodplain and deltaic communities are often characterized by a high degree of primary biological production; by the constant replacement of older soil–vegetation complexes that are destroyed by floods and channel erosion; or by new successional sites which are initiated on ground that is formed by the deposition of sediment during receding flood-stages. Floodplain backwaters provide important breeding refuges and habitats for a variety of lotic fauna, and the absence, or curtailment, of flooding limits opportunities for exchanges between the parent river and floodplain pools and lakes (Shiel, 1976). Many of the alluvial plans are being lost because of flood-control by upstream impoundments: the fringing floodplains of the Nile have virtually disappeared, as has much of the Nigerian plain; and the Mekong in south-east Asia is but one example of a regularly-inundated habitat that is being altered at the present time by flood-control measures. Below the Tsimlyansk Reservoir, USSR, the total area of annual floodland of the Lower Don has been reduced from 95 000 to 30 000 ha and the duration of flooding has been reduced on average from 49 to 12 days (Bronfman, 1977). Between 1879 and 1967 the area of floodplain wetlands over a 145-km sample reach of the Missouri River, USA, was reduced by 67% (Whitley & Campbell, 1974). Today, major floodplain rivers are relatively rare, and many of the special marshland and water-meadow habitats have been lost to agricultural, industrial, or urban, development.

Under natural conditions, flood discharges periodically submerge portions of the river-bank or floodplain. These flows rearrange sedimentary deposits, and can be key events in maintaining riparian habitats. An important downstream manifestation of river impoundment is the loss of pulse-stimulated responses at the water–land interface of the riverine system (Gill, 1971). High discharges can retard the encroachment of true terrestrial species, but many riparian plants have evolved with, and become adapted to, the natural flood-regime. Ambasht (1971), for example, identifies four particular adaptations within tropical riparian zones: (1) growth stimulated by flooding (e.g. *Corchorus olitorius*); (2) growth stimulated by silt deposition (e.g. *Echinochloa macclounii*); (3) the development of vegetative buds, or roots, exposed by soil or bank erosion (e.g. *Acacia albida*); and (4) the development of unilateral root-growths to provide anchorage on unstable substrata (e.g. *Euphorbia hirta*).

Under conditions of flow-constancy, riparian vegetation often becomes dominated by tree species. Lush growths of *Salix* spp. (willows) and *Equisetum* spp. (Horsetails) occur along the Mackenzie River, Canada (Gill, 1971); *Salix* spp. and *Populus* spp. (poplars, etc.) are common along the Peace River, Canada (Blench, 1972) and the Republican River, USA (Northrop, 1965); and *Salix* spp. and *Tamarix* spp. (tamarisks) are now common along the Colorado River, USA (Dolan et al., 1974; Turner & Karpiscak, 1980). In other areas, however, the species adapted to pulse-stimulated habitats are adversely affected by flow-regulation. The Eucalyptus forests (primarily *Eucalyptus camaldulensis*) of the Murray floodplain, Australia, depend upon periodic flooding for seed germination, and regeneration has been curtailed by headwater impoundment (Walker, 1979). Artificial pulses generated by dam releases at the wrong time—in ecological terms—have also been recognized as a cause of forest destruction: *Acacia xanthophloea*, for example, is disappearing from the Pongolo system below Pongolapoort Dam, South Africa, as a result of mistimed high-flows (Furness, 1978).

The direct loss of annual silt and nutrient replenishment, consequent upon upstream impoundment, is thought to have contributed to the gradual loss of fertility of formerly productive floodplain soils (Attwell, 1970; Tinley, 1975). However, the expected ill-effects of the Aswan High Dam upon the fertility of the Nile floodplain have been shown to be unfounded (Kinawy et al., 1973): under natural conditions, 88% of the annual silt-load was transported to the sea, depositing only 12% on the floodplain. Moreover, whilst the reservoir traps the inorganic solids, the fine particulate organic matter is retained in suspension. The net result of the High Dam was to increase productivity of the floodplain as a result of the provision of a reliable, all-year-round water-supply. The naturally high productivity of the 0.5×10^6 ha delta wetlands around Lake Athabaska, Canada, was a pulse-stimulated phenomenon, where perched wetlands were rewatered and fertilized annually by flood-waters. After the construction of the Bennet Dam on the Peace River, and the consequent flow-stabilization, the perched wetlands immediately began succession toward a meadow system dominated by Salix spp. (Geen, 1974). Reports of similar fates of productive wetlands are widespread, and include the Kor River below Dorudsan Dam, Iran (Cornwallis, 1968), and the Mogi Guassu–Rio Grande system, Brazil (Godoy, 1975).

Confined Rivers

Many large rivers flow for relatively long distances through canyons with steep, rocky walls. Such bedrock channels do not develop important alluvial deposits along their margins and, under natural conditions, stable substrates for the establishment of plant communities are rare in them. Any riparian ecosystems within such confined rivers are highly dependent upon the frequency of inundation, and the frequency with which sedimentary deposits are moved. Turner & Karpiscak (1980) highlighted the significance of these two factors in their detailed description of floral changes along the Colorado River since the completion of Glen Canyon Dam. Moreover, they demonstrated that floral and vegetational response relies mainly upon the spread—both longitudinally and vertically—of existing local species, although exotic introductions may assume an important role in the longer term.

Through the Grand Canyon, the Colorado River is confined between valley walls, and prior to impoundment upstream, sedimentary deposits were highly unstable, being destroyed by annual flood events. The bounding talus slopes were frequently undermined and eroded by these floods, so that stable substrata were indeed rare. Prior to dam construction, the valley bottom was devoid of a dense riparian community such as is typical of neighbouring streams. Newly-established plants were subjected to early inundation, and even uprooting, by flood-waters. Subsequent to dam closure at Glen Canyon in 1963, the Upper Colorado River above Lake Mead experienced marked hydrological changes. The elimination of destructive high floods and the stabilization of normal low-flows, has created a moist and stable riparian habitat through the arid environment along the canyon bottom. New sand-and silt-deposits, and stabilized gravel-bars, fans, and talus slopes, all provide more suitable substrates for plants at, and near, the water's edge than formerly existed. The interaction between reduced flood magnitude and frequency, increased diurnal flow-variability, and reduced capacity and competence for sediment transport, has induced a marked ecological response within the riparian zone. Three major habitats had been recognized (Dolan et al., 1974), namely: pre-dam flood-terraces, aeolian deposits, and stabilized talus slopes; pre-dam fluvial sediments; and post-dam fluvial sediments.

Detailed vegetation-changes have been identified by Turner & Karpiscak (1980), who compared photographic evidence taken before and after the closure of Glen Canyon Dam. The zone of pre-dam flood-terraces, aeolian deposits, and talus, is clearly defined by a change in the vegetations' density and composition, because the scouring action of the pre-dam flood-waters produced a sharply-defined lower limit to this community. Dominant plants are *Fallugia paradoxia, Acacia greggii*, and *Prosopis juliflora*, which provide some of the seed-source for the vegetation of the deposits at lower elevations. *A. greggii* seedlings in particular are found below the old flood-line, and the two zones are slowly merging. However, vegetation density on the low flood-terraces is decreasing, probably due to the lack of inundation and consequent reduced moisture-content of the soil. The new stability of the talus slopes, on the other hand, may allow the development of a climax vegetation.

Substrate stabilization is again important within the zone of pre-dam fluvial sediments. Prior to dam construction, few plants became established, although *Pluchea sericea* and *Baccharis sarothroides* occurred in the occasional quiet backwaters, and with *Sporobolus* sp. on aeolian deposits. As a result of flow-regulation, the stable substrate now favours plant colonization, but the lack of water has again inhibited growth. Nevertheless, a new community is slowly developing, including species that are commonly found up-slope, and an old-world exotic—*Alhagi camelorum*, which was first reported 140 km below the dam in 1970—is rapidly spreading upstream to become an important member of the community.

The most dramatic changes have occurred within the zone of the post-dam fluvial sediments lying nearest to the River. Here, a barren skirt along both sides of the River has been transformed into a dynamic double strip of vegetation. Under conditions of relative hydrological stability, plant establishment and growth have been rapid. The present community is characterized by both the spread of existing common species, and the colonization and spread of formerly rare or absent species.

Tamarix chinensis had occurred during the pre-dam period as widespread though isolated individuals; but by 1976 it was present almost continuously, the high density of the plants being related to the daily inundation of the river bars—caused by power-peaking reservoir releases, which provide a dependable source of water during the critical period of seed production.

Cynodon dactylon and *Salix* spp. are also recent colonizers, but are subordinate to *Tamarix chinensis*. *Salix interior* was found throughout the Colorado River system, although restricted to stable substrates near to normal river-level. *Pluchea sericea* was apparently unable to withstand the scouring action of flood-waters during the pre-dam period, but its ability to reproduce vegetatively enabled the plant to colonize open alluvial deposits rapidly after dam construction, and it has now become a prominent member of the riparian community.

Recent introductions—including *Ulmus* sp. introduced in 1976—are, however, expected to continue to expand in importance, so that, together with the spread of *A. camelorum* at higher elevations, and the increased community coverage that will result from the maturation of new plants, changes in the riparian community composition are expected to continue. Turner & Karpiscak (1980) concluded (p. 21) that, 13 years after dam-closure, a new ecological equilibrium had not been reached along the Colorado River within Grand Canyon, and a stable riparian community may require decades to become established. Certainly, the changing riparian habitat is influencing the distribution of mammals and, probably, other animals and birds as well. Ruffner & Carothers (1975), for example, reported the introduction to the riparian habitat of *Ammospermophilus leucurus* (White-tailed Antelope-squirrel) and *Peromyscus maniculatus* (Deer Mouse).

Floodplain Rivers

Welcomme (1979) has examined the fate of major floodplain rivers from the fisheries viewpoint, and his assignment of rivers to two main classes, namely, reservoir- and flood-rivers may be usefully adopted for this discussion of riparian communities. Flood-rivers are characterized by marked seasonal variations of discharge, so that the lateral floodplains become submerged by overspill from the main channel. In some areas, vast temporary or seasonal lakes may form. Extreme variations in water chemistry and primary production occur throughout the cycle of flood and drought, and the floodplain community is necessarily adapted to surviving periodic stress—be it flooding during the wet season, or drought during the dry period when the river returns to its channel. Reservoir-rivers, on the other hand, differ both behaviourally and dynamically, and have communities that tend to resemble the populations of lakes rather than those of seasonally-flowing rivers. The latter rivers have a stable flow throughout the year; they lack major flood-periods and only overflow in years of exceptional rainfall, so that the floodplains are typically terrestrial ecosystems. Under natural conditions, reservoir-rivers may be found in climates that are characterized by all-year-round precipitation, or within drainage basins which are dominated by large natural lakes or swamp systems which serve to store water, and hence to regulate discharges throughout the year. Artificial regulation through river impoundment has transformed many former flood-rivers into the stable 'reservoir' type.

Riparian communities within two particular environments are most severely altered by impoundment, namely semi-arid and subarctic ones, although all areas may be affected to some extent. Within the eastern Danube basin, for example, an area of 2 645 000 ha has been protected from floods (Liepolt, 1972). Delta areas will also experience marked changes. The 2 500 km² Peace River delta, in British Columbia, Canada, comprised a unique ecological system of lakes and rivers with sedge-meadows inhabited by Bison (*Bison bison*); after headwater impoundment, mud-flats developed a cover of unpalatable grasses, dense thickets of willows became established on the higher ground, and the natural foliage available for Bison was restricted (Blench, 1972). The regulation of the higher flow-variability rivers reduces or eliminates the perturbing effect of spring floods, so that plant succession will decrease the productivity of the riparian habitat. Thus, Kellerhals & Gill (1973) considered that, in the absence of the continued initiation of primary succession, the biological productivity of any northern Canadian floodplain, or delta, would quickly diminish. These riparian habitats are characterized by a distinct zone of sedges (*Carex* spp.), cottongrasses (*Eriophorum* spp.), and semi-aquatic grasses (Gramineae), which parallels the shore and abuts sharply against the riparian woodland—a junction maintained by ice-scour during the spring flood.

Within the zone of permafrost, such as in the Mackenzie Delta, Kellerhals and the late Don Gill (1973) envisaged that flow-regulation will increase the areal extent and depth of ground-ice: the inorganic soils will change to organic and waterlogged soils, having reduced temperatures in summer and a smaller depth of root penetration. River impoundment could completely alter the character of the riparian habitat: the elimination of silt-laden floods would prevent the periodic replenishment of nutrients in many floodplain and delta lakes, and reduced flooding and ice-scouring would allow the rapid succession of densely-growing *Salix* spp.

It is, however, within the low-latitude areas, having highly seasonable rains, that the riparian habitats are most sensitive to hydrological change. The riparian zone provides a valuable habitat within which the fauna and flora interact to form a delicate balance, characterized by seasonal changes of density and composition. Yet the riparian zone is often ignored:

> 'Two singularly contemputous actions are repeated by planners of almost every large-dam scheme in Africa. The total ecology of the below-dam region is ignored; and major downdraw releases bear no relation to the original flood regime.'—Tinley (1975 p. 25).

The riparian communities in the Kafue Flats of southern Zambia (Rees, 1978) rely heavily upon the annual flow-regime for the maintenance of a highly productive ecosystem. The wet season is long, and the 4 340 km² Kafue Flats remain inundated from December through May, producing a luxuriant water-meadow grassland. This ecosystem is highly dynamic: plant communities exhibit partial monthly successions and annual variations, dependent upon the annual flow-regime; herbivores utilize the area in large numbers, their grazing being facilitated by the slow but continuous uncovering of nutritious herbage as flood-waters retreat during the dry season. The construction of the Iteshiteshi Dam upstream would confine the flows to the main channel for long periods, soils would become drier, grazing pressures would increase, and the floristic combinations within the Kafue Flats would be markedly altered.

The full magnitude of the effects of upstream impoundment upon the riparian habitats of African floodplain rivers has been further demonstrated by studies of the Zambezi (Attwell, 1970; Davies *et al.*, 1975; Tinley, 1975). Impoundment of the Zambezi, with the creation of Lake Kariba, and the consequent flow-regulation—which has allowed salt water incursion in coastal areas—and reduced silt-loads, has destroyed large areas of mangrove along the coastal floodplain (Davies *et al.*, 1975). Reduced flooding in the delta area has dried out the rich alluvial soils—many of which have become alkaline or saline—and savanna vegetation has invaded (Tinley, 1975). However, it is within the 70-km^2 floodplain of the Mana Pools Game Reserve, 130 km below Kariba Dam, that the full consequences of upstream impoundment for riparian communities has been documented (Attwell, 1970). Under natural conditions, the active migration of the Zambezi River had produced a broad floodplain containing a series of residual pools. The floodplain was regularly inundated to a depth of up to 5 m, and flooding persisted for 3 months or more; silts and nutrients were supplied; and stagnant pools were flushed. At the same time, large herbivores were driven off, so that, aided by water-borne seed dispersal (especially of *Acacia albida*), the vegetation could recover from the intense grazing pressure to which it was subjected during the dry season. Thus, the productivity of the floodplain was replenished annually. The full cycle, from flood to drought, significantly contributed to the high productivity of the floodplain area.

The annual floodplain-regime was related *sensu lato* to the desiccation of the hinterland areas, because the drying out of transient wet-season water-supplies forced the fauna to return to the relative luxuriance of the floodplains' replenished vegetation. Now river regulation has reduced the ecological dynamism of the floodplain and removed the very instability which at one time was a key factor in the ecosystem, resulting in reduced productivity. Attwell (1970) recognized two particular effects of Lake Kariba upon floodplain communities: the effect of flood-control during the wet-season, and the effect of reservoir releases at abnormal times.

Flood-control during the natural wet-season in the above instance has induced a response of the vegetation's composition to one favouring drier conditions. Indeed, Savory (1961) stated that the alluvial flats' vegetation was undergroing an observable decline only 3 years after dam closure. Forbs increased on the more raised part of the floodplain, and the species composition of the grass sward declined, with an increase in unpalatable herbs and grasses—particularly *Vetriveria nigritana* (Kerr, 1968). Floodplain pools became infested with emergent rooted aquatics (mainly Cyperaceae) and the water-fern *Salvinia auriculata*. The elimination of flooding has allowed grazing to persist for longer periods than formerly, and some species (e.g. the Waterbuck, *Kobus ellipsiprymnus*) have moved from their wet-season dispersal areas to spend virtually all the year on the floodplain. Consequently the floodplain rest—or recovery—period will be markedly reduced, and the habitat will be placed under additional stress. The ability to release high discharges at abnormal times of the year has encouraged the rapid increase of dry-season crop-growing and this, together with increased recreational activities, has meant that human pressure now makes a wider, and generally greater, impact on the Mana floodplain.

The Floodplain Fauna

'One cannot, nor would one wish to, deny the legitimate needs of the nation for water-conservation works, but one can deplore the lack of consideration that is usually shown to wild-life conservation problems by the planners of these schemes.'

Frith (1977 p. 32)

It is necessary at this point to consider the full significance of upstream impoundment upon floodplain habitats. Indeed, in most areas a delicate balance exists under natural conditions between the flora and fauna of riparian habitats and the annual flood-regime, which is responsible for the creation of their behavioural and reproductive pattern. For the fauna, these habitats are important for food, cover, and breeding. In northern Canada, for example, the riparian areas are the most productive inland habitats for waterfowl, because of the annual replacement in them of nutrients by flowing water, and of their continued initiation of productive successional vegetation (Kellerhals & Gill, 1973). These habitats are particularly important during drought years and, for the fauna of the prairies to the south, the floodplains and deltas function as a sanctuary and breeding-ground for large numbers of waterfowl; the nests of many ducks and geese are concentrated on low alluvial islands, where they are protected from terrestrial predators. Aquatic mammals also depend upon the annual floods to maintain the riparian habitat in a state of high biological productivity. Kellerhals & Gill suggested that flow-regulation may reduce the habitat available for the Muskrat (*Ondatra zibethicus*) and Beaver (*Castor canadensis*) through the deleterious consequences upon food-supply and cover. Along the Mackenzie River, Moose (*Alces alces*) and Ptarmigan (*Lagopus mutus*) are partly dependent upon the lush shoots that grow on annually-produced sediment ridges, but this pulsating regrowth phenomenon would cease if once the flows were stabilized (Gill, 1971).

The riparian areas are of considerable importance for the conservation of wildlife along the Zambezi, because in the dry season they form the main

grazing-areas of the valley. Attwell (1970) has described the biological consequences of the Kariba impoundment on the downstream riparian areas which
constitute important wildlife refuges. As a result of flood-control, extreme
pressure will be placed upon the floodplain vegetation of the Mana Pools
by Elephant (*Loxodonta africana*), Impala (*Aepyceros melampus*), Buffalo
(*Syncerus caffer*), and Zebra (*Equus burchelli*). The impoverishment of the
habitat will place considerable stress upon the populations of these animals.
Already, due to the severe reduction in available habitat, the Hippopotamus
(*Hippopotamus amphibius*), Crocodile (*Crocodylus niloticus*), and various
waterfowl, have suffered a reduction in population numbers. Moreover, the
mistiming of high dam-releases has disrupted the established behavioural and
reproductive pattern of the fauna, and this may affect population numbers
and food-chains. The eggs of reptiles such as the Crocodile and Water
Leguaan (*Varanus niloticus*), and larvae of amphibians have been destroyed
by flushing from pools and backwaters, and birds nesting adjacent to dry-
season water-levels have also been affected (e.g. the White-headed Plover
(*Lobivanellus albiceps*) and the African Skimmer (*Rhynchops flavirostis*)).

Similar problems have resulted from flow-regulation within the Zambezi
delta area (Tinley, 1975). Before the creation of Lake Kariba, and Zambezi
River annually flooded 1800 km² of the delta, within which the Marromeu
Reserve provides an important conservation area for Waterbuck, Zebra,
Elephant, Hippo, and Buffaflo. However, the grasses which support these
populations may become impoverished by excessive grazing and the desiccation, and salinization, of the soils. The potential effects upon the large
Buffalo herds could be disastrous. The consequences for rare species could
be particularly serious: the Kafue Flats, for example, provide a single, unique
habitat for the Kafue Lechwe (*Kobus leche kafuensis*), a semi-aquatic ante-
lope, and the reduced availability of nutritious herbage throughout the dry
season, as a result of flow-regulation, would necessitate a major reduction
of the Lechwe population (Rees, 1978).

The billabongs of the inland Australian rivers are of extreme importance
to many waterfowl, and extensive breeding occurs during the flood-season.
However, the number of breeding ducks is decreasing: river impoundment,
and the consequent flood-control, has greatly decreased the replenishment
of many swamps, lagoons, and billabongs (Frith, 1977). The reduced frequency of flooding decreases both the frequency and scale of breeding, so that
the Freckled Duck (*Sticronetta naevosa*), Grey Teal (*Anas gibberifrons*),
and Hardhead (*Aythya australis*), in particular, have declined owing to the
restriction of breeding areas. Thus, Frith (1977) concluded (p. 31):

'Conservation of breeding habitat in the inland is not so much a matter of
reserving swamps to provide undisturbed conditions, though no one would
deny the value of this also, as of ensuring that the inland rivers continue to
fluctuate in level frequently, so replenishing the lagoons, billabongs, and

depressions on the plains, and creating the correct breeding habitat for the different species, each of which breeds at a different stage of the flooding.'

Frith's statement, directed specifically at waterfowl is, of course, applicable to the whole riparian ecosystem. It is unfortunate that the conflict which exists between the objectives of river impoundment and the needs of riparian ecosystems is so intense. The elimination, or reduction in frequency, of flooding limits opportunities for exchanges between the river and its former floodplain, so that the very processes which cause channel migration, and the construction of new wetlands, will be inhibited. All too often the generation of secondary human activity—agricultural, urban, or industrial, development—on the flood-free land, as a consequence of dam construction opening up remote areas and providing a stable water-supply, prohibits the release of artificial flood-discharges that are needed to maintain riparian ecosystems. As a result, the world's highly productive wetlands are becoming a rare phenomenon.

VEGETATIONAL RESPONSE TO RIVER IMPOUNDMENT

River impoundments can impart a continuum of changes upon the quantity, timing, and quality, of flows within the river downstream, so that a diverse range of responses can be found in terms of vegetation structure and composition. However, primary responses may be recognized, being common to a majority of impounded rivers, and these relate in particular to reduction in the magnitude and frequency of floods (Fig. 29). For the phytobenthos, discharges of reduced competence can increase substrate stability, at least in gravel-bed, or sand-and-gravel-bed, streams. Year-round growth, assisted in high latitudes by thermal regulation which removes the destructive effects of winter ice, can produce an increased standing crop. Filamentous Chlorophyceae are particularly favoured by river regulation, although direct inoculation from the reservoir may give rise to the establishment of lentic diatoms in low-velocity areas. Indeed, reduced turbidities encourage a wide range of attached Algae and higher plants, but the latter may not become established for tens of years after dam closure—at least in the absence of tributary, or existing in-channel sources, to aid dispersal.

For floodplain habitats, river impoundment and flow-regulation can have disastrous consequences. Periodic inundation is a vital component of wetland ecosystems, and the removal or altered timing of the annual flooding has changed, in many cases, highly productive habitats into unproductive scrub or, with further intervention by Man, into vast areas of monoculture or industrial or urban uses. Within channels, the increased growth of macrophytes can support substantial numbers of invertebrates, but on floodplains, the elimination of the annual flood has devastated the often unique faunal assemblage.

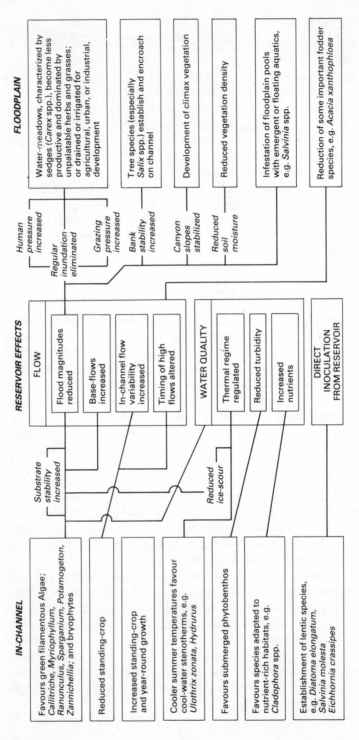

Fig. 29. Primary effects of upstream impoundment upon in-channel and floodplain vegetation.

CHAPTER VII

Macroinvertebrate Response to Upstream Impoundment

'The productivity, diversity and composition of the stream benthic community is extremely important to the total functioning of the stream ecosystem. Besides providing a major source of food for stream fishes, macrobenthos may be the best indicators of past and prevailing ecological conditions.'

Ward (1976b p. 235)

'It appears that whichever way one regulates a southern African stream (at least), some nuisance blackfly species seems capable of exploiting the new conditions!'

Davies (1979 p. 126)

Many factors, both abiotic and biotic, determine lotic habitats, and many macroinvertebrates have become adapted to a specific suite of conditions. Within most natural rivers the pattern of flow and the temperature variations, and various ramifications—particularly substrate particle-size and stability—may be considered the dominant factors controlling macroinvertebrate distributions (Ward & Stanford, 1979b). The life-cycles of many lotic species are related to the natural seasonal variations of discharge, and the biotic communities of streams are significantly affected by flow-velocity because of respiratory, physiological, or feeding, requirements. Several authors have described stream compartmentalization on the basis of velocity—for example, for species of Trichoptera (Highler, 1975; Oswood, 1979). In some of their habitats, most lotic species are limited in their preference of water-depth (Fraser, 1972), and the short-term magnitude and frequency of flow-variations can have a marked influence upon organisms of narrow depth-tolerance.

Many life-cycle phenomena, such as hatching, growth, and emergence, depend on thermal cues (Lehmkuhl, 1972), and the alteration of the thermal regime has been credited as a primary factor influencing community change (Gore, 1980). Substrate heterogeneity is a necessary requirement for the maintenance of a diverse number of niches, and the morphology and particle-size composition of the channel-bed affect the quantity, quality, and velocity, of flows within the substrate—factors of major importance for the lotic fauna (Vaux, 1968). Most aquatic insect adults are rheophilic and select upstream areas to colonize, reproduce, and deposit eggs (Hynes, 1970b). A dam will act as a barrier to aerially colonizing adults as well as to the passive down-

175

stream disperal (drift) of nymphs and larvae (Hynes, 1955; Minckley, 1964). Dam construction will also have an immediate impact upon the flow-regime, water quality, and sediment loads, but the resulting changes of channel morphology may require several years.

Most studies of impounded rivers utilize comparative data derived either from upstream or from adjacent or tributary rivers, because of the lack of reliable pre-dam data. Such comparisons are handicapped by the expected 'normal' downstream variation of the macroinvertebrate fauna along a river, as the channel and network enlarge (Cummins, 1975). Gore (1980) observed the downstream replacement of particular members of a community by their ecological equivalents: for example, on the Tongue River, Montana, USA, at downstream sections, *Acroneuria abnormis* replaced *Claassenia sabulosa*. Nevertheless, marked changes of macroinvertebrate populations have been reported for tens of kilometres downstream from dams (Table XXXVII). The richest zoobenthos in the River Svratka, Czechoslovakia, is found immediately below the Vir Valley Reservoir (Peňáz et al., 1968), and below Possum Kingdom Reservoir on the Brazos River, USA, the benthos is dominated by 38 species of insects (compared with only 15 upstream) having high standing-crops (Poole & Stewart, 1976). In contrast, below Gardiner Dam on the South Sakatchewan River, Canada, the zoobenthos was composed only of a few chironomid (midge) larvae (Lehmkuhl, 1972).

Despite these diverse changes, several responses can be recognized which are common to many impounded rivers. In a review of twenty-three studies from USA, Europe, and South Africa, Stanford & Ward (1979) found that in all but three, species-diversity was reduced by river impoundment, although the majority (13) showed an increase in the overall standing-crop of macroinvertebrates. Many taxa exhibited a consistent response: amphipods were enhanced in eight of the studies and were often introduced into the impounded rivers; Plecoptera (stoneflies) and Ephemeroptera (mayflies) generally declined, or remained similar in numbers, in 13 and 19 of the studies, respectively; but the response of the Trichoptera (caddis-flies) was more varied, with nine studies reporting increases and ten studies reporting decreases. However, even if the relative abundance does not change, the composition of each taxonomic group can be modified considerably. In some cases the effect of impoundment has been to displace species that had been characteristic of upstream areas into channel-reaches below the dam. Thus, the construction of Hungry Horse Dam on the South Fork, Flathead River, USA, changed the character of the main channel from a 6th order to a 4th order stream (Hauer & Stanford, 1982): in terms of water temperature the annual number of degree days was reduced from 2600 to 2050, coarse particulate organic matter (derived from the sloughing of attached Algae by daily flow-fluctuations) increased over fine particles, and, of the Trichoptera, *Arctopsyche grandis* was displaced downstream to dominate over the Hydropsychinae. Similarly, four main-stream impoundments on the Gunnison River, USA, produced a major downstream shift of the thermal regime and partic-

Table XXXVII   Effects of impoundment upon benthic macroinvertebrates.

River, reservoir, location	Macroinvertebrate changes	Source
River Elan, Craig Goch Reservoir, UK	Reduced abundance and diversity	Scullion et al. (1982)
S. Saskatchewan River, Gardiner Dam, Canada	Marked reduction of macroinvertebrates downstream for over 100 km; 19 species of Ephemeroptera were probably eliminated	Lehmkuhl (1972)
Green River, Flaming Gorge Dam, USA	Number of taxonomic groups reduced and density of benthos increased for over 100 km downstream	Pearson et al. (1968)
Brazos River, Possum Kingdom Reservoir, USA	Increased zoobenthos diversity for 80 km below the dam	McClure & Stewart (1976)
S. Platte River, Cheesman Lake, USA	Reduced diversity but increased standing-crop for 32 km	Ward (1976b)
Upper Colorado River, Navajo Dam, USA	Invertebrate densities increased from 820 per m^2 to 6727 per m^2 within a 13 km reach	Mullan et al. (1976)
River Tees, Cow Green Reservoir, UK	Reduced diversity and increased biomass for only 400 m below the dam	Armitgage (1978)
Stevens Creek, Central California, USA	Biomass more than doubled	Briggs (1948)
River Svratka, Vir Valley Reservoir, Czechoslovakia	Numbers increased by up to 3.5 times and biomass by up to 2.8 times in comparison with the natural river	Peňáz et al. (1968)
Mill Creek, Wisconsin, USA	Many species disappeared and the fauna became dominated by a few species: Simulium sp., Chironomidae, and Gammarus sp.	Hilsenhoff (1971)
Tennessee Valley, South Holston Reservoir, USA	Increased numbers attributed to large population of simuliids, chironomids, Gammarus and Hydropsyche	Pfitzer (1954)
Guadalupe River, Canyon Reservoir, USA	Diverse macroinvertebrate community established 24 km downstream 5 years after dam closure	Young et al. (1976)
Clinch River, Norris Dam, Tennessee Valley, USA	Numbers reduced by 30%; Trichoptera and Ephemeroptera replaced by chironomids and gastropods.	Tarzwell (1939)

ulate organic-carbon dynamics which caused major changes in the distribution of hydropsychid caddis-flies over nearly 300 km (Stanford & Ward, 1981).

The changes in the benthic fauna at any location relate to the pattern of reservoir releases, and to the chemical, physical, and biological, quality of the water released; also to the changes of channel morphology, to substrate composition and stability, and to the distribution of aquatic plants. Decreased standing-crops have been associated with the release of low-oxygen water (Isom, 1971), with cold water in summer (Lehmkuhl, 1972), and with unnaturally rapid flow-fluctuations (Radford & Hartland-Rowe, 1971). Increased standing-crops may relate to stabilized discharges and increased winter temperatures (Briggs, 1948), enhanced periphyton (Spence & Hynes, 1971; Armitage, 1976), or improved plankton food-supplies (Chutter, 1963; Cushing, 1963). Thus, different macroinvertebrate responses may be observed below lakes of different type (Table XXXVIII, A). Northern European rivers are characterized by large numbers of Ephemeroptera, dominated by *Baetis* spp., and Plecoptera constitute up to 15% of the total biomass (Henricson & Müller, 1979). In comparison with rivers below natural lakes, the effect of the Messaure Dam, Sweden, has been to reduce the number of species and to alter the species composition. Within the regulated river 84% of the Ephemeroptera, and 81% of the Plecoptera, consisted of only two species each. Below the natural lake the Plecoptera, especially, had a much greater diversity.

Different types of reservoir within a similar geographical region may be responsible for different changes in the zoobenthos (Table XXXVIII, B). Within the North Fork, Colorado, USA, the seasonally variable flow, and thermal, regimes maintain conditions for a diverse fauna with an even distribution of individuals among 49 species. However, the standing crop is poor, and this reflects the variable flow-condition and its ramifications, particularly in terms of substrate stability and water turbidity (Ward, 1976a). Of the three impounded rivers cited in Table XXXVIII, B, the highest diversity is found in Joe Wright Creek, where the fluctuating reservoir-releases may maintain conditions alternately favouring different species, and allowing a degree of niche-overlap that would not be possible under more constant flow-conditions (Ward, 1976b). Significantly, this site also has the lowest standing crop of the impounded streams. Trout Creek and the South Platte River have increased standing-crops, as has the naturally-regulated stream, although species-diversity is commonly low, and this may reflect the constant discharges, stable substrates, and greater cover of aquatic plants.

The effect of flow-regulation is further demonstrated by consideration of some important taxa. Plecoptera are absent from the sites of constant discharge, but are relatively abundant in Joe Wright Creek and the North Fork (although species-composition is different at these two sites). However, other factors are influential, and the abundant trichopterans below Maniton Lake, where *Cheumatopsyche* spp. comprised 50% of the total fauna, may reflect the enhanced food-supply provided by lake plankton (Müller, 1962) from

Table XXXVIII   Summary comparisons of the zoobenthos for different river types.
A: Sweden. (From Henricson & Müller (1979). Reproduced by permission of J.
Henricson and Plenum Publishing Corporation.)

Location and flow-type	River Lule below Messaure Dam Hypolimnial-release	River Lule below Lake Vaikijaure Natural Lake
PLECTOPTERA—		
Number of species	16	21
Dominant species	*Leuctra fusca* *Amphinemura standfussi*	*Capnia atra* *Nemurella picteti*
EPHEMEROPTERA—		
Number of species	16	25
Dominant species	*Metretopus borealis* *Siphlonurus lacustris*	*Baetis* spp.
TRICHOPTERA—		
Dominant species	*Hydroptila tineoides*	—

B: USA. (From Ward & Short (1978). Reproduced by permission of E.
Schweizerbart'sche Verlagsbuchhandlungen.)

Reservoir/river	Chambers Lake, Joe Wright Creek	Maniton Lake, Trout Creek	Cheesman Lake, South Platte River	Spring Brook	North Fork
Flow-type	Irrigation reservoir; fluctuation flow	Epilimnial release; regular flow	Hypolimnial release; regular flow	Natural stream; regular flow	Natural river; variable flow
Relative water temperature:					
Winter	Ice-cover	Ice-cover	Warm	Warm	Cold
Summer	Warm	Warm	Cool	Cool	Warm
Species—					
diversity index	2.8	1.7	2.3	1.8	3.7
Density (per m²)	1259	12,903	1988	2512	406
Biomass (g per m²)	2.5	95.5	15.3	48.3	7.6
Periphyton*	+	+++	+++	++	++
Rooted Plants*	—	+++	—	+++	—

* + to ++ to +++ in order of increasing abundance.

the epilimnial-release dam. The above comparisons clearly demonstrate that
the reorganization of the invertebrate community below dams will relate to
the hydrological characteristics of the impounded river, and to the depth
within the reservoir from which discharge releases are made.

180

The creation of a uniform habitat below dams (in terms of temporal variability), by regulating floods and seasonal discharge-variations, has been isolated as a primary cause of the development of macroinvertebrate communities having a reduced diversity and increased standing-crop. Armitage (1977) concluded that the altered flow-regime below Cow Green Reservoir, UK, was influential in causing an increase in the numbers and biomass of certain organisms, particularly *Lymnaea peregra*, Naididae, Orthocladiinae, *Baetis rhodani*, and *Brachycentrus subnubilus*. Increased flow-constancy within the Fasleelva, Norway, produced a benthic fauna similarly dominated by gastropods, Ephemeroptera, and Trichoptera (Lillehammer & Saltveit, 1979). However, regulations of the discharge regime produces relatively constant and predictable conditions which appear not to favour a diverse community (Patrick, 1970). Normal seasonal flow-changes, and the less predictable changes that take place during storms, favour different organisms as the conditions change. Monopolization of a given resource is less likely to occur, and a diverse fauna can be maintained. In regulated rivers, competitive interactions should intensify, opportunist species may be eliminated, and a reduced community-diversity can be expected.

Water quality and temperature were unaffected by the construction of Soldier Creek Dam, USA, and so changes in the benthic macroinvertebrate fauna of the involved Strawberry River, Utah, USA, were attributed to the direct, and indirect, effects of flow-regulation (Williams & Winget, 1979). Prior to dam closure the mean daily discharges ranged from 0.34 m³ per s. to over 7.0 m³ per s; but subsequently flows have been maintained below 1.5 m³ per s.The extended period of low, uniform, flow changed the macroinvertebrate community (Table XXXIX) by increasing the relative abundance of species which are tolerant of the constant-flow conditions. Predators are often important in maintaining species-diversity, and low predation-pressure can produce a benthic community dominated by a few species with high densities (Patrick, 1970). Ward (1976a) suggested that the absence of predacious plecopterans below Cheesman Dam, Colorado River, USA, may relate to the regulation of the flow regime. Plecoptera were also eliminated from the regulated section of the Strawberry River. However, the secondary effects of flow-regulation appear to exert an important control upon the particular species-composition, involving, as they do, not only substrate stability and sedimentation, but also algal growth.

Increased substrate-stability induced by the regulation of competent flood-discharges, together with reduced flow-turbidity (resulting in increased light-penetration), and sometimes with increased nutrients and higher winter temperatures, can encourage the development of Algae and aquatic plants. Furthermore, Spence & Hynes (1971) found that reduced summer temperatures below deep-release dams could reduce the number of macroinvertebrate grazers and allow the increased growth of periphyton. The periphyton,

Table XXXIX Effects of constant flows on invertebrate densities (per m²) within Strawberry River, Utah, USA. (From Williams & Winget (1979). Reproduced by permission of R. N. Winget and Plenum Publishing Corporation.)

	Regulated section	Natural flow-regime
GASTROPODS	187	0
OLIGOCHAETA	3391	829
EPHEMEROPTERA		
(Number of species)	(4)	(6)
Baetis sp.	41 146	4196
Ephemerella inermis	416	22
PLECOPTERA		
(Number of species)	(0)	(4)
TRICHOPTERA		
(Number of species)	(4)	(6)
Brachycentrus sp.	1083	323
Glossosoma sp.	0	1367
Hydropsyche sp.	72	377
Micrasema sp.	803	43
COLEOPTERA		
Elmidae	366	1119
DIPTERA		
Simuliidae	5767	32
Chironomidae	6320	3637

in particular, offers considerable shelter from the current, traps detrital material, and creates an abundant food-supply, thus providing additional habitat. Species that are otherwise unable to maintain themselves in the stream may be enabled to become established, and dense mats of *Cladophora* spp. below the Tennessee Valley dams may provide a false bottom inhabited by large numbers of organisms (Pfitzer, 1954). Substantial increases in the numbers of *Baetis* spp. below dams have often been associated with enhanced algal growth, which provides an additional food-source and improved niche-diversification for herbivorous organisms (Peñáz et al, 1968; Ward & Short, 1978; Williams & Winget, 1979; Petts & Greenwood, 1981). Increases in the numbers of other taxa have also been attributed to increased algal growth: *Ephemerella inermis* and the dipteran *Antocha monticola* (Williams & Winget, 1979); Chironomidae (Pfitzer, 1954; Peñáz et al., 1968; Williams & Winget, 1979); and amphipods, especially *Gammarus* sp. and *Hyalella* sp. (Pfitzer, 1954; Ward, 1974; Spence & Hynes, 1971). Extensive growths of *Vallisneria aethiopica* below the Akosombo Dam on the Volta, Ghana, have been associated with the increased abundance of *Neritina* sp., a gastropod (Hall & Pople, 1968). Also the occurrence of the ephemeropteran *Tricorythodes* sp. —often the dominant, and in some cases the only species present—has been directly related to the distribution of well-developed submerged angiosperms (Ward, 1976a; Ward & Short, 1978). Particular species may be associated with particular vegetation-types. Within the Gavins

Point tailwater on the Missouri River, USA, the chironomid *Orthocladius obumbratus* was associated with periphyton, and *Nanocladius distinctus* with bryozoans (Walburg *et al.*, 1971). However, the extensive development of periphyton, in particular, can eliminate the important clean rock-surface habitat, and species requiring such surfaces to secure their position may be precluded. Forms which utilize suckers, or friction pads, as holdfasts may disappear from these physically stable habitats, and this may explain the absence of two mayflies, *Ephemerella doddsi* and *Rithrogena doddsi*, from below Cheesman Dam, Colorado, USA (Ward, 1976*a*). Furthermore, the respiratory demand at night of dense periphyton communities may cause oxygen depletion—or even exhaustion—in the surface layers of the substrate. Spence & Hynes (1971) suggested that this caused the elimination of three predatory Plecoptera (*Acroneuria lycorias*, *Paragnetina media*, and *Phasgonophora capitata*) from the Grand River below Shand Dam, Ontario, Canada, due to their sensitivity to a slight diminution in dissolved-oxygen levels.

A biologically-diverse ecosystem (Cummins *et al.*, 1972) requires, for its maintenance, a supply of chemically-diverse organic matter. Only a low diversity of organic matter may exist below most deep-release dams, at least, and the near-absence of terrestrial leaf-litter may explain the restricted taxa. Thus, Ward (1976*a*) suggested that the absence of large-particle detritivores, including several species of Plecoptera, below Cheesman Dam, South Platte River, USA, could relate to the lack of decomposing vascular plants. However, the increased growth of riparian vegetation along regulated rivers could introduce an important allochthonous input, and the common establishment, below dams, of filamentous Algae and aquatic angiosperms, could enhance the autochthonous supply. Indeed, because supplies from upstream are isolated by the dam, local sources become of increased importance in affecting the macroinvertebrate populations. Leaf-packs and their microbial communities provide food and microhabitats for many particle-detritivores, and, upon breakdown, become available to filter, and other fine-detritus, feeders. In rivers receiving constant releases, large accumulations of organic debris have been observed on the channel bed (Petts, 1978); but during decomposition, the development of a high oxygen demand can adversely affect the benthic macroinvertebrate fauna. Within impounded rivers experiencing widely fluctuating discharges, accumulations of organic debris, and leaf-pack formation, may be inhibited.

Flow regulation can also cause channel sedimentation by eliminating high-flows which would periodically flush the substrate of fine sediment. The heterogeneity of substrate particle-size is of critical importance in providing varied microhabitats which can support an abundant and diverse fauna (Hynes, 1970*a*). Thus, Pennak & Gerpen (1947) described a direct relationship between the coarser substrates and higher invertebrate abundance and diversity, although such relationships have only rarely been quantified. Moreover, the spaces between particles in the substrate are of vital importance for many organisms, providing additional living-space and an important refuge against high-flow velocities. Sedimentation can effectively fill the interstices of the substrate, so sealing off this microhabitat, and can severely reduce the periphyton, so that habitat heterogeneity may be greatly reduced (Ward, 1976a).

Nevertheless, diverse faunas have been observed at sites of sedimentation. On Stevens Creek, California, USA, invertebrate biomass more than doubled, although productivity was related to the large numbers of Trichoptera, and the infilling of the interstitial spaces in the rubble substrate by sand and silt caused a marked reduction in the number of Ephemeroptera (Briggs, 1948). In contrast, Ephemeroptera were an important component of the benthos within an aggrading reach below Flaming Gorge Dam (Mullan *et al.*, 1976): the shifting sand substrate inhibited the development of periphyton, but thirty-six invertebrate forms were identified, including Chironomidae, Oligochaeta, and eleven species of Ephemeroptera—of which six were burrowers, debris-inhabiters, or quiet-water dwellers. Burrowers, or organisms favouring fine substrates, often dominate sites of sedimentation in regulated rivers, and increased numbers of nematodes, oligochaetes, and gastropods, in the Strawberry River reflect the build-up of organic sediment in the absence of seasonal 'flushing' flows (Williams & Winget, 1979).

FLOW MANIPULATION

Reservoir operations for hydroelectric power-generation, irrigation supply, or recreational or fishery demands, produce artificial discharge variations. These often involve extreme fluctuations of water depth and flow velocity, having unnatural rates of change, unnatural durations, and unnatural frequencies. Because reservoir releases may reflect an irregular demand, a considerable range of combinations of release duration, magnitude, frequency, and sequence, is possible. Within natural rivers experiencing flows of high variability, a high level of production can be attained—provided that the community present is adapted to the frequency and magnitude of flow fluctuation (Odum, 1969). However, such adaptations require a long time-period, and the combination of severe water-level fluctuations and high content of suspended solids can devastate invertebrate populations (Hynes, 1970b).

Within impounded rivers, fluctuating flows—to which few species can adapt themselves—have produced a benthos of low diversity and density (Powell, 1958; Radford & Hartland-Rowe, 1971; Fisher & LaVoy, 1972; Trotzky & Gregory, 1974). Periphyton and macrophytes may be eliminated, and the fauna may become dominated by those species which can actively migrate into the substrate interstices for protection against rapid increases in flow velocity (Radford & Hartland-Rowe, 1971; Trotzky & Gregory, 1974; Ward & Short, 1978). In many cases, the depletion of invertebrate standing-crops can be dramatic (Table XL), but a few species may favour the condi-

184

Table XL   Observations on the effects of fluctuating discharges upon benthic macroinvertebrates.

Flow character	Composition of benthos	Source and River
Wide daily flow-fluctuations	Macroinvertebrates all but absent: limited to a small Tendipedidae (bloodworm) and a snail (*Helisoma* sp.).	Mullan *et al.* (1976) Colorado River, USA, Glen Canyon Dam.
Daily flow-fluctuations: less than 10 m³ per s to 170 m³ per s	Reduced species-diversity and biomass.	Trotzky & Gregory (1974) Kennebeck River, USA.
Large flow-reductions from 2.8 to 0.3 m³ per s in less than 5 minutes.	Destroyed periphyton, macrophytes, and large numbers of invertebrates.	Kroger (1973) Snake River, USA.
Fluctuating flow below power-dam; high flows associated with high turbidities and substrate instability	Dominated by three species of Ephemeroptera. Trichoptera virtually eliminated.	Radford & Hartland-Rowe (1971) Kananaskis River, Canada.
Turbid-discharge pulses	Biomass reduced to 11% due (primarily) to losses of Trichoptera.	Peňáz *et al.* (1968) Svratka River, Czechoslovakia.
Wide flow-fluctuations with low of O m³ per s.	Elimination of *Baetis* sp.	Henricson & Müller (1979) River Lule, Sweden.
Flow-fluctuations related to irrigation demand produce large depth-variations	Marked increase in the standing-crop of *Simulium chutteri*.	Chutter (1968) Vaal River, South Africa.

tions. For example, the ability of *Amphipsyche meridiana* to colonize small vesicles having a weak internal flow-circulation and therefore requiring respiratory adaptability, allowed the development of large populations with densities of 7300 m²—despite severe water-level fluctuations—below Lake Rawapening, Tuntang River, Central Java (Boon, 1979). Excessive velocity appears to be the primary limiting factor. Thus, the depleted fauna of the main channel below Glen Canyon Dam, experiencing extreme flow-fluctuations, contrasts with the diverse benthic community found in adjacent quiet-water areas, which includes gastropods, Diptera, Trichoptera, annelids, and amphipods (Mullan *et al.*, 1976).

The generation of high turbidities during dam releases can be an important secondary factor causing the devastation of invertebrate communities. Scour releases from reservoirs can transport large quantities of fine sediment, and any sediment introduced by tributaries (silt, sand, fine gravel, and organic debris) may be entrained. Turbid-discharge pulses from the Vir Valley Reservoir reduced the biomass to less than 11% of the original average quantity (Peňáz *et al.*, 1968), and although the density of organisms returned to 'normal' relatively rapidly, the average weight of organisms per unit area was depressed for a considerable distance (Table XLI).

Table XLI   The effects of turbid releases from the Vir Valley Reservoir upon the invertebrate biomass of the Svratka River, Czechoslovakia. (From Peňáz et al., 1968. Reproduced by permission of Academia (Czechoslovakia)).

	Average Numbers (per m²)		Average Weight (g per m²)	
	Normal operations	Flushing flows	Normal operations	Flushing flows
Upstream	1290	1195	10.73	11.54
7.7 km downstream	4530	1705	30.24	4.05
10.5 km "	2390	1345	16.32	3.96
17.4 km "	2720	1510	16.03	5.35
32.9 km "	2480	2020	28.40	6.20

The composition of the macroinvertebrate fauna below pulse-release dams will often reflect the specific habitat requirements of each species. For example, the Green River, Upper Colorado, USA, has become dominated by small chironomids and a few compressed-body forms of Ephemeroptera that are adapted to inhabit substrate spaces below high-velocity currents (Mullan et al., 1976). Three ephemeropteran species (Epeorus longimanus, Ephemerella doddsi, and Rithrogena doddsi)—robust forms with strong legs and an adhesive disk on the ventral side of the abdomen—comprised 65% of the insect fauna of the Kananaskis River, Canada, below the Pocaterra hydroelectic power-dam (Radford & Hartland-Rowe, 1971), although as most ephemeropterans feed on Algae, the erosion or desiccation of the periphyton by the fluctuating flows may limit their numbers. The algal-feeding trichopteran, Glossosoma pterna, was eliminated from the Kananaskis River by fluctuating flows, and other Trichoptera were severely reduced in number—due to the instability of net-spinning forms, such as Panapsyche elsis, to maintain their position. In contrast, Hydroptila sp. appears well-adapted to variable water-flow. Initially, nymphs live deep in the substrate and without cases; but later, when carrying cases, they attach themselves to the substrate and are able to survive temporary drainage (Henricson & Müller, 1979).

Species of Plecoptera show a similarly variable response to the imposition of a fluctuating flow-regime. The herbivorous Nemoura spp. characteristically inhabit leaf-packs, but these niches are eliminated by the rapid flows which flush any organic accumulations, so that such herbivores have been eliminated or markedly reduced in number (Pearson et al., 1968; Radford & Hartland-Rowe, 1971). In contrast, those forms inhabiting clean substrates (i.e. lacking siltation) may increase in relative abundance; thus Alloperla sp., a small slender plecopteran, appears able to tolerate large flow-fluctuations below hydroelectric power-dams (Radford & Hartland-Rowe, 1971; Trotzky & Gregory, 1974).

Artificially high- or low-flows, especially at unseasonal times, can have a major effect upon the composition of the benthos—by eliminating, or favouring, flow-specific species. This is well illustrated by Trotzky & Gregory (1974) for the impounded Kennebec River, Maine, USA. Daily high-flows have eliminated organisms that are adapted to slow-flowing water (e.g. the odonatan *Ophiogomohus* and the trichopteran *Pycnopsyche*), although the highly motile Ephemeroptera nymph *Paraleptophlebia* became more common than formerly. Conversely, low-water flows were a limiting factor for swift-water species, including the trichopteran *Rhyacophila*, the emphemerop-terans *Rithrogena* and *Iron*, and the plecopterans *Acroneuria* and *Paragnetina*. *Rhyacophila* larvae are restricted to locations of fast current, and this reflects both their physiological limitations (high oxygen demand), and the distribution of food-supply. Henricson & Müller (1979) demonstrated that the type of life-cycle may largely determine whether a species can withstand the modified conditions. Plecoptera nymphs, such as *Capnia atra*, *Nemurella picteti*, and *Leuctra hippopus*, having their growth-period between autumn and spring, were adversely affected by the long exposure to short-term flow-fluctuations. In contrast, species emerging in summer and autumn (e.g. *Leuctra fusca* and *Amphinemura standfussi*), experience only a short period of larval growth, and this appears to be an advantage in some impounded rivers.

Locally, the hydraulic effects of flow-fluctuations will be governed by the morphology of the channel—particularly channel shape in cross-section, and bed roughness. Channel shape controls the velocity variation with changing discharge at a cross-section, and such velocity variations influence species-diversity: below a hydro electric power-dam in Maine, USA, velocity fluctua-tions between 0.5 and 0.9 m per s were associated with 19 genera of aquatic insects, in contrast to only four genera at locations where velocity varied from 0.1 to 0.5 m per s (Trotzky & Gregory, 1974). Ward (1976b) suggested that a relatively high diversity may be caused by fluctuating water-levels which create conditions alternately favouring different species, thus allowing an increased degree of niche overlap.

Because of local morphological variations, different niches may also be found within relatively short lengths of river. Indeed, different species have been associated with local changes of flow pattern along the Vaal River, below the Vaal Hartz Diversion Weir, South Africa (Chutter, 1968). In a reach 10 km below the weir, the channel is wide and, as a result, water-depth varies little with changing discharge, and large numbers of Simuliidae and Hydropsychidae dominated the benthos. At narrow sections, however, flow-fluctuations produce large changes of water-depth and, although *Simulium chutteri* successfully exploited the biotope, conditions were unfavourable for the species which were found at the wider sections (*S. adersi*, *S. damnosum*, and *S. mcmahoni*). Chutter suggested that the abundance of *S. chutteri* was a reflection of reduced predation, and reduced competition for food, under the conditions of wide water-depth fluctuations.

Reduced Flows

Sudden flow-reduction, or even cessation, can effectively de-water large areas of channel substrate, producing a depleted fauna of low species-diversity and with few individuals (Trotzky & Gregory, 1974). A flow reduction from 5.38 m³ per s to 0.71 m³ per s on the Tongue River, Montana, USA, for example, reduced flow-width from 30 m to 7 m, and considerably reduced the habitable substrate surface-area (Gore, 1980). Substrate exposure can destroy the primary producers—the Algae and aquatic higher plants—and, until they re-establish themselves, the invertebrates will not return in their former numbers (Waters, 1964). Kroger (1973) concluded that the effects of a single large draw-down on productivity are drastic, and that the net effects of multiple draw-downs may devastate macroinvertebrate populations. Fluctuating water-levels below Jackson Lake Dam, Snake River, USA, reduced discharges from 2.8 m³ per s to 0.3 m³ per s in less than 5 minutes; no invertebrate drift occurred with draw-down, but the stream-bed was exposed, the periphyton was destroyed, and the benthic organisms were left stranded.

In some tropical rivers the practice of flow cessation is used as a control measure to eradicate, or to prevent, the development of pest species (especially Simuliidae). Extreme control measures, for example on the Vaal and Orange Rivers, South Africa, involve flow manipulation including the total cessation of discharge at weekends (Davies, 1979). However, such drastic measures could devastate not only the *Simulium* populations but also downstream ecosystems.

Within rivers that are not experiencing complete discharge cessation, the effects of flow fluctuations will depend upon the frequency of submergence and exposure, the rate of water-level change, and the duration of particular flow-levels. Fisher & LaVoy (1972) have demonstrated that water-level fluctuations, resulting from conventional hydroelectric power-generation, may prevent the establishment of normal benthic invertebrate communities of lower density and diversity than communities in continuously-flooded areas. However, the benthic community may tolerate brief periods of exposure without significant change: substrate exposed for only 13% of the time on the Connecticut River acquired a fauna of comparable density, biomass, numbers, and number of taxa, to that of continuously submerged substrates, but higher levels of exposure were associated with a pronounced reduction in these parameters. Frequent short-term fluctuations may also decimate the fauna by increasing losses through drift which, at least below the dam, are not replaced by drifting organisms from upstream, or from the tributaries.

Artificial Flow-fluctuations and Invertebrate Drift

Drifting of invertebrates in streams is a constant phenomenon throughout both day and night, and involves most organisms that are represented in the benthos. The normal drift exhibits a marked diurnal periodicity (Elliot, 1967), with low numbers during daylight hours and great increases at night, often superimposed upon seasonal fluctuations (Pearson & Franklin, 1968). Variations of drifting organisms are related to many factors, including the density of organisms in the benthos, the stage in their life-history, their activity and behaviour, and the flow-velocity to which they are exposed. Nevertheless, three general classes of drift have been identified (Waters, 1965): constant, behavioural, and catastrophic. Constant drift is caused by the passive entrance of invertebrates into the flow as a result of mechanical removal from the substrate. Greater activity on the surface of the channel-bed at night, for example due to increased foraging (Bishop & Hynes, 1969), increases the probability of an organism being swept away, and may explain, at least in part, the commonly observed night-time drift peak (Waters, 1962). Some species may actively enter the drift as a behavioural response to reduced light-intensity (Waters, 1962), to changes in current velocity and water-depth (Minshall & Winger, 1968), or to competition for available food and space, where production is in excess of the carrying capacity of the channel substrates (Waters, 1966). High drift-rates are commonly associated with an unusually severe 'catastrophic' physical disturbance, such as flooding or ice-floes.

Drifting is, generally, a local transfer of organisms, and detached individuals may drift for only a short period of time, and travel only a short distance, if conditions permit a rapid return to the substrate (Waters, 1965; Elliot, 1967). Artificial water-level fluctuations can affect the drift by altering the population density on the substrate, and by affecting the amount of light penetrating to the substrate (Pearson & Franklin, 1968). Enhanced drifting, particularly the increased frequency of catastrophic drift, can effectively reduce the numbers of benthic organisms immediately below dams; but numbers and species-diversity at downstream locations may be increased as a result of the deposition of drifting organisms. Pearson & Franklin (1968) observed a drift over 50 km on the Green River below Flaming Gorge Dam, Utah, USA, and postulated that organisms drifted in a saltational manner over a number of nights, while flow-fluctuations on the Tongue River, USA, were found to displace many species downstream for a distance of 40 to 120 km (Gore, 1977).

Normally, most aquatic invertebrates avoid light, but rapid changes of stage or turbidity, associated with artificial releases, can cause drift to increase during daylight hours. For example, below Flaming Gorge Dam, Utah, USA, an increase in suspended solids concentrations from 25 to 700 ppm, without any change of water-level, was immediately accompanied by a marked rise in daylight drift-rates of *Baetis* sp. nymphs, and the expected night-time peak was doubled (Pearson & Franklin, 1968). Extreme daily increases in flow also caused sudden increases in the daytime catastrophic drift, which was particularly associated with *Baetis* and Simuliidae. Brooker & Hemsworth (1978) reported that a stage increase of less than 15 cm over a 3-hours' period, associated with a discharge increase from 1.3 to 4.3 m^3 per s, caused a massive increase in the total number of drifting invertebrates (Fig. 30), but also that two types of response could be recognized. The natural 'control' site showed the characteristic diurnal periodicity of drift, and this reflected the dominant contributions by Ephemeroptera and Coleoptera. When the

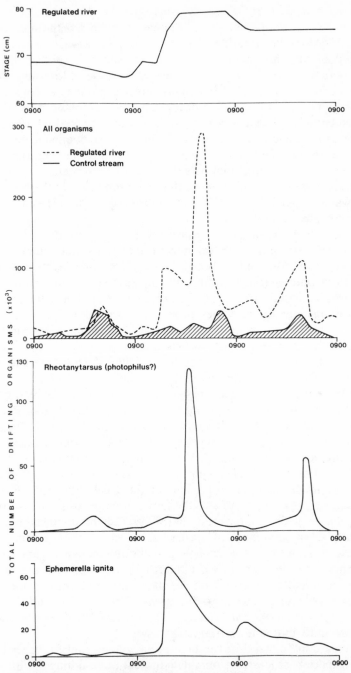

Fig. 30. Effects on invertebrate drift of a rapid increase in water depth during a release, River Wye, UK. (From Brooker & Hemsworth, 1978. Reproduced by permission of Dr W. Junk Publishers.)

release reached the sample site, 16 km below the dam, the stage rise was associated with a simultaneous and dramatic increase in the numbers, and density, of drifting invertebrates. Eighty per cent of the drift at this time was composed of the chiromonid *Rheotanytarsus* sp. This species showed no diurnal periodicity of drifting, but appeared to respond to a flow stimulus, and the numbers declined exponentially under conditions of constant high-flow. The initial response during daylight hours was followed by a massive enhancement of the natural night-time drift, which peaked ten times higher than the previous day; *Ephemerella ignita* was the dominant contributor. During the second day of the release, drift densities had returned to pre-release levels, but the total numbers of drifting invertebrates were still elevated. Despite these dramatic effects, the total proportion of the benthos that drifted was low, being only 0.013% on the first day of the release (cf. 0.006% under the control conditions).

A sudden drop in water-level can cause a 'drought reaction' with large numbers of invertebrates actively moving into the water-column to avoid being marooned (Minshall & Winger, 1968). As water-levels rise, benthic organisms migrate laterally to utilize the increased living-space and to avoid the areas of increased current velocity; but a sudden fall in water-level forces the return movement, and causes an increase in the amount of behavioural drift. A reduction in living-space may cause crowding, jostling, and an increase in invertebrate activity which encourages entry into the current. Indeed, drastic reductions in discharge can produce the most marked changes in drift-rate. The sudden exposure of 25% of the channel-bed below Flaming Gorge Dam, after 5 months of higher flows, induced a massive night-time drift of *Baetis* sp., other Ephemeroptera, Trichoptera, and Plecoptera (Pearson & Franklin, 1968). Flow reduction occurred just prior to the autumn emergence of many aquatic insects and the total invertebrate density was high; during the falling stage of the discharge hydrograph, nymphs and larvae were observed actively migrating towards deeper water by crawling or swimming and becoming incorporated in the local drift. However, mass emigration from the substrate into the flow can occur as a result of reduced current, that is, without a reduction in the size of habitat (Beckett & Miller, 1982). Studies of the effects of artificial flow-reduction have demonstrated an inverse relationship between flow and the number of organisms in the drift (Minshall & Winger, 1968), although in terms of taxa the picture is more complicated, with several species showing an opposing tendency. The most susceptible taxa appear to be *Baetis*, *Ephemerella*, and *Cinygmula* (Ephemeroptera); *Nemoura* (Plecoptera); *Rhyacophila* (Trichoptera); and Simuliidae and Chironomidae (Diptera).

An increase in animal drift can be especially important for fishes because some fish species, at least, are stimulated to eat more during the first 2 days of a flood-release, although feeding subsequently declines to below normal levels (Chaston, 1968). Artificial drift-stimulation in regulated rivers could be used to enhance sources of fish-food, and to stimulate feeding. Periodic reduction of water-levels during daylight hours can induce a behavioural response—overriding normal responses—by causing invertebrates to enter the drift during periods of high light-intensity and when fish are actively feeding. Thus, Mundie (1974) suggested that pulse-releases could be used during daylight periods to enhance drift density in order to increase fish food-supply. However, as yet, the direct effects upon fish species are not fully understood, and other consequences of pulse-releases for turbidity, water quality, phytobenthos, and sediment transport, must be considered.

EPILIMNIAL AND HYPOLIMNIAL RELEASES

Clear differences have been demonstrated between the chemical, physical, and biological, quality of epilimnial and hypolomnial releases, and these differences are reflected in the adjustments of the invertebrate community. Major differences can relate to the contrasting thermal regimes, to the release of water from anoxic hypolimnia, or to the discharge of lake plankton which can provide a reliable food-source. Thus, contrasting invertebrate faunas have been found below two reservoirs in the UK (Table XLII). Downstream from the deep-release Clywedog Dam, both species-diversity and standing crop are low and contrast with the data for the River Vyrnwy, which has more than twice the number of individuals and species. The River Elan is dominated by the ephemeropteran *Baetis rhodani* and three species of Plecoptera. Both in relative and absolute terms *Leuctra inermis*, *Isoperla grammatica*, and *B. rhodani*, are reduced in number below Lake Vyrnwy, and although numbers of *Amphinemura sulcicollis* are three times as great, the major faunal difference is the relative increase in the numbers of Diptera (Chironomidae), Trichoptera, and Oligochaeta. Similar faunas to that below Lake Vyrnwy have been found below epilimnial or mixed-release dams in Scandinavia (Henricson & Müller, 1979), the USA (Fraley, 1979), and on the River Tees, UK (Armitage, 1978).

The most commonly identified factor effecting an increased biomass below epilimnial-release dams is the improved volume and reliability of the planktonic food-supply from the reservoir. Large numbers of filter-feeding organisms, particularly net-spinning caddis-larvae (Trichoptera), and Simuliidae, have been reported from reservoir outflows (Spence & Hynes, 1971; Henricson & Müller, 1979), and filter-feeders are common in rivers draining natural lakes (Ulfstrand, 1968). *Hydropsyche* spp. (Trichoptera) comprised 67% of all invertebrates collected from below the shallow Ennis Reservoir on the Madison River, USA (Fraley, 1979), and these species have been found, numerically, to dominate the fauna within several regulated rivers (Hynes, 1970b; Ward, 1976a). However, filter-feeders occur in relatively low numbers in the River Tees, UK, immediately below the dam, despite a supply of zooplankton from the reservoir (Armitage & Capper, 1976). Also, filter-feeders did not dominate the benthos below the hypolimnial-release Cheesman Dam despite available plankton (Ward, 1975).

Table XLII   Contrasting macroinvertebrate populations below a deep-release, flow-regulation reservoir (Clywedog), and intermediate-release, water-supply dam (Vyrnwy), Wales, UK (data supplied by the Severn–Trent Water Authority).

	River Vyrnwy (Lake Vyrnwy)		River Severn (Clywedog Dam)
Distance below dam (km)	2.4		2.4
Flow character	Stable		Variable
August temperature (°C)	17		7
Number of species	67		32
Total numbers per standard sample (individuals)	3391		1422
Relative abundance of major taxa (%):			
Plecoptera	18.8		33.2
Amphinemura sulcicollis	12.6	*Leuctra inermis*	11
		Isoperla grammatica	9.3
		Amphinemura sulcicollis	9.0
Ephemeroptera	20.9		37.1
Ecdyonurus dispar	12.0	*Baetis rhodani*	31.0
Baetis rhodani	6.8		
Trichoptera	24.0		8.9
Sericostoma personatum	6.8		
Diptera (chironomidae)	20.7		7.0
Oligochaeta	6.6*		3.7

* Increases to 20% at some downstream sites

The poor plankton-drift from hypolimnial-release dams in Sweden eliminated the early summer maxima of *Simulium* sp. that are typical of many natural rivers (Müller, 1962). Deep-release dams may not provide a reliable enough food supply for the development of a fauna that is dependent upon micro-seston (Ward, 1975), although other factors may be influential. Indeed, Trichoptera comprised nearly 80% of the standing-crop below the Stevens Creek flood-storage basin, USA—a response to reduced flow-fluctuations and increased minimum temperatures (Briggs, 1948). Large numbers of *Hydropsyche*, in particular, have been associated with warm waters (Coutant, 1962). Within the River Tees, UK, the rarity of simuliid larvae was related to competition for settlement by very high densities of *Hydra* (Armitage, 1977). Furthermore, Cushing (1963) demonstrated that the plankton food-supply can be utilized rapidly, so that its effects upon the benthos may be restricted to only a short distance below the dam.

The seston may also be supplied from an autochthonous source—the periphyton—within the channel. Ward & Short (1978) suggested that the different distributions of *Cheumatopsyche* (below epilimnial-release dams) and *Hydropsyche* (below hypolimnial-release dams) may relate to the size of seston particles. Low numbers of *Hydropsyche* below Cow Green Reservoir were sugggested to reflect low concentrations of algal filaments in the seston, while the summer abundance of the net-spinning caddis larva *Brachycentrus subnubilus*, 600 m below the dam, was related to increased seston loads derived from channel sources (Armitage, 1977). The abundance of filter-feeders below dams may also result from the exclusion of predators (Ward, 1976a).

Ward (1974) identified fourteen characteristics of the South Platte River, below the hypolimnial-release Cheesman Lake, USA, which should have created conditions conducive to a high diversity and biomass of benthic macroinvertebrates, although many streams below hypolimnial-release dams are characterized by reduced macroinvertebrate diversities (Pearson *et al.*, 1968; Hilsenhoff, 1971; Isom, 1971; Spence & Hynes, 1971; Fisher & LaVoy, 1972; Lehmkuhl, 1972; Ward, 1974; Ward & Short, 1978; Scullion *et al.*, 1982). In an extreme case, a natural, diverse fauna of Ephemeroptera, Plecoptera, Trichoptera, Diptera, and Coleoptera, may be replaced by a grossly simplified fauna dominated by Chironomidae and Simuliidae, with amphipods and molluscs. Hilsenhoff (1971) reported such a change below a dam on Mill Creek, Wisconsin, USA, and a similar impoverishment of the zoobenthos, characterized by a marked depletion of Insecta, has been described by Gore (1977) on the Tongue River, Montana, USA: the impoverished insect fauna was dominated by molluscs (*Physa* and *Sphaerium*) and the riffle beetle *Stenelmis*. The latter inhabited dense mats of *Cladophora* spp. and had densities of 2250 per m^2. Other rivers receiving hypolimnial releases may have a relatively more diverse and more abundant fauna, although the species diversity will, characteristically, be reduced in comparison with natural rivers. Spence & Hynes (1971), for example, reported a reduction in species-diversity below Shand Dam on the Grand River, Canada, but the numbers of certain species were considerably increased and this response to impoundment is comparable with that produced by mild organic enrichment.

The above-mentioned different responses of the Mill Creek and Grand River faunas (Table XLIII) are reflected in other studies, and may be viewed as two points on a continuum of possible faunal adjustments. Several effects were common to both rivers: the Plecoptera were eliminated or considerably reduced in number, Diptera (especially the chironomids and simuliids) were increased and became dominant, and amphipods and molluscs became important. Pfitzer (1954), Pearson *et al.* (1968), Peňáz *et al.* (1968), and Ward (1974), also reported that plecopterans were rare below hypolimnial-release dams, although Briggs (1948), Henricson & Müller (1979), and Scullion *et al.* (1982), noted that species of Plecoptera formed an important part of the benthos; for example, *Leuctra fusca*, a detritivore, was relatively abundant in both the regulated River Elan, UK, and River Lule, Sweden. Increased populations of chironomids and simuliids are commonly found downstream from hypolimnial-release dams (Pfitzer, 1954; Peňáz *et al.*, 1968; Lehmkuhl, 1972; Ward 1974; Young *et al.*, 1976;) and Pearson *et al.*, (1968) reported densities of 27, 675 m^2 and 20, 261 m^2 for the Chironomidae and Simuliidae, respectively, on the Green River, USA, below Flaming Gorge Dam. Amphipods (especially *Gammarus* sp. and *Hyalella* sp.) and gastropods

Table XLIII Summary of changes in the physical etc. factors and macroinvertebrate components below two hypolimnial-release dams. (Based on Hilsenhoff, 1971* (A) and Spence & Hynes, 1971†(B) .)

	(A) Mill Creek, Wisconsin, USA	(B) Grand River, Ontario, Canada
Reservoir/Dam type	Recreation reservoir	Flood-control dam
Floods	'Scouring floods' prevented	Reduced flood magnitudes
Temperature (°C)		
Winter	Warmed to *ca* 4°C; ice formation prevented	Similar to natural (*ca* 0°C)
Summer	Reduced by more than 3°C	Reduced by more than 6°C
Water quality	Markedly higher N and P concentrations	Slight N increase
Other	Sedimentation at riffle sites	Autumn discharge of moribund plankton produces high seston-loads
	Slight increase in macrophytes	Substrate always covered by periphyton
PLECOPTERA	Eliminated	Eliminated
EPHEMEROPTERA	Numbers markedly reduced	Numbers of *Baetis* and *Caenis* increased, but abundance and number of species of *Stenonema* were reduced
TRICHOPTERA	Virtually eliminated	Number of species reduced but Hydropsychiidae were abundant
DIPTERA	Numbers of Chironomidae and *Simulium vittatum* increased markedly	Numbers of Chironomidae and Simuliidae (esp. *Simulium vittatum*—absent upstream) were increased
COLEOPTERA	Eliminated	Number of *Optioservus* sp. increased and *Stenelmis* sp. decreased but abundance and diversity little changed.
AMPHIPODA	Increased abundance (esp. *Gammarus* sp.)	Increased abundance (esp. *Hyalella* sp.)

* Data reproduced by permission of the Entomological Society of America.
† Data reproduced by permission of the *Journal of the Fisheries Research Board of Canada.*

(especially *Lymnaea* and *Physa*) are also commonly found (Pfitzer, 1954; Ward, 1976*a*). Other taxa display very different responses.

The population of Trichoptera is frequently reduced in diversity, although in terms of abundance, some rivers are characterized by a rarity of trichopterans whereas others have enriched standing-crops (Briggs, 1948; Pfitzer, 1954). The Coleoptera, also, may be numerically deficient (e.g. Scullion *et al.*, 1982), or dominated by one or two important species (Ward, 1974; Gore, 1977). However, the most dramatic difference between rivers receiving hypolimnial releases is in the composition of the Order Ephemeroptera, which were completely eliminated from below Gardiner Dam, South Saskatchewan River (Lehmkuhl, 1972), whereas eight species were found on the Grand River (Spence & Hynes, 1971); Ward (1974) reported abundant Ephemeroptera in the regulated South Platte, although the number of species was lower than that of a natural river. Ephemerid mayflies are commonly reduced in number: for example, *Ephemerella g. grandis*, an important mayfly in numerical terms in natural channels, was absent below Cheesman Dam on the South Platte River (Ward, 1976a). The baetids show a less consistent response, being reduced in several cases (Gore, 1977; Scullion *et al.*, 1982), but increased in others (Peñáz *et al.*, 1968; Ward, 1976a), while Pearson *et al.* (1968) reported exceptional densities of *Baetis* spp. (40 124 per m^2) on the regulated Green River, USA.

Mill Creek and the Grand River are similar in that both regulated rivers have experienced reduced flood magnitudes and increased substrate stability. Although the Mill Creek reservoir has a recreational usage, and would be maintained near capacity, the surface-area is approximately 2% of the drainage-area, so that floods may be attenuated (see p. 35). This is supported by the observation that 'scouring floods' had been prevented (Hilsenhoff, 1971). However, the controlled drawdown of the Belwood Lake, on the Grand River, in late summer and autumn, to provide flood-control storage, has enhanced seston loads (primarily moribund plankton), and produced only negligible thermal regulation in winter (Spence & Hynes, 1971). Thus, in detail, the precise hydrological effects appear to exert an important control upon the macroinvertebrate populations. Indeed, comparisons of the fauna of rivers receiving both epilimnial and hypolimnial discharges reveal certain similarities which may reflect the overriding control of discharge, irrespective of release-depth. Nevertheless, the water quality differences of the two types of release appear also to play an important role.

Water Quality

The release of water from an anoxic hypolimnion can have adverse consequences for the benthic fauna, not least if low dissolved-oxygen levels are transmitted to the receiving stream (Isom, 1971). Thus, the limited benthos below Norris Dam, Tennessee Valley, USA, has been attributed to the markedly reduced oxygen levels in the releases, which are often less than 4 ppm, and sometimes as low as 0.5 ppm (Pfitzer, 1962). Discharges from anoxic hypolimnia are often associated with high levels of hydrogen sulphide, and below Canyon Reservoir, Guadalupe River, USA, high H_2S levels were suggested as the cause of the low species-diversity, dominated by Chironomidae, with a few species, tolerant of the altered environment, reaching unusually high population densities owing to reduced competition (Young *et al.*, 1976). Relatively high concentrations of iron and manganese, also associated with anoxic hypolimnial releases, have been identified as a major cause of faunal depletion below Craig Goch Reservoir, UK (Scullion *et al.*, 1982). The reduced abundance and diversity of the zoobenthos on the River Elan,

Wales, UK, was related, principally, to the accumulation of iron and manganese deposits on, and within, the substrate sediments, and this probably accounts for the absence or scarcity of *Glossosoma conformis* and the Simuliidae.

Reservoirs can act as nutrient sinks, so that the releases from them may have relatively high concentrations of many chemical constituents. Wetzel (1975) predicted that Ca/Na ratios can increase in rivers receiving hypolimnial releases. In the autumn and winter months, particularly, the hypolimnion can contain ionic calcium concentrations 45% higher than in epilimnial areas of the reservoir, whilst sodium concentrations remain relatively constant. Gore (1980) attributed the abundance of physid snails below the Tongue River Reservoir, Montana, USA, partly to the relatively high Ca/Na ratios in the releases, as these favour the reproductive success, and the osmotic and ionic regulation abilities, of freshwater molluscs. Many gastropods and amphipods are limited by water-quality characteristics, and favourable conditions are commonly found below deep-release dams. Thus, Ward (1976a) found the gastropod *Lymnaea auricularia* only within a reach of the South Platte River below Cheesman Dam, Colorado, USA, associated with water of high alkalinity.

Thermal Regime

Water temperature in rivers characteristically exhibits a high degree of seasonal, and diurnal, variability, producing a winter ice-cover in high-altitude, or high-latitude, areas. Thermal regulation by reservoirs will dampen these natural variations. For example, stream temperatures below Cheesman Dam, Colorado, USA, were higher than normal in winter, lower in summer, fluctuated less seasonally, and exhibited a seasonally-displaced maximum (Ward, 1974). In temperate regions, at least, the faunas are adapted to the normal diurnal temperature-fluctuations, so that a constant water-temperature would be disadvantageous (Hubbs, 1972). Seasonal thermal regulation can have similar consequences, because different physiological and behavioural components have different temperature optima. Thermal modifications below dams may be influential in effecting the alteration of benthic macroinvertebrate communities both within rivers receiving hypolimnial releases (Pearson *et al.*, 1968; Peňáz *et al.*, 1968; Spence & Hynes, 1971; Lehmkuhl, 1972; Ward, 1974; Gore, 1977) and eplimnial releases (Ward & Short, 1978; Fraley, 1979). Indeed, Armitage (1977) explained the lack of effect of Cow Green Reservoir, UK upon the downstream zoobenthic diversity, by the simultaneous draw-off of surface and deep water, which reduced the seasonal temperature-range by only 1–2°C. Suppression of the seasonal temperature-range from 0–15°C by hypolimnial releases from Cheesman Dam, in contrast, probably caused the reduced species-diversity within the South Platte River (Ward, 1976c).

The phenomena of depressed summer, and elevated winter, temperatures may have particularly adverse effects on emergence cues and triggers for egg-hatching, or diapause-breaking, in many aquatic insects. Thermal constancy, and the seasonal temperature-pattern below deep-release dams, may not provide the thermal signals essential for the completion of the life-cycles of certain species, so that only species which are able to complete their life-cycle under the modified regime would be able to occupy the impounded river. Several species of Ephemeroptera and Plecoptera appear to be particularly sensitive. Within the South Saskatchewan River above Gardiner Dam, Canada, Lehmkuhl (1972) found the eggs of ten species of Ephemeroptera which have a long period of diapause (arrested development), and hatch only in response to an appropriate environmental stimulus in June, July, or August, depending on the species. Three specific requirements, characteristic of the natural sites, were not met below the dam: firstly, near-freezing water temperature to break egg-diapause; secondly, a rapid rise in temperature, following freezing, to induce hatching; and thirdly, minimum temperatures of about 18°C over several months to stimulate nymph maturation. Even mayfly species having eggs which hatch soon after they are laid (no diapause) are adversely affected by thermal regulation, because maturation can require about 1 year, and a specific number of degree days in spring appear to be needed for the nymphs to emerge as adults. This condition was met in the natural sections of the South Saskatchewan River, but spring temperatures were reduced below the dam and several species, such as *Ephemera simulans* and *Baetisca bajkovi*, were eliminated.

Ward (1976c), and Ward & Stanford (1979b), have further elucidated the potential effects of thermal modification which can result in the selective elimination of macroinvertebrate species in streams below deep-release dams (Fig. 31), although many of the interrelationships are, as yet, hypothetical. Ward isolated five specific effects of thermal regulation which can induce selective species-elimination, namely: delayed maxima, depressed maxima, elevated minima, seasonal constancy, and diurnal constancy. The consequences of these effects relate, in part only, to unsuccessful reproduction, because the establishment of competitive disadvantages, and reduced predation, will also be reflected by reduced species-diversity. Nevertheless, depressed water-temperatures associated with high releases into the Green River, USA, were observed to delay hatching and to cause the near-complete failure of the summer generation of *Baetis* sp. for 11.7 km below Flaming Gorge Dam (Pearson *et al.*, 1968). Insect larvae and nymphs were eliminated from the Tongue River, USA, because eggs deposited within a 25-km reach below the hypolimnial-release dam did not experience sufficient temperature fluctuations to break diapause (Gore, 1977); the insect eggs probably remained in diapause until death.

Lehmkuhl (1972) presented data to demonstrate that three temperature requirements of the ephemeropteran *Ephoron album* were not met below Gardiner Dam (Fig. 32): exposure of the eggs to temperatures at or near 0°C to break diapause, rapid spring temperature-rise to 10–13°C to induce

198

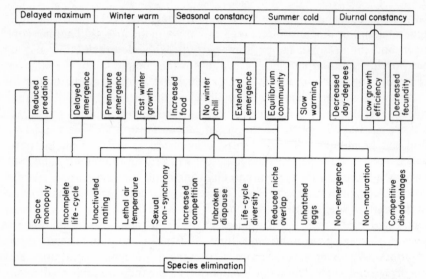

Fig. 31. Thermal modifications below deep-release dams: interrelationships hypothesized as influential in the selective elimination of zoobenthic species. (From Ward & Stanford, 1979b. Reproduced by permission of J. V. Ward and Plenum Publishing Corporation.)

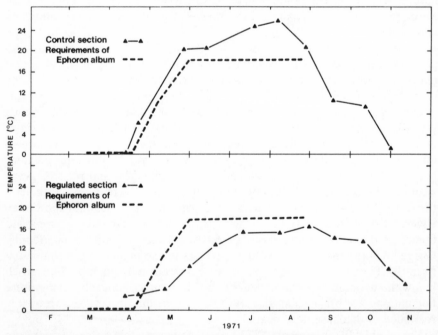

Fig. 32. Seasonal temperature pattern of natural and regulated sections of the South Saskatchewan River and the temperature requiremenrts of *Ephoron album*. (From Lehmkuhl, 1972. Reproduced by permission of the Journal of the Fisheries Research Board of Canada.)

hatching, and at least 2½ months at, or about, 18°C for the nymphs to reach maturity. Premature emergence, induced by unseasonally high water-temperatures in winter, may eliminate some species, because the air temperature could be below the tolerance of the aerial adults, below the threshold for activating mating, or because of non-optimum emergence of males and females (Nebeker, 1971). The emergence failure of the plecopteran *Pternarcys californica*, below both epilimnial (Fraley, 1979) and hypolimnial (Ward, 1976c) release dams, has been attributed to the thermal constancy and lack of thermal cues. The relative abundance, below dams, of groups lacking aerial adults—such as amphipods, gastropods, and oligochaetes—may reflect the non-impact of changed water-temperature–air-temperature relationships during winter (Ward & Stanford, 1979b).

Epilimnial-release dams will influence the macroinvertebrate fauna particularly by elevating summer temperatures, but also by enhancing thermal constancy. Thermal enrichment for 22 km below Ennis Reservoir on the Madison River, USA, stimulated an increase in invertebrate numbers but reduced diversity—probably due to the earlier emergence of some insects—and caused a shift of invertebrate dominance to warm-water forms (Coutant, 1962). Upstream of Ennis Reservoir, *Ephemerella grandis*, *Arctopsyche inermis*, *Glossosoma* spp., and *Physa* spp., had frequencies of occurrence at least four times as great as in the River below the dam—all four were shown to be cold-water preference forms. The downstream fauna, in contrast, was characterized by warm-water preference forms: *Tricorythodes minutus*, *Hydroptila* sp., *Asellus* sp., *Leucotrichia pictipes*, and *Zumatrichia notosa*. *Hydropsyche* sp. and *Cheumatopsyche* sp. were also classified as warm-water-preference forms, and the former has been found also at warm-water sites on unregulated rivers.

Large populations of amphipods, especially *Gammarus*, are favoured by reduced summer temperatures (Ward, 1974), and the absence of Plecoptera below epilimnial-release dams may also be associated with elevated summer temperatures (Ward & Short, 1978). However, several species appear able to maintain viable populations within both hypolimnial and epilimnial releases. *Ephemerella inermis*, for example, is a species that is capable of tolerating a wide range of thermal conditions, and has been found upstream and downstream of hypolimnial (Ward, 1974) and epilimnial (Fraley, 1979) release dams. Nevertheless, thermally-cued invertebrates can be reduced or eliminated, and only those forms with broad tolerance-levels persist. Considerable geographical differences may be expected, however, as the effects of winter temperature-changes will be less significant where the natural winter-flows are relatively warm, and where the natural community is not adjusted to natural extremes.

DOWNSTREAM VARIATIONS OF BENTHOS

Macroinvertebrate diversity and biomass generally increase and decrease, respectively, downstream from hypolimnial-release dams (Pearson *et al.*, 1968), but detailed comparisons between rivers are problematical because of the common coarse sampling-framework used, and the usual small-scale variability in population density and composition that can be found —particularly within stressed systems. Rapid changes in the macroinvertebrate fauna can occur within a few hundred metres—and certainly within the first few kilometres—below a dam (Hilsenhoff, 1971; Ward, 1974; Armitage, 1977), and different changes can occur within different channel-forms (Chutter, 1968; Scullion *et al.*, 1982).

Changes of the thermal regime, water quality, and flow variability, downstream from dams, produce marked changes in community composition and standing-crop. Downstream, atmospheric conditions and tributary discharges combine to return the river to its natural state, although changes of density and diversity may be detected for a considerable distance. On the Kananaskis River, Canada, for example, the productivity of the River below a hydroelectric power-dam is low, with less than 500 individuals per unit area (1 m²) all-year-round, in contrast to similar natural rivers having more than 1000 individuals per unit area for at least 6 months (Radford & Hartland-Rowe, 1971). Downstream of most hypolimnial-release rivers, the invertebrate fauna have been reported to be impoverished (Spence & Hynes, 1971; Lehmkuhl, 1972; Gore, 1977), though the pattern of change downstream will be affected by four primary variables: thermal regime, supply of organic seston, substrate stability (with associated development of periphyton and aquatic angiosperms), and unregulated tributary discharges.

The changing importance of these variables for the stream benthos has been demonstrated by Ward (1974, 1976*a*) for the South Platte River below Cheesman Dam, Colorado, USA. For 32 km below the dam, the character of the River is dominated by the reservoir releases, the biomass is elevated (by about 30 g per m²), and species-diversity is relatively low. Downstream from the first major tributary confluence the biomass drops to 8.3 g per m², but the species-diversity approaches that of the natural river. Discharges upstream from the confluences vary with the dam releases (1.5–12.5 m³ per s), and contrast with those of between 4 and 66 m³ per s downstream. Despite the control on discharges, exerted by the dam within the 32-km reach to the tributary confluence, the diversity and standing-crop display a marked variability (Table XLIV).

Throughout the reach the substrate is stable and has a relatively high cover of periphyton and aquatic angiosperms, in comparison with the unregulated 'control' river. Aquatic angiosperms are absent from the first kilometre below the dam, but macroalgae develop all-year-round, due to the elevated winter water-temperatures and ice-free conditions. The density of organisms per m² increases from 1988.1 immediately below the dam, to 3357.1 five kilometres downstream, and to 4786 upstream from the tributary confluence; the biomass also increases within the first 5 km from 15.3 g per m² to 38.4 g per m², and then decreases, albeit only slightly, to 30.9 g per m² at the end of

Table XLIV   Downstream changes in the benthos below a hypolimnial-release dam (A) and a mixed-release reservoir (B).

A. South Platte River, Cheesman Dam, Colorado, USA (Based on Ward, 1974, 1976a*)	Distance below dam				Control channel
	250 m	2400 m	5000 m	32 km	
TOTAL NUMBER OF SPECIES	19	22	30	64	49
EPHEMEROPTERA (per m²)					
Baetis sp.	760.4	1204.2	1523.5	331.4	49.1
Ephemerella inermis	98.1	527.9	193.0	838.6	19.4
Tricorythodes fallax(?)	–	7.8	27.4	11.8	–
Rithrogena doddsi	–	–	–	110.4	103.6
Biomass (g per m²)	9.9			6.2	1.3
PLECOPTERA (per m²)					
Isoperla patricia	–	15.1	27.5	31.6	+
Pteronarcella badia	–	–	–	7.1	15.8
Biomass (g per m²)	0.3			1.2	4.0
TRICHOPTERA (per m²)	89.5	94.4	56.1	1233.8	108.2
DIPTERA (per m²)	1001.9	1135.8	1346.2	1954.4	62.9
AMPHIPODA (per m²)					
Gammarus lacustris	25.3	13.0	62.2	+	–
GASTROPODA (per m²)					
Physa anatina	–	+	8.2	–	–

B. River Tees, Cow Green Reservoir, UK (Based on Armitage, 1977†)	Distance below dam				Control channel
	2 m	240 m	500 m	900 m	
TOTAL NUMBER OF SPECIES/TAXA	45	56	68	67	58
EPHEMEROPTERA‡					
Baetis rhodani	27	162	161	95	131
Ephemerella ingita	54	134	73	25	6
Rithrogena semicolorata	+	2	6	57	141
Ecyonurus dispar	+	3	19	60	47
PLECOPTERA‡					
Leuctra inermis	–	35	35	38	57
L. fusca	7	14	28	53	19
TRICHOPTERA					
Brachycentrus subnubilus	2	44	175	153	28
DIPTERA‡					
Orthocladiinae	61	476	369	89	23
AMPHIPODA‡					
Ancyclus fluviatis	+	314	139	14	7
Gammarus pulex	106	230	5	2	1
GASTROPODA‡					
Lymnaea peregra	510	473	59	28	+

* Reproduced by permission of E. Schweizerbart'sche Verlagsbuchhandlung.
† Reproduced by permission of Dr W. Junk Publishers.
‡ Mean numbers per standard sample × 10.

the truly regulated reach. The number of species also increases rapidly within the first 5 km below the dam. The distribution of a few species may relate to the improved periphyton cover (e.g. *Rithrogena doddsi*—negatively related; *Baetis* sp.—positively related), and to the development of angiosperms (e.g. *Tricorythodes* sp.—positively related); but the major changes were associated with the thermal regime, which recovered rapidly in response to air temperature variations under conditions of low-flow.

A 'constant' thermal regime, with delayed and depressed seasonal extremes, has been shown to affect benthic macroinvertebrates adversely. Thus, within the thermally regulated zone, species-diversity is reduced and the fauna is dominated by Diptera and cold-water species (e.g. *Ephemerella inermis*). However, within the first 2.4 km, the mean temperature increases by 1.4°C; the seasonal peaks in April and August become clearly detectable; and the pattern of diurnal variation is re-established. By 32 km, the thermal regime shows no discernible change from the pre-impoundment situation. While the benthos of the sites within 5 km of the dam reflect the changing thermal conditions, more remote sites, unaffected by additional tributary inputs, reflect the constant flows and increased substrate stability by having a relatively diverse fauna and large standing-crop.

The dramatic increase in the density of Trichoptera probably reflects both the favourable flow-conditions and the enhanced organic seston derived from periphyton within the channel upstream. Spence & Hynes (1971) attributed the downstream increase in the density of *Hyalella azteca*, *Caenis* sp., *Baetis* spp., *Optioservus* sp., and Chironomidae, to the greater availability of detritus and the more uniform flow on the Grand River below Shand Dam, Canada, than above. Similar patterns can be found also below epilimnial, or mixed-release, dams—for example along the River Tees, UK (Table XLIV, B). Armitage (1977) found that marked changes of the benthos occurred within a few hundred metres of the dam; the initially reduced diversity persisted for only 500 m; faunal numbers of individuals were greatest only 240 m downstream; and the numbers of filter-feeders (especially *Brachycentrus subnubulus*) were associated with the improved supply of food material under regulated flow conditions.

Pool-riffle Variations

At the small scale involved, different changes of the benthic macroinvertebrates may occur at different morphological locations along the channel, particularly at pools and riffles. Most studies have restricted their attention to riffle locations, but Scullion *et al.* (1982), for a comparison of the unregulated River Wye and regulated Elan, Wales, UK, have shown that markedly different responses may be observed. Within natural rivers, riffle-sites have been observed to support a greater invertebrate abundance than pools, although the number of taxa may be similar (Egglishaw & Mackay, 1967; Armitage *et al.*, 1974; Scullion *et al.*, 1982; Logan & Brooker, 1983).

However, pool and riffle sites with similar invertebrate densities, and a higher number of taxa at the riffle sites, have also been found (Minshall & Minshall, 1977). Riffles are commonly associated with relatively high densities of ephemeropterans (especially *Baetis* sp.), trichopterans, and simuliids; while pools may be dominated by oligochaetes, chironomids, and coleopterans (Egglishaw & Mackay, 1967; Armitage, 1976; Hynes *et al.*, 1976; Minshall & Minshall, 1977; Scullion *et al.*, 1982; Logan & Brooker, 1983). High riffle-densities of particular species have been related to current requirements associated with respiratory physiology (e.g. *Baetis rhodani* and *Rhyacophila dorsalis*), or feeding (e.g. Simuliidae); the flushing of 'fines' (e.g. *Glosssosoma conformis*), or biotic factors, such as prey-density of predatory species (e.g. *Perla bipunctata*) (Jaag & Ambühl, 1964; Rabeni & Minshall, 1977; Wallace & Merritt, 1980; Scullion *et al.*, 1982).

Within the unregulated River Wye, Wales, UK, the riffle and pool sites are almost equally rich in taxa, but many species are either exclusive to, or abundant in, either one or the other. In contrast, the regulated Elan sites are numerically deficient and, although the riffle contained more species than the pool, invertebrate density was similar at both (Table XLV). Moreover, of the species eliminated from the Elan, the majority were found on the Wye riffles (e.g. *Ecdyonurus dispar, Perla bipunctata, Glossosoma conformis,*

Table XLV   Arithmetic mean densities (per m²) of major taxonomic groups in the River Wye (unregulated) and River Elan (regulated) Wales, UK, during July 1980. (From Scullion *et al.*, 1982. Reproduced by permission of Blackwell Scientific Publications.)

	Wye		Elan	
	Riffle	Pool	Riffle	Pool
Ephemeroptera	3946	1058	48	12
Plecoptera	904	162	1306	550
Trichoptera	592	120	332	22
Diptera				
(Chironomidae)	7642	13 374	8056	9450
(Simuliidae)	11 474	228	256	14
Coleoptera	946	1636	30	50
Hydracarina	1766	242	106	—
Platyhelminthes	282	2	532	16
Oligochaeta	4724	1820	4050	1162
Others	490	206	490	38
Total numbers	32 766	18 846	15 206	11 314

Hydropsyche siltatai, and *Planaria torva*). Whereas the Wye riffles contained a high proportion of ephemeropterans (12%) and simuliids (49%), the Elan riffle was dominated numerically (80%) by oligochaetes and chironomids. Chironomids dominated both the Wye pool (71%) and the Elan pool (over 80%), and oligochaetes were the next most abundant, comprising 10% of the numbers in each. The dominance of chironomids and oligochaetes, with relatively low densities of ephemeropterans, trichopterans, and simuliids, at the Elan riffle, suggests a tendency for the invertebrate community to evolve characteristics more normally associated with pools. On the Elan, this probably reflects the reduced flow-velocites over the riffles, and the associated siltation which here is related, principally, to the deposition of iron- and manganese-rich materials.

TIME-SCALE FOR INVERTEBRATE RESPONSE

Invertebrates, generally, are both sensitive to the environmental conditions in which they live, and able to colonize newly-available habitats by upstream migration of adults and downstream drift of nymphs and larvae. Thus, Young et al. (1976) considered that the successful reorganization of the benthic macroinvertebrate community to fill the new set of niches made available by changes of the environment, required only 5 years on the Guadalupe River, Texas, USA, after the closure of Canyon Reservoir. Nevertheless, important changes in the density, and diversity, of the benthic invertebrates, may occur during, and for a few years after, dam closure.

Ward & Short (1978) grouped macroinvertebrates into four types, on the basis of observed responses to river impoundment:

(1) Tolerant organisms having a ubiquitous distribution, but forming large populations under certain types of river regulation.
(2) Organisms present in unregulated streams, but favoured by certain types of regulation.
(3) Intolerant organisms present in unregulated streams, but reduced in, or eliminated from, regulated streams.
(4) Indicator species not normally present in unregulated streams, but favoured by regulation.

The interaction of these groups of organisms will produce characteristic distributions of macroinvertebrates downstream from dams, as unregulated tributary runoff restores the river to its natural condition. Thus, downstream from Cheesman Lake, South Platte River, USA, Ward (1974) recognized three typical macroinvertebrate distributions: species uncommon below the dam but increasingly prevalent downstream (e.g. *Ephemerella grandis*, *Rhithrogena* sp.); species reaching much greater abundance below the dam and decreasing downstream (e.g. *Simulium articum*, *Baetis* sp.); and species apparently restricted to the area below the dam (e.g. *Gammarus lacustris*, *Lymnaea auricularia*).

Considerable reorganization of the invertebrate community existing during and shortly after dam construction may, however, occur as the water quality, and thermal and discharge regimes, stabilize. For example, the impoundment of Mill Creek, Wisconsin, USA, had a dramatic effect upon the benthos

(Table XLVI): many species were eliminated, and the fauna became dominated by chironomid larvae and amphipods, while other benthos changed
rapidly during, and after, impoundment (Hilsenhoff, 1971). Four general
response-groups may be recognized: species that were common prior to
impoundment but became reduced or eliminated after it (e.g. *Hydropsyche*
and Elmidae); species that were common under pre-dam conditions, eliminated during dam construction or shortly after, but returned subsequently
in limited numbers (e.g. *Baetis brunneicolor*); species that seemed largely
unaffected by impoundment but experienced population explosions during,
and immediately after, dam closure (e.g. *Chironomus* spp.); and species
that showed an increase in abundance after dam completion (e.g. *Simulium
vittatum*).

Table XLVI Short-term macroinvertebrate changes during and shortly after dam
construction at a riffle 300 m below the dam, Mill Creek, USA. (From Hilsenhoff,
1971. Reproduced by permission of the Entomological Society of America.)

	Before	During	7–15 months after	19–27 months after
EPHEMEROPTERA, Baetidae				
Baetis brunneicolor	21	0	0	10
DIPTERA, Chironomidae				
Chironomus spp.	0	508	345	3
Micropsectra spp.	5	80	239	41
Orthocladius spp.	74	54	66	210
Stictochironomus spp.	0	6	25	0
DIPTERA, Simuliidae				
Simulium vittatum	173	336	901	742
DIPTERA, Tipulidae				
Dicranota spp.	1	20	24	4
AMPHIPODA, Gammaridae				
Gammarus pseudolimnaeus	34	–	50	132
TRICHOPTERA, Hydropsychidae				
Hydropsyche betteni	177	81	0	2
COLEOPTERA, Elmidae	142	17	0	0
Summer total phosphorous (ppm)	0.08	–	1.24	0.58
Summer total nitrogen (ppm)	0.77	–	3.16	1.37

(Values = number per standard sample).

The patterns of benthos variation during, and after, dam construction
probably relate to extreme changes of water chemistry, associated with the
initial submergence of soil and vegetation by the reservoir. Concentrations
of total nitrogen and total phosphorus in the discharges increased drama-
tically upon dam-closure, but fell away after about 2 years. High nutrient
levels were associated with copious amounts of slime Bacteria on the
substrate, and extensive aquatic macrophytes; but these also declined in
abundance to become only slightly more prevalent than before dam construc-

tion. A reorganization of the macroinvertebrate community may, therefore, be anticipated as the river 'matures' after reservoir-filling. Changes in the zoobenthic community below Soldier Creek Dam, Strawberry River, USA, occurred over a 4-years' period after dam-closure (Williams & Winget, 1979) in association with the reduced, and more uniform, flows, and the development of dense beds of aquatic macrophytes and periphyton. Downstream reaches may also respond to the changing conditions below the dam, because of the changing abundance and composition of the drift. Gore (1980) has suggested that changes in the river regime below a dam can act as a barrier to upstream insect migration. Certainly the dam, reservoir, and hypolimnial-release-dominated reach, may markedly reduce the available reproductive channel-area, and reduce inoculation by downstream drift.

Channel Change and Invertebrate Communities

Many of the factors that influence the benthic macroinvertebrates will change immediately on dam closure, but others, such as channel morphology, and substrate composition and stability, will change only slowly. Consequently, a progressive response of the zoobenthos could occur over a period of many years. Channel changes in an impounded Dartmoor River, England, UK, effectively maintained flow variability within the channel, and this provided for a diverse fauna (Petts & Greenwood, 1981). Moreover, comparison with the regulated River Meavy below Burrator Reservoir, where similar channel changes had previously been reported (Gregory, 1976b), revealed a clear degree of conformity between the impounded rivers. Different changes of channel morphology have been reported along impounded rivers in response to changing relationships between discharge, sediment load, and boundary resistance. The character of the channel substrate, in particular, will be influential in determining the composition, and abundance, of the benthos.

The heterogeneity of the channel-bed sediments, in terms of size, is of critical importance in providing varied microhabitats which can support an abundant and diverse fauna (Hynes, 1970a) and Pennak & Gerpen (1947) described a direct relationship between the sediment size and invertebrate abundance and diversity. However, such relationships have been quantified only rarely, and many authors (e.g. Williams & Hynes, 1974; Williams, 1980; Scullion et al., 1982) have failed to discern significant relationships between invertebrate abundance and substrate size and heterogeneity. Indeed, other related factors, such as the frequency of substrate turnover, and the rates of erosional or depositional processes, may impose a dominant control on the benthic community (Petts & Greenwood, in press)—particularly as, within many rivers, the character of the sediment loads is more a function of source availability, and the processes of sediment production, than of channel hydraulics. Thus, below the Navajo Dam, Upper Colorado River, USA, distinct macroinvertebrate communities were associated with channel reaches dominated by erosional and depositional processes (Mullan et al., 1976). Some 13 km below the dam, invertebrate densities were elevated to 6727 per m², and the abundance and diversity of invertebrates related to the benthos of erosional streams, being dominated by Ephemeroptera, Diptera, Hydropsychidae, gastropods, and amphipods. Thirty kilometres downstream, however, sedimentation processes dominated, turbidity was high, and the benthos was devastated: the invertebrate density was reduced to between 32 and 192 per m², and a trout fishery was eliminated.

As channel changes progress, processes of accelerated erosion or sedimentation will be replaced by more stable conditions, and this change may be associated with a change of the macroinvertebrate community. Mullan et al. (1976) also reported that, for the Navajo tailwater, invertebrate densities progressively increased during the 5-years' period after dam closure, from 828 per m² to 6727 per m², and this change may relate to the development of improved substrate stability associated with the degradation process and channel armouring. Indeed, despite major, daily flow-fluctuations, an enhanced benthos has been observed below hydroelectric power-dams (Pearson et al., 1968), and this may also relate to improved substrate stability (Pfitzer, 1954), and to the removal of 'fines' from interstitial spaces. Channel-reaches experiencing sedimentation can also change to a more stable or even 'erosional' type of river, when once morphological changes have become adjusted to provide for the transport of introduced sediment by the regulated flows.

The effect of this change upon the benthic macroinvertebrates has been investigated within the River Rheidol, Wales, UK, below Nant-y-Môch Reservoir (Petts & Greenwood, in press). Tributary sediments introduced to the regulated mainstream provide a point-source downstream from which sedimentation processes have produced channel change. A sequence of four different morphological and sedimentological units were identified downstream from the tributary confluence (Table XLVII and see pp. 134–137): aggradation immediately below the confluence has produced a channel-form of reduced dimensions, adjusted to the regulated flows, and the substrate is coarse and stable (1); in the next reach downstream (2) channel change is less advanced, the channel is braided, the substrate is composed of smaller particle-sizes, and is unstable; the coarse sediments are deposited within reach 2 and only the finer materials are transported into reach 3, which is characterized by progressive sedimentation, and represents the downstream limit of channel change induced by point-source sediment injection; below the limit of sedimentation (4) the existing substrate is stabilized by flow-regulation, and mosses and Algae cover much of the channel-bed. At the time of survey, channel depth within these units averaged between 14 and 22 cm.

The stabilized, and morphologically unaltered, reach has the largest number of species and the greatest species-diversity, whereas the adjusted channel-reach (1) has a fauna which is comparable in many ways to that of

Table XLVII Channel change and invertebrate distributions within the Afon Rheidol, below Nant-y-Môch Reservoir. (After Petts & Greenwood, in press.)

Reach	Control channel	1	2	3	4
Number of samples	10	8	17	18	13
Mean substrate particle-size (mm)	54	75	40	13	Gravel and boulders
Proportion less than 2 mm (%)	18.1	6.6	11.3	30.1	19.6
Channel width (m)	14	8	16	22	14
Substrate stability	Unstable	Stable	Unstable	Unstable	Stable
Proportion of sites with moss (%)	0	10	18	0	63
Average number of individuals (per m²)	336	402	177	313	437
Number of species	10	13	12	21	23
Species-diversity	0.917	1.239	1.833	2.250	2.298
DIPTERA (number per m²)	275	309	91	76	153
Chironomidae	258	116	45	50	114
Simuliidae	16	189	42	2	19
Dicranota sp.	0	4	4	20	7
EPHEMEROPTERA (number per m²)	10	5	21	11	14
PLECOPTERA (number per m²)	6.5	41.5	29	48	73
Diura bicaudata	5	1.5	5	0	3
Protonemura meyeri	0	7	13	3	41
Leuctra hippopus	1.5	33	11	45	29
TRICHOPTERA (number per m²)	39	41	42	116	163
Polycentropus sp.	1	8	+	+	8
Rhyacophilidae	18	10	5	7	28
Cased Caddis	20	23	36	89	116
OLIGOCHAETA (number per m²)	4	2	12	31	8

the natural channel, and this similarity was confirmed by the Kulezynski index. As the processes of sedimentation adjust channel morphology to the imposed flow-conditions, reach 1 will extend downstream, reaches 2 and 3 will shift downstream also, and reach 4 will be progressively eliminated; the macroinvertebrate distribution is expected to follow a similar pattern of change. Channel change, in many cases, can require tens of years to establish a stable morphology, so that a continuous readjustment of the macroinverte-brate community, over many years, may result from river impoundment.

CHAPTER VIII

Fish and Fisheries

The large-scale development of water resources has produced sudden and dramatic changes within riverine habitats. Indeed, dam construction appears to have had a greater impact upon riverine fishes than any other human activity. In 2 years after the completion of Lake Kainji, for example, catches of fish from the River Niger, Nigeria, were reduced by 30% (Lelek & El-Zarka, 1973). One immediate consequence of river impoundment is the conversion of naturally lotic environments to lentic habitats. The impoundment of relatively fast-flowing rivers may totally preclude riverine fishes, which are dependent upon flowing water for all their ecological requirements (Fraser, 1972), and species that are able to live only in running water can be eliminated (Zhadin & Gerd, 1963). Such changes have involved entire river systems: the John Day Dam, for example, completed in 1968, inundated the last free-flowing portion of the Columbia River, USA, between the tidal limit and the mouth of the Snake River, forming an uninterrupted series of impoundments for 300 km (Beiningen & Ebel, 1970). On the Snake River itself—one of the major North American salmon rivers—a total of eight dams have created a further 522 km of lentic habitat (Raymond, 1979).

Many important 'commercial' fishes are diadromous, that is, they migrate between the river-system and the sea, either for breeding or for feeding. Anadromous species, such as Acipenseridae (Sturgeon) and salmonids, which breed in fresh water, have been particularly affected by impoundments. Reservoirs will inundate vast spawning-grounds, and great dams provide barriers to upstream and downstream migrations, but the effects of a single impoundment upon the discharge regime, water quality, and habitat structure, of rivers, may be transmitted for considerable distances downstream. Much research has been directed towards the fish and fisheries potential of man-made lakes (e.g. FAO, 1968; Hall, 1971), but major changes of the fish fauna downstream from impoundments have also been identified; for example, for 105 km below Flaming Gorge Dam, on the Green River, USA (Vanicek et al., 1970).

Fish species can become extinct as a result of natural evolutionary events, but perturbations induced by river impoundment have markedly increased the rates of extinction. Some of the native species which have disappeared from impounded rivers are presented in Table XLVIII. *Salmo salar* (Atlantic

209

Table XLVIII   Selected examples of fish-fauna changes consequent upon river impoundment.

Location	Native species disappeared	Introductions	Source
Australia	Plectroplites ambiguus	Salmo spp.	Walker et al. (1979)
	Tandanus tandanus	Perca fluviatilis Carassius auratus Tinca tinca	
Scandinavia	Salmo solar Salmo trutta	Perca fluviatilis Acerina cernua	Henricson & Müller (1979)
	Thymallus thymallus	Rutilus rutilus Esox lucius	Lillehammer & Saltveit (1979)
Central Europe	Barbus spp. Perca fluviatilis Esox lucius	Thymallus thymallus Salmo trutta Cottus gobio	Peñáz et al. (1968) Lehmann (1927)
Western Europe		Salmo trutta Cottus gobio	Armitage (1979)
	Salmo salar Lampetra planeri Petromyzon marinus Alosa alosa	Salmo trutta Barbus barbus Anguilla anguilla	Décamps et al. (1979)
USSR	Acipenseridae	Perca fluviatilis Acerina cernua Rutilus rutilus	Zhadin & Gerd (1963)
India	Hilsa ilisha Puntius dubius		Sreenivasan (1977)
USA	Ptychocheilus lucius		Minckley & Deacon (1968)
	Micropterus treculi Ictalurus punctatus	Salmo gairdneri Notemigonus crysoleucas	Edwards (1978)
	Hybopsis aestivalis Cichlasoma cyanoguttatum Notropis volucellues	Poecilia latipinna Pimephales vigilax	

Salmon), for example, disappeared from the Dordogne River, France, soon after the first dams were built on the lower reaches between 1842 and 1904 (Décamps et al., 1979). The extinctions have often been associated, not with the increased abundance of other native species, but with the introduction of exotics. Valuable fishes, such as Salmo salar and the Acipenseridae (Sturgeon family), have often been replaced by less valuable and slower-growing species: for example, Rutilus rutilus (Roach), and Perca fluviatilis (Perch). The changes contrast for different rivers, however, and other reports suggest that Salmo spp. (trout etc.) and Thymallus thymallus (Grayling) may replace less valuable species such as Esox lucius (Pike), Plectroplites ambiguus, and Tandanus tandanus. Cottus gobio, Anguilla anguilla (Eel), and Salmo trutta

(Trout), in particular, appear to favour conditions below some reservoirs.

The fish faunas of impounded rivers will be characterized by extinctions of some faunal elements, but changes in the faunal composition, biomass, and diversity, will also be apparent. For example, on the Guadalupe River, Texas, USA, the total number of fish species above Canyon Reservoir (22 species) is greater than that found below the dam (18 species), and the percentage of introduced exotic species—13.6% upstream, and 27.8% downstream—reflects a reduced fish species-diversity in the downstream channel (Edwards, 1978). The most noticeable impact of the dam, however, is the gross reduction in biomass of the fishes: above the reservoir the aggregate weight of fishes captured per collection averaged 848 g, compared with that of only 413 g for the downstream reach. The extinctions were related to two particular consequences of river impoundment, namely: inundation of spawning grounds, and the construction of barriers to migration. From a study of the threatened fishes of Oklahoma, Hubbs & Pigg (1976) suggested that 55% of the man-induced species depletions had been caused by the loss of free-flowing river habitat—primarily spawning riffles—resulting from inundation by reservoirs, and a further 19% of the depletions were caused by the construction of dams, acting as barriers to fish movement.

Impoundments have also altered the physico-chemical regimes of many rivers, thereby adversely affecting the reproductive ability of those native fishes which require specific minimum temperatures, and/or floods, as triggering mechanisms for spawning, or for the subsequent survival of eggs and young (Cadwallader, 1978). However, the effect of impoundments are certainly not always detrimental, as the reduction in suspended loads (a source of great mortality of developing eggs), the regulation of floods and prevention of drought (causes of egg and young fish losses), and thermal regulation (important to many species for reproduction), may prove beneficial. For example, the immediate effects of Shasta Dam, Sacramento River, California, USA, were to create headwater conditions for at least 80 km below the dam (Moffett, 1949). This change had several beneficial effects on salmon and trout production: the regulated thermal regime provided water temperatures within, or near, the optimum range; higher water temperatures in the late autumn and early winter accelerated the development of salmon eggs and fry, and increased the autochthonous production of fish-food; lower summer-temperatures reduced predator and competitive fish species; and losses of eggs and young fish were reduced by flow regulation, and by reduced discharge turbidity. Indeed, tailwaters immediately below dams often provide excellent fisheries (Eschmeyer & Smith, 1943; Hall, 1951; Pfitzer, 1954; Parsons, 1957).

Tailwater fishing can provide a better rate of catch, and size of fish, than the reservoir itself (Bennett, 1970). Walburg et al. (1971) suggested that the probable causes for fish concentrations below dams include the availability of food, supplied from the reservoir; suitable water temperature and discharges; and the creation of a barrier to further upstream movement. The recruitment to tailwater fish-stocks will be supplied, at least partially, from the reservoir itself *via* the discharge valves. Moreover, the growth of many fishes in tailwaters, for example below Gavins Point Dam, Missouri River, USA, can be superior to the growth attained in the reservoir. The tailwater fishes of the Volga hydroelectric power-dam were dominated by predators (Sharonov, 1963), but this may reflect the extreme range of dam releases, producing water-level fluctuations of 8.5 m, which would limit benthic production and reduce the growth of benthic feeding-fish (Walburg *et al.*, 1971).

For fishes which spawn on inundated floodplains during the annual flood-season, the formation of shallow reservoirs with large surface-areas may enhance the spawning conditions. Such a change on the Kafue River, Zambia, will benefit important species of *Tilapia*, which form a major part of the commercial catch, through improving growth-rate and the survival of juveniles (Dudley, 1974). Under natural conditions, the Kafue River regularly inundated its floodplain for more than 8 months of each year. The Kafue Gorge Dam reduced both the frequency and duration of floodplain inundation downstream, but the shallow lake causes the floodplain upstream to be inundated more rapidly at the start of the flood-season, and to drain more slowly during the dry period, than formerly. Migratory fish-species within rivers having extensive, branching, stream-networks, may find alternative spawning-grounds. In the UK, many of the rivers having a high water-resource potential are those that produce the main weight of migratory fish (Blezzard *et al.*, 1971). Although Llyn Celyn, for example, submerged more than 6 km of first-class spawning-ground on the Welsh Dee, *Salmo* spp. have adopted small unregulated tributaries downstream of the dam for spawning, and river impoundment has proved not to be detrimental to salmonids (Armitage, 1979). However, the implications for anadromous fishes cannot be assessed simply in terms of the numbers of fish reaching the spawning-grounds or returning to the ocean; for long-term assessments, the 'health' of these fish must also be considered.

Assessments of impacts on fish, shortly after impoundment, may not provide an indication of the long-term effects, because the fauna will require time to readjust, and to recover after the initial impact of dam-closure and reservoir-filling. Post-impoundment studies of Lake Kainji, Nigeria, for example, suggested that the diverse, abundant, and commercially important Mormyridae—highly specialized for lotic habitats—of the Niger system were radically affected by impoundment, being reduced from about 20% (Banks *et al.*, 1965; Motwani & Kanwai, 1970) to less than 5% (Lelek & El-Zarka, 1973; Lewis, 1974) of the fish caught. Some species were exterminated, others adjusted rapidly to the change to lacustrine conditions, while yet others may recover but at a slower rate (Blake, 1977). For example, *Hippopotamyrus pictus* and *Macusenius senegalensis* were of similar abundance in 1971 (35% each), but changed in contrasting ways: *H. pictus* increased in number to comprise over 70% of the total mormyrids caught, whereas *M.*

senegalensis declined to 0.5%. Moreover, despite the sudden decline in total abundance during the first post-impoundment year, an increasing trend was subsequently apparent; and as the species-composition adjusted to the changed conditions, the abundance of the Mormyridae recovered towards pre-impoundment levels.

EFFECTS OF PHYSIO-CHEMICAL CHANGES

Discharge Modification

The modification of downstream river-flow characteristics by an impoundment can have a variety of effects upon fish species: food production, stimuli for migration, the success of migration and spawning, the survival of eggs and juveniles, spatial requirements, and species composition, can all be adversely affected (Fraser, 1972). However, discharge is a complex parameter involving the width, depth, and velocity, of flow within a cross-section, short-term (daily) and long-term (annual) variations, and the timing of flows of different magnitude. Changes of any one, or any combination, of these parameters may significantly affect the resident fish-fauna. Edwards (1978), for example, related the increase in abundance of *Gambusia affinis*, within the impounded Guadalupe River, to the influence of stabilized water-flows and the elimination of periodic flood-discharges. One major habitat alteration that occurs as a result of upstream impoundment, within arid areas in particular, is dewatering—the reduction of flows due to the loss of water by transmission into the channel-boundary deposits, and abstraction for irrigation. The reduction of usable habitat associated with dewatering has a significant impact upon specialized channel-sites, such as spawning areas (Smith, 1976). Thus, Zakharyan (1972) demonstrated that flow reduction downstream of irrigation dams dewatered spawning areas along the Kura River, USSR, and adversely affected the reproduction of the Acipenseridae.

Short-term Flow-fluctuations

Although the abnormal discharge regime, and associated thermal and chemical conditions, that are apt to be imposed by impoundments, have significant effects upon the aquatic fauna, the rate of flow increase or decrease is arguably the most important factor. Rapid rises of discharge can erode spawning-gravels and remove benthic invertebrates, which provide a primary food-source for salmon parr and trout. The removal of food-sources may be particularly significant if only a few tributaries are available to act as sources for recolonization.

The operational procedures of most hydroelectric power-dams result in a rapidly variable downstream flow-regime, and such variations can have significant consequences for the fish fauna. A daily 2 m to 3 m fluctuation of Colorado river-levels below Glen Canyon Dam may have contributed to the decline in endemic fishes (Holden & Stalnaker, 1975). The native fishes have largely been replaced by the introduced *Salmo gairdneri*. Spawning of most of the native species is restricted to tributaries, and the juveniles are seldom found in the main-stream. Holden (1979) suggested that the ephemeral nature of the habitat, caused by the rapid, daily fluctuations, is the major factor preventing juvenile fishes from utilizing the Colorado main-stream. Walker *et al.* (1979) related the disappearance of *Tandanus tandanus*—which favours sluggish, warm waters, and spawns in shallow water—within the Murray River, Australia, to short-term fluctuations in water-level, caused by reservoir releases in response to downstream water-user requirements.

The fluctuations of water-level due to irrigation or power demand, could have disastrous effects on fish, such as *Hypseleotris klenzingeri*, which spawn in shallow water over grasses and twigs (Lake, 1975). Hamilton & Buell (1976) identified four causes of loss in salmon and trout production for predicted increases in the daily discharge-range for the Campbell River, Canada, from 31–122 m^3 per s to 28–263 m^3 per s. Rapid increases in depth and velocity of the water would exceed the limits tolerated by adults, and would therefore result in a reduction of the spawning area. Spawners would be displaced by abrupt changes of flow, and the spawning behaviour would be inhibited. Juveniles would be swept downstream by high-flows, and sudden reductions in flow would leave juveniles stranded. Finally, low-velocity areas for feeding, particularly by juveniles, would be reduced.

Floodplain Inundation Control

The prevention of floodplain inundation by flow-regulation, downstream from an impoundment, will deprive many fish species of spawning grounds and valuable food supplies. The construction of the Pa Mong Dam, Thailand, initiated a dam-building programme which will eliminate flooding from the 49 560 km^2 Mekong delta, and most of the 5000 km^2 floodplain of the Senegal River, West Africa, will soon become dry because of headwater impoundment (Welcomme, 1979). Dam construction for industrial uses within the Rio Mogi Guassu, Brazil, has also resulted in the progressive loss of floodplain wetlands (Godoy, 1975). The loss of floodplain area for feeding and breeding can have serious effects on fish populations, and the reaction of the fish communities of the Chari, Niger, and Senegal, Rivers, Africa, to flood failures, provoked by natural climatic variations such as the Sahelian drought (1970–74), confirm the highly detrimental effects of suppressing annual floods (Welcomme, 1979). Several semi-migratory species have been affected (Table XLIX). The decline of these species may relate to flood-control, or to sporadic flooding caused by dam releases. Below the V. I. Lenin Volga Hydroelectric Power Station, USSR, irregular flooding has stranded species which spawn on drowned vegetation, whereas main-stream spawners such as *Stizosterion (= Lucioperca) lucioperca* and the Acipenseridae, are not so affected (Eliseev & Chikova, 1974). Whitley & Campbell (1974) attributed the steady decline of the fish-catch from the Missouri River, USA, primarily to the loss of fish habitats following the construction of dams. Because of flow-regulation, the amount of floodable wetlands over a 145-km sample reach of the River has declined from 15,167 km^2 in 1879 to 6414 km^2 in 1967, and during this period annual fish-catches dropped from 680 t to 122 t.

Table XLIX Fish species affected by the control of floodplain inundation.

River, dam, and location	Affected fish	Source
Murray River, Hume Dam, Australia	Reduced inundation of backwater systems will reduce the spawning area for *Maccullochella peeli*	Lake (1975)
Missouri River, USA	*Petalurus punctatus* and *Ichthyobus* spp. eliminated, due to loss of floodplain wetlands, and replaced by *Cyprinus carpio*	Whitley & Campbell (1974)
Pongolo River, J. G. Strydom Dam, S. Africa	Spawning failure of *Hydrocynus vittatus* and *Labeo* spp. due to the failure of annual flood to link the river and floodplain water-bodies	Jubb (1972)
River Dnieper, Kakhovka Dam, USSR	Primary spawning-habitat—floodplain water-meadows—eliminated by river regulation, for *Rutilus rutilus heckeli, Abramis* sp., and *Cyprinus carpio*	Zalumi (1970)
River Volga, V. I. Lenin Dam, USSR	Phytophilous fishes—*Abramis brama, A. ballerus*, and *Rutilus rutilus*—reduced, due to failure and irregularity of floods	Chikova (1974)

The Pongolo Floodplain Pans, South Africa, comprise 25 major bodies of water, together with smaller pans, and have a normal total surface-areas of 2650 ha. The pans depend, almost exclusively, upon an annual inundation by the Pongolo River, which may flood 10 000 ha. The occurrence and timing of these floods are the prime factors associated with the stimulation of upstream migration of mature adult fish, which can then enter the pans to spawn. Jubb (1972) noted that the failure of the annual flood, since the construction of the J. G. Strydom Dam, has resulted in spawning failure by *Hydrocynus vittatus* and *Labeo* spp. Moreover, the spawning of these species was observed in the pans consequent upon controlled artificial releases from the dam, which inundated two major pans—suggesting that considered flow-regulation can accommodate a well-established stock of native fishes. Davies (1979), in a review of the effects of the Pongolapoort Dam, presented evidence to suggest that the careful management of reservoir-releases could sustain high levels of fish productivity: firstly, by providing a preliminary flow to induce adult fish migration, in readiness to enter the pans during the main flood, and, secondly, by filling the pans above supply-level, so that discharge then takes place back to the river as the flood recedes, attracting fish into the pans for spawning. This is supported by the successful utilization by fishes of an artificial flood-release from Lake Kariba for spawning on the Mana Pools floodplain (Kenmuir, 1976).

Discharge as a Stimulus to Migration:

Migratory fishes are often encouraged to move upstream by rising flows, and little movement occurs when discharge is uniformly low (Hynes, 1970b). Extreme floods have been observed to provide a barrier to upstream migrations (Pritchard, 1936), but many migrations occur at times which are coincident with increasing, or seasonally high, discharge. Even in rivers that are experiencing regular migrations throughout the year, such as the Sacramento River, California, USA, significant peaks of migration have been observed (Fraser, 1972). Observations on the different reactions of *Oncorhynchus gorbuscha* (Pacific or Pink Salmon) to flows in three streams in northwestern USA (Davidson *et al.*, 1943), suggested that, in some streams, the migration of the Salmon can be delayed by flow-regulation. Such delays may cause the premature use of energy-reserves, and the development of stresses, causing death or reduced reproductive success (Brett, 1957). Because of this apparent simple relationship, Baxter (1977), and others, suggested that reservoir releases could be manipulated so as to maintain populations of valuable fish-stocks. However, Alabaster (1970) noted that there is no consistent relationship between discharge and fish migration, either within or between rivers, and that no particular flow is associated with fish movement throughout the year. A sudden decrease in the numbers of Acipenseridae within the Kura River, USSR, below the Mingechaur Dam, was related, principally, to both the increased transparency of the water and to flow-regulation, because the main spawning-run was adapted to the period of high, turbid flows (Abdurakhmanov, 1958). Thus, it is doubtful whether flow, as such, is the sole stimulating factor.

In a study of the migration of *Salmo salar* (Atlantic Salmon), Hayes (1953) noted that, in comparison with natural rains, the effect of artificial freshets produced by reservoir releases should be considered as small. Alabaster (1970) suggested that short-term changes in flow, together with accompanying changes in the concentration of dissolved substances, may provide the stimulus. However, freshet releases from the Stocks and Tryweryn Reservoirs, UK, did not significantly affect Salmon migration (N.W.W.A., 1972), and it was concluded that the water released from the reservoirs as a freshet flow did not have the same influence upon the fish and their migratory behaviour as a natural flood.

Water Quality

The quality of reservoir tailwaters will differ from that of the natural river because the thermal, and chemical, character of the dam-releases will be determined by the limnology of the lake itself. Surface-release reservoirs act as nutrient traps and heat exporters, whilst deep-release reservoirs may export nutrients and provide a heat-store. Changes of water-quality below dams can affect fish species and populations in three ways: by exceeding tolerance limits; by inhibiting the normal sequence of reproduction, development, and survival; and by altering the competitive balance and predator–prey relationships.

The thermal influences of impoundment have had the most significant effect upon local fish faunas, eliminating many temperature-specific species (Trautman & Gartman, 1974). Water-temperature changes have been commonly identified as a cause of the reduction in native species, particularly as a result of reduced spawning-success (Table L). Species with specific temperature requirements may disappear if they are unable to tolerate the imposed conditions and, even if adults are able to withstand the imposed stress, reproduction may be adversely affected. Cold-water releases from the high dams of the Colorado River, for example, have resulted in the decline in native fish abundance (Holden & Stalnaker, 1975). Cold discharges from Glen Canyon Dam, together with little solar warming because of high canyon walls and the absence of large, warm-water tributaries, have made the River unfavourable for most native fishes. Spawning temperatures, especially for rare forms, seldom occur, and the endemics are being replaced by introduced species that are better adapted to the imposed conditions. The native *Catostomus latipinnis* and *C. discobulus* were the dominant species, but, after dam completion, introduced species came to outnumber the native species by 19 to 10, and four endemics—*Ptychocheilus lucius, Gila elegans, G. cypha,* and *Xyrauchen texanus*—were considered to be endangered. Indeed, the Colorado River in the Grand Canyon remains too cold for most native fishes for over 400 km below Glen Canyon Dam. Prior to the filling of Flaming Gorge Reservoir, the estimated maximum net gain in weight of Rainbow Trout (*Salmo gairdneri*), of 10.4×10^3kg per km^2, was associated with temperatures of 20–36°C, but after filling this dropped to 0.2×10^3 kg per km^2 at a reduced water temperature-range of 22–31°C (Mullan *et al.*, 1976).

In many cases, water temperature may provide the dominant limiting condition. Despite the relative abundance of insect food below Flaming Gorge Reservoir, *S. gairdneri* were concentrated not in areas of high velocity, associated with the typically aggressive drift-feeding, but in areas of least velocity, where Algae were most abundant. This suggests that the fish were forced by low temperatures, resulting in low metabolism, to seek sheltered

Table L Observations on the relationship between water temperature and fishes within impounded rivers.

General Location	River	Effects of thermal regulation	Source
Canada	Kananaskis	Lowered summer temperatures have virtually eliminated trout.	Radford (1972)
USA	Madison	Increased summer temperatures are threatening trout.	Vincent (1977)
USA	Green	Successful reproduction of *Ptychocheilus lucius* has been eliminated by imposed thermal regime.	Holden & Crist (1979)
USA	Guadalupe	*Etheostoma spectabile* extended its normal winter and spring breeding-season to all parts of the year as a result of cold hypolimnial releases; this has produced a ten-fold increase in its abundance.	Edwards (1978)
USA	Caney Fork	Temperatures 1–1.5% too low for *Polyodon spathula*.	Pasch *et al.* (1980)
USSR	Kura	Sturgeon reproduction has been inhibited by changed temperature regime.	Zakharyan (1972)
Central Europe	Svratka, and Elder	Altered temperature regime, and decreased mean temperature, detrimental to *Barbus* sp., *Esox lucius*, and *Perca fluviatilis*.	Peňáz *et al.* (1968) and Lehmann (1927)
Austrlia	Goulburn	10–15°C reduction of water temperature resulted in loss of the formerly abundant native *Maccullochella peeli*	Williams (1967)
	Murray	*Macquaria ambigua* became totally extinct due to temperatures being maintained below spawning levels. Also, temperatures now are rarely high enough to induce spawning by *Plectroplites ambiguus*.	Williams (1967)

positions out of the main currents and away from their major food-source (Mullan *et al.*, 1976). Similarly, in Tennessee, USA, apparently good spawning habitat for *Polyodon* sp. exists within the Caney Fork River below Center Hill Dam, but no eggs or larvae were observed, despite the movement of large fish into the channel over a prolonged period: the temperatures of the reservoir releases were 1–1.5°C colder during the spawning period than the lowest temperature at which *Polyodon* larvae have been collected (Pasch *et al.*, 1980).

For many studies of fish changes it is difficult to isolate a single causal variable, because other parameters will also have an effect—especially turbidity and water chemistry. Indeed, the Lower Colorado was formerly a warm, turbid, often violent, river which was given also to drastic changes in volume and tubidity, and characterized by coarse-fish species. The change from a turbid and warm-water, to a clear and cold-water, habitat has been associated with a rise in the dominance of *Salmo* spp., which have replaced some twenty native and introduced species (Mullan *et al.*, 1976). Nevertheless, in general, epilimnial-release dams produce a 'warm' tailwater more suitable for coarse-fish species, whereas hypolimnial releases give rise to 'cold' tailwaters, which are more appropriate to 'game' fishes. Below hypolimnial-release dams, trout may be able to survive at low altitudes and low latitudes, because of the release of cold, clear, water during summer. Thus, the range of Brown and Rainbow Trout (*Salmo trutta* and *S. gairdneri*) has been extended throughout the Murray–Darling system in Australia (Cadwallader, 1978).

Water-chemistry changes can also be significant for riverine fish. Chemical changes relate to the retention-time of the reservoir and to the level of outflow at the dam—so that, downstream, changes of water chemistry will vary both between dams, and seasonally at any one dam, as the reservoir stratifies and overturns. Downstream from an impoundment, nutrient availability may be decreased, because of metabolism within the reservoir, and this will be reflected by reduced loads of nitrates, phosphates, and other dissolved 'elements'. Such changes would effectively lower the primary productivity of the river, accompanied by changes in the energy transfer of higher trophic levels, as the various species adjust to the altered food-supply. However, with less convertible energy available, the productivity of the community as a whole must be reduced. A change of the water pH, and increases in ammonia or hydrogen sulphide, due to anoxic conditions in the hypolimnion, may cause a further unnatural stress to be placed upon organisms inhabiting the water downstream. Even if the concentration of any one toxic substance is maintained below a critical level, the combined action of many subtle changes can cause disruptions of the local stream fauna. The release of anoxic water from the hypolimnion has caused extensive fish mortality below some Czechoslovakian dams (Brádka & Rehacková, 1964), and problems of oxygen supersaturation for fry mortality have been widely identified (e.g. Peňáz *et al.*, 1968; Beiningen & Ebel, 1970). However, little direct evidence exists for an evaluation of the influence of these biochemical changes upon riverine fishes. Low oxygen-levels may, in fact, only have a local effect, because dissolved oxygen can be rapidly replaced by turbulence at the discharge point or within the downstream channel.

INDIRECT CONSEQUENCES

Hydrological, seston, and water-quality, changes induced by reservoir creation will have a number of important indirect effects upon lotic fish faunas. Changes in autochthonous food production and allochthonous food supplies, riparian and aquatic vegetation, channel-substrate composition, and channel morphology, consequent upon dam closures, will cause adjustments within fish populations. However, these attributes of lotic systems may require a long-time period before a complete adjustment of the fish population to the imposed flow-characteristics is achieved. Readjustments of fish species composition and fish biomass will also require a long time-period. Such readjustments may be slow and progressive, but they are more likely to be episodic, occurring relatively suddenly when once a critical level of tolerance, or competition, is exceeded. With respect to the latter, deliberate, or unintentional, introductions of exotic species are often important; but even deliberate introduction may fail to establish viable populations. Thus, although many warm-water fishes were eliminated from the Clinch River by thermally-regulated releases from Norris Dam, Tennessee, USA, the introduced cold-water species, especially *Salmo gairdneri*, failed to develop because the depleted benthic invertebrate fauna did not provide the necessary food-supply (Tarzwell, 1939). Primarily because of a failure to appreciate the magnitude and spatial significance of these long-term changes, which often appear to be very subtle in the short-term, and because of the difficulty of isolating discrete factors, the effects of these changes upon riverine fishes are only poorly understood, and have often been ignored within the literature. Nevertheless, these 'delayed' impacts can impose important controls upon fish faunas.

Food Supply Alteration

The fish fauna is influenced by changes in the plankton, invertebrate drift, and bottom faunas. Although specific data are sparse, the food preference of different species, in relation to the dam-induced changes in the supply of different food materials, may play a fundamental role in the determination of stable fish populations within impounded rivers. The formation of Lake Kainji, River Niger, led to a decrease in fishes of the family Mormyridae primarily because of the change of food availability (Lewis, 1974): the food supply of bottom-feeding insectivores was severely depleted. The biomass of benthos, which contributes a substantial part of the fish-food, may be detrimentally affected, for example, by the increased salinity of the lower Volga and Caspian Sea, USSR, as a result of flow-regulation. The level of the latter fell by nearly 3 m between 1930 and 1962 as a result of the controlling of 50% of the Volga River's inflows (Micklin, 1979). Moreover, dam construction on the Sefid River made no allowance for the requirements of spawning Sturgeon (*Acipenser* sp.) from the Caspian Sea (Vladykov, 1964). Water flow was reduced, shallow-river water became overheated and, although the increased water temperatures were tolerated by the adult Acipenseridae, the catastrophic depletion of important food items, such as aquatic insects and crustaceans, resulted in severe mortalities.

Changes of the flow regime—involving reduced seasonal, and daily, flow-variability—may alter the natural drift of food organisms, and affect the feeding habits and possibly the reproduction of fish. For example, artificial freshets can contribute to food supplies for salmonids by dislodging benthic organisms (Fraser, 1972). It is well established that salmonid fish, in particular, utilize drift as a food-source, and Mundie (1974) suggested that drift density during daylight periods might be artificially enhanced by controlled releases, in order to increase fishes' food-supply. During two small experimental releases, the percentage of benthos in the drift at any one time has been increased from 0.006% to 0.013% and 0.010% (Brooker & Hemsworth, 1978). However, only some species responded—those whose pattern of behaviour is related to flow variations—whilst others, displaying a day–night periodicity, retain this diurnal behaviour (See pp. 187–191).

River impoundment, however, does not always impact negatively upon food supplies and, thus, fish communities. Supplies of plankton from reservoirs can provide an improved food-source, and annual plankton losses from Gavins Point Dam, Nebraska, USA, provide an important food supply for Paddlefish (*Polyodon spathula*) concentrations in tailwaters (Boehmer, 1973). Peňáz et al. (1968) observed that the changes in the characteristics of the phyto- and zoo-benthos, as well as the plankton and drifted organisms, which were induced by the formation of the Vir River Valley Reservoir, Czechoslovakia, had a favourable effect upon all sizes and ages of fish, because all the major food-components were enriched. Subsequent to impoundment, the zooplanktonic component of the drift in the river below the dam has been increased during the spring, summer, and autumn, periods, coinciding with the period of increased feeding activity of fishes—especially *Salmo trutta*, which is the most important species of the Svratka River, both from the economic and sporting viewpoints. Juveniles are particularly favoured by the increased drift of nematodes, oligochaetes, and chironomids. An increase in the populations of some small native fishes was observed in the section below the dam, also as a consequence of an improved food-supply, and these form another important component of the natural food of *S. trutta*. However, within the Missouri River, USA, Whitley & Campbell (1974) detected that the food supply was enriched for only about 2 km below a dam. This suggests that improved conditions for fish production, in terms of the enriched water from the reservoir which, in passing through the sluices, carries with it zooplankton, insects, and fish, may only be of local significance. Disorientated and injured fish, passed through turbines into stilling basins, may also provide a food-source for predators (Begg, 1973); but this, once again, will be of only local significance.

Morphological Changes

The regulation of flows and the abstraction of the sediment-load will cause a readjustment of channel morphology, and these changes describe a continuum of responses from narrower, deeper channels to broader, shallower ones, depending upon the local conditions and vegetational reactions (cf. Chapter 6). The dimensions of the channel in cross-section are important for fishes for two reasons: firstly, because the width of the channel controls the area of usable habitat; and secondly, because the channel-shape controls the hydraulic geometry—specifically, the variation of velocity with changing discharge. The area of channel-bed inundated by a particular discharge, and controlled by the configuration of the channel in cross-section, can be important in terms of the area available for fish-spawning and food production, for temperature change, and for satisfying the spatial requirements of some species (Fraser, 1972). Territorial requirements, however, are perhaps more related to the velocity of the current, and to the velocity distribution across the channel. Juvenile Salmon (*Salmo salar*) and Brown Trout (*S. trutta*), for example, have been found to occupy, and defend, well-defined territories, which become smaller with increased velocities (Kalleberg, 1958). At reduced velocities, individual fish will enlarge the area of their territories and the resident population will be reduced. Thus, even where the area of habitat remains unaltered below dams, a change of the channel shape may alter the velocity characteristics of the flows and cause a change of the potential carrying capacity of the stream.

Changes of the channel-bed and substrate composition may have severe implications for the habitats available for lotic fauna. Together, these factors influence the velocity of flows, by affecting channel roughness and the flow-resistance, and the shelter that is is available for lotic organisms. The velocity of the water influences shelter by creating turbulence and white-water areas, which in turn control the visibility for predators, and also the re-aeration of the water. Shelter requirements also involve the existence of deep pools for large fish, and a heterogeneous bed for smaller organisms. The requirements of Trout, and their food organisms, can be critical if channel changes involve the formation of a monotonous, uniform bed. The filling of pools would deprive riverine fishes of necessary hiding-places, and channel sedimentation may induce a marked change in insect populations and a loss of fish-food. Such changes have occurred within the Colorado River, where the accumulation of sediment has caused the replacement of Trichoptera, and nymphs of several species of Plecoptera and Ephemeroptera, by dipteran larvae (Saunders & Smith, 1962).

An accumulation of sediments, within river channels below dams, occurs where tributaries inject larger volumes of sediment than can be transported by the regulated flows. However, even in the absence of such obvious sediment sources, important changes of the substrate materials will occur. Immediately downstream from a reservoir, the surface of the bed-material may become eroded, so that an incomplete 'armoured' layer of relatively coarse material is produced. Very coarse gravels may provide unsuitable substrate for spawning. Thus, below Shasta Dam, Sacramento River, California, degradation, and the increase in size of the bed-sediments, has been a major factor responsible for the decreased spawning stock of fall migration Chinook Salmon (*Oncorhynchus tshawytscha*) from over 250 000 to less than 110 000 (Milhous, 1982). Substrate gravels may also become compacted and the interstices, protected by an armoured layer, may become filled with finer material. Changes in the particle-size composition of bed sediments may result in a reduction of the intragravel flow-velocities, deleteriously affecting fish-stocks. Fish species spawning on gravels generally require a flow of well-oxygenated water through the substrate, in which the eggs are incubating, for successful reproduction. Loose gravels are used predominantly by fishes in depositing and covering the eggs. It is essential that the eggs are not smothered by sediments, or dislodged, and water must percolate freely to aerate the eggs for 2 to 3 months to allow full development and hatching. The lack of hatching and survival, particularly of trout forms, in many tailwaters, has been ascribed, primarily, to fluctuations in flow; but the alteration of gravel size or permeability, associated with river-bed degradation or sedimentation, could be a common limiting factor.

Riparian and Aquatic Vegetation

Shelter and food production for many fish species is strongly related to the type and density of riparian and aquatic vegetation, but the encroachment of vegetation into the channel, as a result of flood control, may adversely affect spawning-grounds. Fraser (1972) observed the encroachment of willows and other vegetation onto *Oncorhynchus tshawytscha* spawning-beds in several Californian rivers where flood-flows have been controlled. Such encroachment may render the spawning areas unusable by the Chinook Salmon, at least, even though the discharges during the spawning season may be entirely satisfactory. Furthermore, Fraser observed that, within other regulated Californian rivers, stabilized low-flows contributed to favourable conditions for luxuriant growths of *Eichhornia crassipes* (Water-hyacinth) in some years, providing an impairment to Salmon migration. In other rivers, the establishment of macrophytes may provide ideal spawning-grounds for some species: abundant water-weeds and reeds along the Murrumbidgee–Darling system, for example, provide spawning grounds for *Perca fluviatilis* and *Tinca tinca* (Cadwallader, 1977).

Introduction of Exotics

Reservoirs may create habitats which can be detrimental to native warm-water species, but the artificial environment often provides conditions that are suitable for exotic species, which may be introduced either deliberately or accidentally. Below many dams, the artificial environment can provide for viable trout fisheries, although fish-stocks are usualy maintained by stocking rather than by natural reproduction (Hoffman & Kilambi, 1970). In the Murray River, Australia, introduced species such as *Carassius auratus*, *Perca fluviatilis*, *Tinca tinca*, and *Salmo* spp., are becoming increasingly dominant at the expense of native species, such as *Plectroplites ambiguus* and *Tandanus tandanus* (Walker *et al.*, 1979). The introduction of exotic species has induced the local extinction of many native species which, although reduced in population, might otherwise have maintained a reproductive stock under the regulated flow-regime. Indeed predation, or competition, by exotic species—frequently introduced deliberately to improve the sporting potential or to satisfy an 'economic' demand—has had an immediate impact upon the native fishes of rivers (Kimsey, 1957; Long & Krema, 1969). Competition from introduced species has influenced the decline of the numbers of native fish in the Upper Colorado basin (Holden & Stalnaker, 1975): *Ptychocheilus lucius* and *Xyrauchen texanus* have become rare simultaneously with the establishment of the exotic *Ictalurus punctatus*. The addition of an abundant carnivore to the fish fauna has created additional competition for space and food, whilst the introduction of small cyprinids (especially *Notropis lutrensis*) has created more competition with juveniles of native species (Holden & Stalnaker, 1975).

Unintentional introductions may be nonetheless significant. The natural intrusion of piscivorous or competitive species may take several years, but the frequent provision of water-transfer tunnels, or canals, may provide for the increased dispersal of these species. Cambray & Jubb (1977), for example, reported the passage of five species of fishes through the 82.45-km-long Orange–Fish Tunnel from the Hendrik Verwoerd Dam, and their successful introduction, albeit unintentionally, to the headwaters of the Great Fish River, South Africa. Such water-transfers break down natural geographical barriers and permit the dispersal of fish that are indigenous to one river into a neighbouring stream-system. Juveniles of some indigenous fish-species can survive being transported by centrifugal pumps from rivers into dams, or *vice versa* (Jubb, 1976), and can survive, in limited numbers, a journey through hydroelectric turbines. In the Kura River, USSR, some limnophilic fishes (e.g. *Abramis brama*) have survived passage through the turbines and diversion conduits to form large populations below the Mingechaur Dam (Abdurakhmanov, 1958). The Lake Tanganyika Sardine (*Limnothrissa miodon*) was introduced into Lake Kariba to exploit the zooplankton resources, but was soon observed in the Zambezi downstream (Kenmuir, 1975). The *L. miodon* alevins appear to have survived being released *via* the tailrace discharge through hydroelectric turbines driven by a head of water of some 90 m. Although *L. miodon* is essentially a lacustrine species, favouring deep water, the regulation of the Mid-Zambezi below Lake Kariba has, apparently, provided suitable conditions for the establishment of a size-able stock. Such observations indicate that deliberate species-manipulation within a reservoir should not be undertaken without due consideration for the tailwater fisheries.

Considerable efforts have often been expended in the introduction of new, valuable, fish species that are able to tolerate lentic conditions, and at the same time to compete successfully against, and often to suppress, the multi-plication of native species of lower economic value. In many cases, however, the introductions have not been exotic species, but species from downstream, low-velocity, reaches of the same stream-network. Thus, the Tsimlyansk Reservoir was stocked with species from the lower Don, and the fish popula-tion of the Kakhovka Reservoir, on the Dnieper River, was established by introductions from adjacent water-bodies (Zhadin & Gerd, 1963). Character-istically, the development of fisheries within man-made lakes involve two complementary operations: namely, the introduction of new, valuable, fish species, able to exist in standing water, and the eradication of non-valuable 'rough fish', which are primarily indigenous species. Pre-impoundment fish-eradication programmes in the Colorado River Basin were conducted at the Flaming Gorge Reservoir and Navajo Dam sites (Holden, 1979). Subse-quently, large numbers of an exotic chub, *Gila atraria*, and *Gila robusta*, became abundant in Flaming Gorge Reservoir and Navajo Reservoir, respec-tively. Holden concluded that both within, and downstream from, reservoirs, eradication programmes create conditions that may prove favourable for opportunist exotic species and, in many cases, give unwanted exotics a head start on deliberately introduced game-fishes.

INTERRUPTED FISH MIGRATIONS

The effect of dams upon the population of diadromous and semi-diadro-mous fishes has received considerable attention, particularly with reference to the reproductive biology of anadromous species—a concern stimulated by the need for the maintenance of economic fisheries and for the conservation of endemic species. It has been shown that river impoundment has a profound, and usually an adverse, effect on the reproduction of native fishes, because of the inundation of spawning-grounds and the obstruction to upstream migration. Some native species may be found in a reservoir for a few years after dam closure, but if the suitable lotic spawning-grounds are inaccessible, extinction will result because of a lack of recruitment. Adult Paddlefish (*Polyodon spathula*) of the Missouri River can survive in reser-voirs, but migrate upstream to spawn in tributaries (Pflieger, 1975); further dam construction would isolate these spawning areas. In northern Iran the construction of a large earthfill dam on the River Lar will destroy 40 km of stream habitat that is ideal for the spawning of *Salmo* spp. (Coad, 1980), and on the Indus River, the construction of the Gulum Mahommed Dam has deprived the migratory *Hillsa ilisha* of 60% of their previous spawning-areas (Welcomme, 1979). The valuable Acipenseridae have been particularly threatened by hydroelectric power dams on the Volga, Don, and Caucasian rivers, USSR, and since the Mingechaur Dam was constructed, Acipenseridae have not entered the Kura River, and can no longer pass from the Caspian Sea to their spawning-grounds in the middle and upper reaches of the Volga (Zhadin & Gerd, 1963).

The 'barrier-effect' imposed by a dam obviously depends upon the height of a structure and the motile capabilities of particular species. A series of small weirs on the River Murray have little effect upon the native fish species which may undertake rapid, and long, migrations during periods of high-flows (e.g. migrations of 100 km in 163 days: Reynolds, 1976), but drastic reductions of these species, notably the *Macquaria ambigua*, upstream of the Yarrawonga Weir, suggests that this larger structure provides a physical barrier to fish migrations (Walker, 1979). Major dams introduce considerable obstacles to migratory fishes, but relatively small structures may also prove to be impassable, even during high-flows, and block vital movements of certain species. Irrespective of the size of the dam, delayed migrations, related to fish movement through the reservoir, can also adversely affect the survival and reproduction of migratory species (Raymond, 1969; Bentley & Raymond, 1976). Thus, impoundments have introduced major problems for many species: *Oncorhynchus* sp. (Raymond, 1968), *Salmo* sp. (Henricson & Müller, 1979), *Polyodon* sp. (Branson, 1974), *Anguilla* sp. (Jackson, 1966), Acipenseridae (Zhadin & Gerd, 1963), *Thymallus* sp. (Bishop & Bell, 1978), and *Alosa* sp. and *Morone* sp. (Nichols, 1968). The Columbia River once supported a great run of anadromous salmon and trout into more than 80% of the basin, but less than 30% of the original salmon habitat remains accessible to sea-run spawners, due to the concentration of main-stream dams (Robinson, 1978). Indeed, this impact has resulted in diminishing catches of *Oncorhynchus* sp. in the Columbia River, from a peak of 22 440 t in 1911 to less than 7000 t today (Trefethen, 1972). The 800 km-long migrations of *Plectroplites ambiguus* within the Murray River have been stopped by flood-control structures (Butcher, 1967).

Some species, however, appear to be able to circumvent or scale the obstruction imposed by high dams. Prior to the construction of Lake Kariba on the Zambezi River, Jubb (1960) predicted that catadromous eels, *Anguilla nebulosa labiata*, would disappear from the lake and upstream channel network, claiming that the extinction of the eels would be caused by the failure of small eels, returning from the breeding-grounds at sea, to negotiate the 128 m-high dam. However, Balon (1975) observed an abundant population of *Anguilla* sp. at depths of 25–40 m in Lake Kariba, because the juvenile eels were capable of surmounting the dam-wall.

Juvenile eels, of the species *Anguilla reinhardtii* and *A. australis*, and the local gudgeon *Gobiomorphus coxii*, have also been observed to move up a vertical section of the Tallowa Dam wall on the Shoalhaven River, Australia (Bishop & Bell, 1978). However, even if upstream migration is maintained, any delay to this migration, caused perhaps by hydro-chemical changes, can have severe consequences. Shikhshabekov (1971) for example, demonstrated that a reduction in the reproduction of semi-diadromous fishes may occur under the conditions of regulated discharges produced by the Arakum Lakes, USSR, if the arrival of the spawners is delayed in relation to critical spawning temperatures and hydrological conditions. Moreover, under such conditions, Shikhshabekov indicated that several species of fishes reabsorbed their eggs, and this would severely affect not only the period of reproduction, but also might cause delays to the next sexual cycle.

Gas Supersaturation

During high-flow years, water may spill over the crest of a dam and become saturated with atmospheric gases to a level that can be lethal to fishes. Indeed, even though large numbers of smolts (migratory young salmon) can pass dams in spillweir discharges, mortality can result from prolonged exposure to lethal concentrations of dissolved gases. The severity of gas-bubble disease, and its consequences, depend upon the magnitude of supersaturation, the duration of fish exposure, water temperature, and the general physical condition of the individual fish. Nevertheless, the impact of gas supersaturation can be severe. The supersaturation of dissolved nitrogen in the Columbia River, during heavy spillweir discharges, caused the mortality of juvenile salmonids (*Oncorhynchus kisutch* and *O. tshawytscha*) from gas-bubble disease, and also appeared to explain the mortalities of adult *O. nerka* and *Salmo gairdneri* (Beiningen & Ebel, 1970). Supersaturation by dissolved nitrogen has also occurred in turbine discharges from hydroelectric power-dams at low generating levels. Turbine discharges caused gas-bubble diseases, resulting in major fish-kills, in the Saint John River, New Brunswick, Canada, below the Mactaquac hydrelectric power-dam (MacDonald & Hyatt, 1973): two hundred adult *Salmo salar* were killed—some 10% of the annual up-river 'run'—*Anguilla rostrata* were also severely affected, and dead, or dying, fish were observed for nearly 2 km downstream from the tailrace.

The mechanisms involved in creating dissolved-nitrogen supersaturation have been elucidated by Ebel & Raymond (1976) for a study of the Snake and Columbia Rivers, USA. Intensive research was stimulated by substantial mortalities of both adult and juvenile salmonids, caused by high spillweir-flows, which produced abnormally high (123–143%) supersaturation below John Day Dam in 1968. Differences in mortality between species may be related to two factors: firstly, the differences in depth that each species travels and, secondly, the timing of the migration. Live-cage studies in the Columbia River (Ebel, 1969; Beiningen & Ebel, 1970), indicated that whilst less than 30% mortality occurred in populations of juvenile *O. kisutch* (Coho Salmon) and *O. tshawytscha* (Chinook Salmon) held in a deep (> 6 m) cage, mortality reached 100% in fish held in a shallow (> 1 m) cage, with nitrogen levels between 118 and 145%. Moreover, fewer of the deeper-migrating *Salmo gairdneri* succumbed (15% mortality)—compared with the 25% mortality of *O. tshawytscha*—to lethal concentrations of dissolved atmospheric gas between two dams on the Snake River (Raymond, 1979). The largest percentage of downstream migrants are commonly found in the top 2 m of water, so that the average hydrostatic compensation achieved is about 7.5% of saturation—insufficient to compensate for saturation levels as high as 135–140%, which occur over wide areas during high-flow years (Ebel & Raymond, 1976). Laboratory studies (Dawley *et al.*, 1976) have demonstrated that mortalities may vary with depth for saturation levels below 120% but, at higher levels, substantial mortalities can occur irrespective of other factors.

Those species which migrate prior to the period of peak discharge may avoid being stressed by gas-bubble disease. Thus, the early migrations of *O. tshawytscha* during low-flows of the Snake River are affected mainly by turbine passage, but the later *S. gairdneri* migration coincides with higher flows and spillage, and is subjected to 10 additional days of exposure to lethal gas-concentrations (127–140%) (Raymond, 1979). Prior to 1968, the main migrations of juvenile *O. tshawytscha* from the Snake River usually entered the Columbia River before flows in the Columbia had peaked, and were not subjected to the highest levels of gas supersaturation. The arrival of peak migrations to the Dalles Dam, however, has latterly been delayed by 10–20 days due to the closure of three large upstream dams, so that fish exposure to supersaturated water has been increased (Ebel & Raymond, 1976). The migration rate is particularly important in determining the length of time that fish are exposed to supersaturation. Moreover, whilst juvenile salmonids subjected to sublethal periods of exposure to supersaturation can recover if returned to normally saturated water, adults do not recover and generally die from the direct or indirect effects of exposure to supersaturated water (Ebel & Raymond, 1976). It should also be noted that stressed individuals are particularly vulnerable to predation.

Downstream Migrations

Impoundments present problems to fishes migrating downstream, of such magnitude that survival may be as low as 5% (Raymond, 1979). Such catastrophic mortalities can occur during low-flow years as a result of four factors: limited over-dam spillage, reduced flow velocities through reservoirs, passage through turbines, and increased predation in the stilling basin below the dam. For hydroelectric power-dams, mortality resulting from passage through turbines is especially significant, and turbine losses of juveniles of between 10% and 40% have been reported widely (Schoeneman & Junge, 1961; Ebel & Raymond, 1976). Such factors are in addition to the imposed changes of discharge and water quality, particularly gas supersaturation, which affect all fishes within the riverine section below dams.

The travel-time for downstream fish migration is directly related to river velocity, so that the quicker the flow is, the higher will be the rate of fish migration. Raymond (1979) has reviewed a decade of study on migration rates and travel-times for various discharges (Table LI). This demonstrates that the reduced water-velocities within a reservoir may reduce the migration rate to one-third of that through free-flowing stretches of river. Smolts travel the 636 km of free-flowing Salmon River, USA, from the spawning grounds to the estuary, in about 26 days under low-flow conditions, but consequent to the impoundment of 45 km of the River, low-flow migration required 65 days. The effect of these delays would increase the possible mortality from exposure to poor-quality water—water that is supersaturated with atmospheric gases during high-flow years—or from predation and disease at times of low-flow. Indeed, the 'migratory urge' in young salmonids may only last for a limited period, so that delay may cause this urge to wane, and migration may accordingly cease (Pyefinch, 1966). Significant delays during periods of low-flow may induce juveniles to remain in reservoirs for extended periods before completing migrations.

Studies made on salmonids in the Brownlee Reservoir, Columbia River, USA (Collins, 1976), showed that while adult fish were able to migrate successfully through the 90-km-long reservoir, the juveniles found conditions too severe. Strong thermal stratification in the reservoir is associated with surface temperatures which reach levels that are lethal to young salmon, whilst the cooler, subsurface water becomes deficient in oxygen. Reservoirs can act as a storage for the accumulation of pollutants to levels that may become toxic to juvenile fishes exposed for an increased time-period as a result of delays in migration. Thus, Dominy (1973) speculated that the considerable volume of pollution within the impounded Saint John River, New Brunswick, Canada, may have caused the severe reduction there of Atlantic Salmon (*Salmo salar*). The large reservoirs concentrate much of the organic and toxic waste, water storage has reduced the re-aeration of the water, and severe oxygen-depletion has resulted: the ecosystem has become unsuitable for salmonids or for any other species of desirable sport-fish. Delayed migrations may also affect the limited smoltification period (Raymond, 1979), during which juveniles (smolts) are motivated to migrate downstream and are physiologically capable of adjusting from a freshwater to a salt water environment. Any delay may result in the loss of this capability, so that, for Steelheads (sea-run *Salmo gairdneri*), the survival rate may drop from almost 100% to less than 20% during the adjustment from fresh to salt-water (Adams *et al.*, 1975). Further smolt losses will result from predation. Predators such as *Ptychocheilus oregonensis* and seagulls cause up to a 33% loss of fish released into the tailwaters of the Columbia–Snake River dams (Long & Krema, 1969). Smolt losses of 86%, 71%, and 79%, occurred in three successive years during passage through Loch Tummel, UK, due to delayed migrations, loss of migratory urge, and increased predation by *Esox lucius*, *Salmo trutta*, and birds (Pyefinch, 1966).

Hydroelectric power-dams create an additional cause of mortality as a result of fish passage through turbines. Extensive tests have been conducted

Table LI   Migration rates of juvenile *Oncorhynchus tshawytscha* and *Salmo gairdneri* through free-flowing and impounded sections of the Snake and Columbia Rivers, USA, for varying river-flows. (After Raymond, 1979.)

Discharge (m³ per s)	Low flow		Moderate flow		High flow	
Snake River	1000	1500	1500	3000	3000	5000
Columbia River	4000	5000	5000	10 000	10 000	14 000
Migration rates (km per day)						
Free-flowing	24		40		54	
Impounded	8		13		24	

on the downstream passage of fishes through hydraulic turbines utilizing both model and prototype installations (Long & Krema, 1969). The tests have sought dual objectives: to establish design criteria for turbines that will provide optimum fish-passage, and to formulate operational procedures for existing turbines to provide for maximum survival of fish under prevailing conditions. Although designs have been proposed for collecting fish entering turbine intakes (Park & Farr, 1972), no satisfactory device has been developed to ensure the safe downstream passage of juveniles. Under natural conditions, young downstream migrants moving to the sea are not exposed to major pressure-changes. Upon encountering dams, however, fish are involuntarily swept through spillways, turbines, and outlet works, and are influenced by extreme turbulence and pressure changes which can inflict severe stresses on the fish, causing injury or mortality.

The passage of fish over spillweirs at relatively low-head dams (Schoeneman & Junge, 1961), or at high dams with a free-fall discharge into a deep plunge-pool (Schoeneman, 1959), appears to be relatively safe. Under conditions prevailing at dams, most observations indicate that injury and mortality are not apparent, except in areas of extreme hydraulic turbulence, or in areas of pressure at, or near, the point of vaporization. However, fishes passing through turbines with small clearances in relation to the fish size may experience mechanical injury. Experimentation by Cramer & Oligher (1964) revealed that turbine mortality could fluctuate between 9% and 54%, depending upon tailwater levels, turbine type, and operational conditions. Moreover, an examination of the mortalities revealed that only 20%, or less, suffered injuries associated with water pressure and high-speed flow, the vast majority—over two-thirds—had been injured by mechanical means. During high-flow periods, large numbers of smolts can pass hydroelectric power-dams in the overflows. Within the Saint John River, New Brunswick, Canada, *Salmo salar* smolts begin their migration to the sea in the spring of each year, at a time when waterflows are high and water spills over the dam-crest; smolts pass unharmed over the dam, and, in 1971, losses of naturally reared Salmon from upstream tributaries were less than 10% (Dominy, 1973). Thus, in the absence of gas supersaturation, fish can safely pass over dams, and, through careful design, setting, and operation, successful fish-passage may be achieved through high-head turbines.

Maintenance of Anadromous Fisheries

The need for anadromous fishes to spawn in fresh water has made them particularly vulnerable to river impoundment for a variety of reasons (Table LII). River impoundment will adversely affect the rate of fish migration, and the availability of suitable channel-reaches for spawning, so that the number of valuable fish, such as *Salmo* spp. and Acipenseridae, will be reduced. Further losses may result from predation (particularly by introduced species), competition, reduced food-supplies, pollution, or the failure of regulated releases to stimulate upstream migration. However, plans for improving the management of fisheries for anadromous species have involved considerations of three main operations: fishway construction, fish collection and trucking, and the creation of artificial spawning-grounds. These primary operations may be supplemented by hatchery construction for specialized breeding, controlled fishing of 'undesirable' species, and 'planting' of new species of fish and food organisms (Tyurin, 1966).

Fishways

In order that fish may circumvent the blockages to migrations that are imposed by large dams, 'fishways' have commonly been constructed. These allow fishes to bypass the dam, and so provide continuity between fish populations along the river. Fish 'ladders' generally consist of a long sequence of weirs and pools, starting from the tailrace below the dam and ascending in small increments until reaching the water-level of the reservoir. Such fishways often provide adequate access for salmon and trout over low-head structures (Long & Krema, 1969). Cadwallader (1977) reported that large numbers of native fish had been recorded as passing the Euston–Robinvale fish-ladder on the Murray River, Australia, which by-passes a low-head weir and lock. Fish passage facilities at Whitehead on the Yukon River, Canada, have ensured that runs of *Oncorhynchus tshawytscha* and other fishes are able to pass the 14-m-high dam (Geen, 1974).

Four canals constructed to enable spawners to migrate between the Caspian Sea and the Arakum Lakes, Dagestan, USSR, have provided for the successful passage of *Rutilus* and cyprinids (Shikhshabekov, 1971). A 500-m-long fishway with 57 steps and five pools, which allow fish to rest after each 20-m ascent, on the Tuloma River, USSR, is used annually by some 4500 *Salmo salar* in addition to other *Salmo* spp. and *Thymallus thymallus* (Zhadin & Gerd, 1963). Welcomme (1979) describes several successfully installed fish-passes at dams in the tropics. At Cachoeira de Emes on the Mogi Guassu River, Brazil, a fish-ladder has operated successfully since 1936: species of Characoidei leap up a series of shallow steps, whilst the siluroids, which do not jump, move through a series of special tunnels. Successful migrations of cyprinids have also been achieved by the construction of several fishways at dams on the Tigris and Euphrates rivers, in Iraq.

Many fish-ladders, however, have proved to be costly failures. Acipenseridae and *Coregonus lavaretus baeri* have been unable to negotiate fishways on the Volga and Volkov Rivers, USSR, respectively (Zhadin & Gerd, 1963). Large numbers of *Oncorhynchus tshawytscha, O. nerka,* and *Salmo gairdneri,* together with *Cyprinus carpio* and *Alosa sapidissima,* failed to enter the ladders by passing the John Day and Dalles dams on the Columbia River, USA (Beiningen & Ebel, 1970). *Salminus maxillosus* failed to negotiate by-pass ladders at dams on the Carcarana tributary of the Paraná River, Argentina (Bonetto *et al.,* 1971), and fish-ladders installed at the Markala Dam on the Niger in Mali also failed to provide a facility for the passage of migratory species (Daget, 1960).

Table LII   The effects of river impoundment upon anadromous fishes (modified after Mundie, 1979).

STRUCTURAL EFFECTS	PHYSICO-CHEMICAL EFFECTS
(A) *Construction of dam*	Discharge, sediment load, water quality,
(i) Obstruction to upstream migration	food supply, channel form, riparian
(ii) Loss of fry during passage over or	and aquatic vegegtation:
through dams	
(iii) Gas supersaturation	(i) Unacceptable changes of seasonal
(iv) Accumulation of predators below	discharge pattern
dams	(ii) Unnatural short-term flow fluctuations
(B) *Creation of lake*	(iii) Unacceptable changes of seasonal
(i) Flooding of spawning-grounds	water-quality pattern (particularly
(ii) Delays to migration of spawners	temperature)
through reservoirs to spawning-	(iv) Elimination of natural trigger-
grounds	mechanisms for migration (related to
(iii) Delays to migration of fry	discharge and water quality)
downstream through reservoirs	(v) Unnatural 'toxic' pulses of poor-quality
(iv) Reluctance of fry to move through	water
stratified water-masses	(vi) Removal of flushing capacity for the
	dilution of toxic wastes
	(vii) Suitability of substrate for spawning
	(viii) Survival of eggs in gravels
	(ix) Amount of drifting food organisms
	(x) Spatial and territorial requirements
	(xi) Shelter and susceptibility of young fish
	to predation from birds

Not all anadromous fishes are able to negotiate all fishways during upstream passage. The primary problem appears to be the ability of the fish-collection system to intercept, or to attract, migrants—by providing entrances at proper locations and with suitable hydraulic conditions—rather than the ability of migrants to negotiate a ladder when once it has been found and entered. It has been demonstrated, for example, that the form of the stand-ing-wave downstream of weirs is essential for salmon to be able to generate enough speed to clear the crest of the weir (Welcomme, 1979), and failures have often been associated with improperly adjusted attraction-flows (Bein-ingen & Ebel, 1970). Detailed experimentation at Bonneville Dam on the Columbia River has related the behaviour of adult migrants, during upstream migration, to fishway dimensions, slope, and flow-rate—particularly with regard to the response of fishes to orifices of different size and shape, and to passage through pipes (Trefethen, 1972; Collins, 1976). Maximum critical velocities were shown to be appropriate for different species, but the design and placement of fishway entrances were shown to be of primary importance.

Even if fish-passage is achieved, undue delays may be incurred—such that a significant fraction of the spawners would never reach their spawning-grounds, or would spawn with reduced effectiveness. Indeed, the failure of the Markala ladder, River Niger, has been attributed to its insufficient capacity in view of the very large numbers of fish waiting to pass the dam; and although some fish did pass the barrage successfully, many were delayed and their stimulus for migration was lost (Daget, 1960). Fish can spend a substantial period of time passing fishways (up to 57 hours to pass Bonneville Dam, Columbia River: Monan & Liscom cited in Ebel & Raymond, 1976); they are forced to utilize restricted depths when entering the fishway, and delay at fishways has contributed to the severity of salmon losses from gas-bubble disease (Beiningen & Ebel, 1970).

The design and efficiency of fishways past high-head dams is dependent upon a detailed knowledge of the swimming capabilities and behaviour of migrating fish, particularly in relation to the discharge and water quality requirements for movement to be induced: for many species this knowledge is unavailable. Thus, the inclusion of a fish-ladder at the Tallowa Dam on the Shoalhaven River, Australia, could not be justified during the planning phase (Bishop & Bell, 1978). The effectivness of fish passage was uncertain under the proposed conditions of fluctuating water-levels in the reservoir, and there was no evidence that the indigenous species would utilize a fish-ladder over a dam with a proposed wall-height of 40–50 m.

'In view of the costly nature of fish-ladders, their incorporation in relatively high-level dams cannot be justified unless there is reasonable evidence that they prove effective.' (Bishop & Bell, 1978 p. 548).

Collection and Transportation Facilities

In an attempt to avoid turbine mortalities and the hazards of delayed migration imposed by slow passage through reservoirs and fishways, particu-larly where a long series of impoundments are involved, upstream and down-stream transportation by truck or barge has been employed, together with specialized fish-collection facilities (Long & Krema, 1969). For example, the location and size of the Mactaquac Dam on the Saint John River, New Brunswick, Canada, necessitated the introduction of a procedure for the collection, handling, and mechanical transportation, of fish upstream (Dominy, 1973). Large numbers of fish are annually transported over-land to the headwater reaches in order to facilitate natural spawning. The collection and trucking of juveniles for downstream release may also have advantages, by reducing losses from turbine passage, predation, and gas supersaturation, etc. A 300-km trucking system on the Columbia River, for example, increased survival rates, in comparison with those of natural migrations, by up to 350% during periods of high spillage, and by 64% for periods before major spillweir discharge (Trefethen, 1972). Importantly, the transported fish successfully returned to the River and migrated upstream to spawning areas—indicating that transportation did not adversely affect their 'homing' ability.

Artificial Spawning-grounds

The loss of spawning-grounds is primarily related to the isolation and inundation of headwater reaches by the dam and reservoir, but considerable losses may also result from channel changes below the dam. Indeed, the survival of a river's anadromous fish-run is highly dependent on the condition of the river channel itself. Although continuously changing, the structure of the channel controls the conditions conducive to spawning, resting, rearing, and food production. Sedimentation can be particularly detrimental to fish: adult fish cannot successfully spawn in sand-sized sediments; the filling of deep pools deprives fish of shelter, resting-places, and cooler temperatures; the narrowing of stream channels can reduce living-space; and food production provided by bottom-dwelling insects can be eliminated.

Rivers that have successful, natural, anadromous fish-runs commonly have slopes of 2 m per km to 10 m per km, channels composed of well-defined pools and riffles, and major pools at frequent intervals to provide resting areas and shelter. Riffles may provide suitable spawning-areas only if flow velocity, water depth, and substrate composition, meet specific requirements: velocities of 0.4 to 1.0 m per s, depths of 0.25 to 1.0 m, and well-sorted, permeable, gravel substrates of a predominant size within the range of 5 to 50 mm, provide the most suitable conditions. Food production is also related to substrate composition, although riparian plants contribute to food supply as well as providing shelter and shade.

In many rivers where impoundment is likely to produce major losses of breeding grounds, artificial spawning-channels are being used with increasing frequency, and success, to ensure that fish populations are maintained (Geen, 1974). Special fish-cultural measures are necessary, in particular, to maintain the reserves of valuable anadromous fishes. These measures entail the constant release of increasing numbers of juveniles into the artificial reservoirs and downstream river-reaches. On the Volga, sturgeon-breeding farms have been established, and artificial spawning-structures have also been created for the cyprinoid fishes (Zhadin & Gerd, 1963). In the Don basin, fish-breeding plants have restored the stock of sturgeon (*Acipenser* sp.) whose numbers over the last decade have increased fifteen-fold (Bronfman, 1977).

A detailed examination of salmon management through controlled channel-sites has been provided by Mundie (1979). Artificial spawning-channels provide both regulated flow and optimum substrate-size for the eggs of the particular species that is being reared. The survival of eggs and alevins under such controlled, and protected, conditions is often substantially above that observed within natural rivers, giving from two to nine times the egg-to-fry survival of a natural river. The Weaver Creek Spawning Channel, Harrison Creek, Canada, has been utilized annually by between 18 000 and 27 000 spawners in recent years, and has given an egg-to-fry survival of 61–81%: this is nine times more efficient per unit area for the production-returning adults than natural spawning-grounds, producing 148 952 and 194 744 returning *Oncorhynchus* in two consecutive years (Mundie, 1979). Similarly, the Tehama–Colusa Fish Facility of the Sacramento River, California, USA—constructed, in part, to replace the spawning-grounds lost by the construction of the 29 740-surface-acres, 183-m-high, Shasta Lake and Dam on the headwaters—annually provides an artificial spawning-area for approximately 40 000 adult salmon, which are captured and selected at a fish-trap on the main river. Under optimum spawning conditions of controlled water velocity, depth, and temperature, clean gravel, and elimination of predators, over 30 million juvenile migrants are released annually (U.S. Fish and Wildlife Service, pers. comm.). The maximum use of such facilities, however, requires detailed knowledge of the needs of particular species (e.g. channel-bed slope, substrate composition, and water quality); also cleaning techniques to remove dead salmon, algal growths, and silt, may be an expensive and substantial undertaking. Nevertheless, artificial spawning-channels have proved so successful that they are also being used to increase salmon populations in areas that are unaffected by impoundments (MacKinnon *et al.*, 1961).

FISHERIES OF IMPOUNDED RIVERS

River impoundment can have a considerable impact to the detriment of native riverine fish-populations, and particularly of those species that are highly specialized to the lotic conditions. Research workers have attempted to define simple bivariate cause-and-effect relationships, in order to provide predictive models for fisheries management; but most deleterious impacts of reservoirs and dams upon fish species result from the combination of various changes to a number of attributes of the lotic habitats. Relatively subtle changes of a single attribute may appear harmless when viewed in isolation, but when considered as a part of the whole system, the addition of many subtle changes may produce conditions which adversely affect particular species. Moreover, many of the changes will occur slowly, so that very gradual habitat-change may produce a slow, but progressive, displacement of indigenous fishes by an exotic, or natural, competitor or predator. Alternatively, the progressive change of some attributes (such as temperature, water quality, substrate composition, etc.) may reach a limiting condition and cause the sudden loss of a particular species. Species rarely become extinct for a single, isolated reason. Species depletion may be caused by one group of attributes, whilst the elimination of the reduced population may be affected by a second, not necessarily related, series of circumstances. In general, species alterations may be associated with three impacts: the establishment of lethal conditions for one, or more, life-history stage or stages; the alteration of environmental conditions in favour of a competitor, prey, or predator; or the introduction of exotics either intentionally or accidentally.

Despite the variety of specific impacts that can result from river impoundment, Zalumi (1970) classified changes of fish fauna into eight categories, reflecting the different habitat requirements of species, for the lower Dnieper, USSR:

(A) Species preferring fast-flowing and cold water, with breeding grounds and main areas of habitat upstream of the dam, are effectively eliminated from the downstream river, e.g. *Acipenser ruthenus, Barbus barbus borysthenicum*, and *Lota lota*.

(B) Species preferring fast-flowing water, with main breeding-grounds above the dam, but feeding in the lower river, are consequently reduced in abundance, e.g. *Leuciscus idus, L. cephalus*, and *Aspius aspius*.

(C) Anadromous fishes (e.g. Acipenseridae) are virtually eliminated.

(D) Some freshwater species, rare before regulation, are eliminated.

(E) Semi-migratory fishes, whose conditions of natural reproduction have not been sharply altered, because of adaptations to the new conditions, maintain viable populations (e.g. *Rutilus rutilus heckeli, Abramis brama, Lucioperca lucioperca, and Cyprinus carpio*).

(F) Non-anadromous lake–river fishes associated with accessory waterbodies linked to the river, under natural flows, by seasonal floods, are markedly reduced in number (e.g. *Rutilus rutilus, Tinca tinca, Carassius carassius, Silurus glanis*, and *Esox lucius*).

(G) Native species favoured by the regulated conditions are increased in number (e.g. *Pungitius platygaster, Alosa caspia nordmanni*, and *Clupeonella delicatula delicatula*).

(H) Marine species are increased in abundance in the lower river.

The first three groups, comprising species which either disappeared or were reduced in abundance, are associated with reduced flow-variability, lower velocities, and gravel-bed siltation; but the phytophilous fishes, spawning on floating or submerged plants—category (E)—are generally less affected. Nevertheless, for most rivers, major fish-changes below dams should be expected.

An evaluation of the ecological consequences of a proposed hydroelectric power-dam on the Fraser River, Canada (Geen, 1975), led to considerable concern within the government agencies responsible for the management of resident and anadromous fishes. Within the Fraser River, five species of the salmon genus *Oncorhynchus* support large commercial, sport, and food, fisheries. Geen estimated that river impoundment would produce signficant habitat changes within a 75-km reach below the dam. Salmon would be adversely affected for several reasons:

(1) The dam would provide a major barrier to the upstream migration of an estimated 750 000 adult salmon per year.
(2) Reduced summer discharges might delay the initiation of the migration of adult salmon, reducing egg-survival because of the late arrival on spawning grounds.
(3) Any major delay to downstream migration could lead to a loss of the migratory tendency of the young salmon, due to physiological changes.
(4) Decreased turbidity would increase early mortality by predation.
(5) Increased travel-time, due to reduced flows and lower summer water-temperatures, would increase predation.
(6) Increased autumn temperatures would accelerate development and induce premature fry-emergence.
(7) Gas supersaturation can lead to injury and mortality of both adult and young salmonids.
(8) Losses would occur by passage through power turbines.

Some beneficial effects were also recognized for example:
(1) Higher winter discharges than formerly might enhance the survival of eggs.
(2) Increased winter flows might reduce the proportion of 'fines' in the spawning-gravels, leading to improved percolation and enhanced egg and alevin survival.

Nevertheless, Geen concluded that the dam would eliminate a major fishery based upon anadromous salmonids. Satisfactory methods for transporting large numbers of young and adults over the dam did not exist, and alternatives to natural production did not seem viable on either biological or economic grounds. The transportation of fish by fishways, locks, or trucking, is the solution most commonly offered, but the magnitude of the facilities of operations required often limit their feasibility or prohibit their adoption—particularly in view of the lack of confidence in their potential success. The maintenance of viable fisheries based upon valuable anadromous species will most probably have to rely upon the successful development of collection facilities and artificial spawning-grounds, together with the appropriate regulation of flows to maintain attractive lotic habitats, and to provide a trigger mechanism for the initiation of upstream migrations.

CHAPTER IX

Management Problems and Prospects

'What is urgently needed is the formulation of long-term development policies, on a sustaining basis, that reflect changing water supply and demand patterns, consistent with efficient use, and better understanding of the social and environmental implications, so that adverse impacts can be minimized. In fact, it can be successfully argued that the time has come when the emphasis should shift to comprehensive land and water planning, treating land and water as an integrated and interacting unit, rather than water planning *per se*.'

Biswas (1978 p. xvii)

In recent years, improved measurement techniques have been employed to determine the effects of dam construction, and the results obtained have generated a new awareness of the sensitivity of rivers to imposed stresses. However, our knowledge is as yet incomplete and, unavoidably, proofs of causality are tenuous. Changes within the river downstream are an inevitable consequence of impoundment, but science cannot precisely predict the consequences of human activity; the methods of the physical and biological sciences can only assess the probability that a particular impact will result. The management of impounded rivers requires the pragmatic application of scientific knowledge to cope with complex, dynamic, and evolving, systems which are only partially understood. This knowledge provides a basis for value-judgements leading to a decision as to the desirability of the new or predicted state. It is the difficulty of resolving environmental changes into two discrete types, namely, acceptable change and impacts or hazards, that is at the root of many managerial problems.

Further information is required concerning the degree and direction of change, and the manner in which change has occurred, in order to predict, or to determine, the ultimate effects of damming a river. 'Stress ecology' (Barrett *et al.*, 1976) is a relatively new focus of research which attempts to measure, and evaluate, the impact of natural or foreign perturbations on the structure and function of environmental systems and, potentially, on the management of these systems for Man's and Nature's benefit and survival (Barrett, 1981). Advances in stress ecology were catalysed by the Jerusalem symposium in 1978 (cf. Barrett & Rosenberg, 1981). What has been called collectively the 'catchment ecosystem' actually consists of a multiplicity of ecosystems and provides a management framework for the application of stress ecology to studies of impounded rivers, for two reasons: firstly, because Man is viewed as a fundamental, rather than an additional, component; and, secondly, because attention is directed not to the main-stream in isolation, but to the relationship between the main-stream, its tributaries, and their catchments.

A symposium on the 'Recovery of Damaged Ecosystems', held at Blacksberg, Virginia, USA, in 1975, provided compelling evidence of the need for immediate action to develop the scientific data, technology, and international and national implementation programmes, to restore damaged ecosystems (Cairns et al., 1977). During the past decade, considerable advances have been made in our ability to restore impounded rivers. However, the symposium also highlighted the need for improved public awareness, and for improved communication between scientists, decision-makers, and the public. Failure to accept that the impacts of dams are highly individualistic, that different responses can be induced by the application of the same stimulus, that the spatial extent of the effects can be considerable, and that a long period of time may be required for the full effects to become apparent, has been a major stumbling-block in public perception which, when once overcome, will provide for an improved understanding of river management problems.

PROBLEMS

River impoundment will generate a sequence of changes within, and along, the channel downstream, but these changes are predictable only in general terms. In detail, the effects of impoundment are manifested by a complex interaction of the different components of the lotic system, producing a variable pattern of changes along the river. The primary impacts, abiotic effects, and linkages between the abiotic and biotic components, are illustrated in Fig. 33. The creation of a lake introduces a new storage capability within the catchment hydrological cycle, allowing increased evaporation losses (so that the annual runoff has been reduced by perhaps 20%, the annual range of discharge has been compressed often to only 15% of the natural variability, and the mean annual flood has been reduced by up to 60%). Where the stored water is released to supply irrigation demands, the natural flow-pattern may be reversed, with a summer peak replacing the natural winter peak. Chemical and thermal regulations modify the quality of the discharges, which are controlled by the level of dam release. Epilimnial releases are depleted in nutrients and are thermally enhanced. On the other hand, hypolimnial releases are associated with enriched nutrient-loads; with elevated winter, and depressed summer, temperatures; and with high loads of ferrous iron but low oxygen levels (often less than 0.5 mg per litre) resulting from the generation of anoxic conditions in the lake.

240

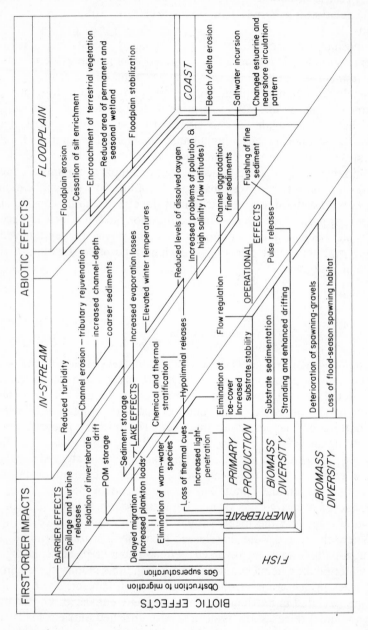

Fig. 33. Summary of impacts of river impoundment.

In some exogenous rivers, or rivers draining saline soils or rocks, flow-regulation can increase salt concentrations to above 500 ppm. Annual plankton loads may be increased by 150 times and 200 times for phyto-plankton and zooplankton, respectively. The release of sediment-free water has caused channel-bed degradation by over 2 m and accelerated erosion rates of more than 500 mm per yr; as a result, the substrate sediments have increased in average size. In contrast, flow regulation has induced channel aggradation below tributary sediment sources at over 100 mm per yr, eleva-ting bed-levels by over 5 m, reducing channel capacity by up to 85%, and producing finer channel substrates than formerly. Regular clear-water releases at locations of stable substrate, have allowed all-year-round growth of periphyton and macrophytes, so enhancing autochthonous primary production. Diatoms and Cyanophyceae are commonly reduced, and are replaced by bryophytes and filamentous Chlorophyceae dominated, in many cases, by *Cladophora glomerata* and *Ulothrix zonata*. *Potamogeton* spp. often become important below dams, and in the tropics floating macrophytes, especially *Eichhornia crassipes* (Water-hyacinth) and the water-fern *Salvinia molesta*, have developed widely to 'nuisance' numbers. On the floodplain, as the soils dry out, the vegetation composition undergoes a reduction in diversity and density, and the reduced productivity leads to increased grazing-pressure and, subsequently, to a reduced fauna. However, a dense, rich but narrow riparian zone may form, dominated by woody species, especially *Salix* spp. The benthic invertebrate community usually experiences a reduction in diversity, although the numbers of some species, including ephemeropterans, dipterans, and some amphipods, as well as gastropods and oligochaetes, may become numerically abundant. Indeed, Simuliidae and Chironomidae may achieve extreme densities of over 25 000 per m^2 and 20 000 per m^2, respectively.

Many endemic fish species have been eliminated in such circumstances, not least because of the loss of seasonal floodplain spawning-grounds, and although reduced summer temperatures have allowed cold-water trout fisheries to be established, valuable anadromous fisheries have been deleteri-ously affected: the weight of fish catches from some impounded rivers has been reduced to less than one-third of the pre-dam levels. The regulation of the hydrological regime and the resulting changes of channel morphology, depressed oxygen-levels, altered seasonal variations of water temperature, enriched phytoplankton loads, and changed macroinvertebrate communities, have been recorded for more than 100 km below some dams. Moreover, extreme 'controlled' stage fluctuations of more than 4 m have been detected for tens of kilometres downstream of flow-regulation reservoirs.

This scenario provides a general description of the magnitude of change which can result from river impoundment. The impacts upon lotic systems have been viewed as stresses, and classified according to whether they produce *favourable* or *unfavourable* changes (Selye, 1974; Odum *et al.*, 1979; Barrett, 1981). Such a classification involves a value-judgement based upon a range of social, economic, and political, criteria beyond the scope of this discussion, but these judgements must be made with due consideration for the full range of environmental consequences. However, even before value-judgements can be debated, with regard to the beneficial or detrimental nature of the consequences of damming a river, there are problems of quantifying impact magnitude, which can refer to the degree of change of individual attributes, or to the cumulative changes of all attributes, at a single location or along the length of river affected. Impact measurement is complicated by the hierarchical organization of ecosystems: effects of a perturbation may vary markedly at the population, community, and system, levels, so that, for example, warm water released into a stream may increase primary productivity but eliminate a population of trout (Odum et al., 1979).

Further problems are introduced because the attributes of river systems vary under natural conditions, so that a 'stable' natural state is hard to define, and long periods of time are often required before changes can be detected. The difficulties of measuring or predicting environmental impacts increase as one moves from a short-term, local-scale approach, concerned with individuals or populations, to a long-term, catchment-scale approach, concerned with communities or ecosystems (Winkle et al., 1976). These difficulties are manifested in impact studies by increasing the cost, time, and degree of uncertainty, of any predictions made.

System Stability and Resilience

The structure of an established river-system has evolved in response to the characteristics of the natural stresses—floods, droughts, thermal extremes, substrate turnover, high turbidities, or floodplain inundation—and it has structural and functional characteristics which are adjusted or adapted to these stresses. Undisturbed systems are continually in a transient state: channels scour and fill, insect populations can range over extremes, and fish populations expand and contract. The 'stability' of the natural river system refers to the persistence of a pattern of fluctuations over time, which produces a constant average condition. Mutualistic interactions bring about system stability by the creation of numerous recovery mechanisms and feedback loops, which introduce a resilience to change—that is, an ability to restore a system's structure and function after disturbance, to absorb changes, and to maintain its initial 'stable' condition (Westman, 1978). 'Stress' has been defined as any force that pushes the functioning of a system or subsystem beyond its ability to restore its former structure and function (Meier, 1972). Stresses induced by river impoundment involve a displacement of the system and its components from the normal operating range. In order to have appropriate and effective management policies and practices for stressed river-systems, or river-systems that are liable to stress, we must have a full understanding, not only of these stresses, but also of the resilience and stability-properties of these systems.

The balance between resilience and stability is clearly a product of the evolutionary history of each system in the face of the range of random fluctuations which they have experienced (Holling, 1973). The various components of the river-system, and the various combinations of these components, will display differing degrees of resilience in the face of disturbance, so that each river-system may respond differently to river impoundment. However, not all stresses cause immediate change—for example, a stress may be lethal for some organisms whereas others will be affected over a longer period of time if the stress inhibits activity, impairs growth, or reduces the chance of survival. A system subjected to a stress can be maintained by a minor internal reorganization if the function performed by affected organisms could be performed by other unaffected organisms within the system (Cairns & Dickson, 1977). However, such an 'accommodation' response will realize an energy-cost (Lugo, 1978), and alter the balance in the use of energy from maintenance and production to repair and recovery (Odum, 1981). Thus, the problem of adult salmonids reaching spawning-grounds is not only related to the obstruction to migration imposed by the dam and lake, but also to factors influencing fish-health, which is related in turn to exposure to elevated pollution-levels or elevated temperatures, causing delayed entry and/or slowed travel-rate.

Degrees of resilience are related to the characteristic amount, and variability, of energy- and mass-transfers within the natural river-catchment ecosystem or series of ecosystems. In rivers where the characteristic processes are highly variable, the components require great resilience, because intrinsic stresses occur frequently. Process variability, in space and time, results in a population which can, simultaneously, retain genetic and behavioural types that maintain their existence in low populations, together with others that can capitalize on chance opportunities for dramatic increase. Such a system will have a low stability but a high resilience. Rivers that are therefore characterized by naturally-regulated processes are less variable and less resilient than unregulated ones, because intrinsic stresses occur in the former relatively infrequently. These regulated habitat conditions give rise to populations which are more constant than those of unregulated ones, but are less able to absorb chance hydrological extremes, i.e. the system has a high stability but low resilience. Thus, in the estuaries of southern Chesapeake Bay, USA, the taxonomically diverse and spatially complex macrobenthos of the shallow-water, stable, high-salinity zone—not easily disturbed by normal river floods—was decimated by exceptional freshwater discharges caused by Tropical Storm Agnes (Boesch et al., 1976). In contrast, estuaries characterized by large seasonal variations in freshwater discharge or salinity, for example in monsoonal regions, have a relatively low-diversity community within which some individuals are seasonally eliminated (Sandison & Hill, 1966).

Copeland (1970) suggested that high-diversity communities are more markedly affected by stresses than those of low diversity. This hypothesis has been supported by experimental studies (e.g. Long, 1974; Goodman, 1975), which invariably show that complex systems are resilient only within a comparatively small domain of conditions and, thus, are dynamically fragile (Boesch & Rosenberg, 1981). The physical characteristics of channel morphology, similarly, have been shown to be adjusted to the variability of the process conditions (Harvey, 1969), and to the factors influencing the rate of recovery of channel morphology following alteration by an extreme event (Wolman & Gerson, 1978; Newson, 1980b).

The magnitude and frequency of the constructive or restorative processes during the intervening intervals between 'effective' erosional flood events, are particularly important. Vegetation, specifically the composition and density of the plant cover, plays an important role in determining the magnitude and frequency of the restorative processes. The rate of vegetation establishment in a particular region, given suitable temperature and other conditions, is primarily dependent on moisture availability, so that within temperate regions, river channels that have been widened by rare flood events, have often regained their original form relatively rapidly. Thus, although Tropical Storm Agnes produced erosion and scour of the Western Run, Maryland, causing an increase of channel-width by up to 160%, the channel was well on its way to recovery towards pre-Agnes dimensions within a single year (Costa, 1974).

The behaviour of river systems is profoundly affected by random events, but these must be viewed in relation to the state of the system as it evolves. Functional reorganization within an ecosystem, in order to maintain stability, might so change the deterministic conditions that the resilience is lost or reduced, and a chance and rare event, that could previously be absorbed by the system, can trigger a sudden dramatic change and loss of structural integrity (Holling, 1973). With reference to channel morphology, Schumm (1973) conceived the existence of intrinsic thresholds, such that whilst external variables remain relatively constant, the progressive change of the system through time renders it unstable, and failure will occur. Adjustment or failure will not occur, however, until the system has evolved to a critical condition. Changes below dams will not only relate to stress (extrinsic) thresholds associated with first-order impacts, and the resilience of the system components, but also to the progressive change of resilience in relation to internal thresholds, which may occur in response to attempts by the system to absorb the imposed stress. River impoundment may be expected to increase stability, but reduce resilience, within the downstream lotic system, so that a rare hydrological event or low-quality release may act as a catalyst, causing the crossing of an internal threshold and the initiation of a complex sequence of adjustments.

Faunal diversity is characteristically reduced below dams, but systems exhibiting a lack of species diversity may be unstable (Cairns, 1974), with minor biotic or abiotic fluctuations producing radical changes in the community structure (Gore, 1977). Benthic invertebrate populations dominated by only a few species may experience great population fluctuations. When the population of the dominant organism changes, as occurs during emergence, the entire trophic balance is disturbed (Boles, 1980b), and fish which depend heavily on benthic secondary production for food may find such supplies too unreliable for survival. Environmental scientists must be concerned with the evaluation of the stability and resilience of lotic systems, and with the identification of both intrinsic and stress thresholds, which, if crossed, could lead to irreversible change.

Complex Response

The sustained impacts of river impoundment, singularly and collectively, produce effects on stream systems, which, in most instances, are of sufficient intensity to cause changes in the character of these systems. Changes below dams relate to four main factors: (1) the number, kinds, and severity (magnitude and frequency) of stress; (2) the stability, resilience of, and threshold characteristics within, the natural system; (3) the secondary stresses upon the biotic component imposed by changes of channel structure; and (4) the presence of epicentres for recolonization of biota. The vulnerability of the system will depend upon the character of the stress in relation to the normal experiences of the natural system. A range of responses may be recognized, from the impoundment of a naturally-regulated river having natural lakes, large floodplain storage or important ground-water supplies, small sediment loads, and an important level of autotrophic production—which may experience relatively little change—to the regulation of a highly variable flow-regime river with pulse-stimulated lotic, riparian, and floodplain, habitats, and transporting large quantities of sediment and allochthonous organic material, which can undergo dramatic change.

A similar range of responses may be identified below impoundments located at different positions along a river. The location of the dam along the river continuum is significant, because the resilience of the system changes downstream (Neuhold, 1981). Natural rivers are characterized by a unidirectional spiralling of nutrients and a gradient of physical conditions—with resulting biotic responses—from headwater to mouth. Typically, the continua describing the downstream variation of individual parameters have non-linear patterns. Biotic diversity, for example, is found to be relatively high in the middle reaches, due to the relatively high degree of both spatial and temporal heterogeneity.

Ward & Stanford (1983) developed a simple, theoretical framework to visualize the effects of impoundment upon the river continuum (Fig. 34) based upon the serial discontinuity concept which describes, hypothetically, the disruptions to physical parameters in relation to biological phenomena. It is theorized that an impoundment will disrupt the continuum, causing a longitudinal shift of a given parameter. The discontinuity distance (DD) has a length (X) which is the displacement of the parameter in stream-order units

246

or kilometres, and may be in the downstream (X_{pos}) or upstream (X_{neg}) direction. The difference in absolute parameter intensity (PI) may be elevated (Y_{pos}), depressed (Y_{neg}), or unchanged (Y_0). Ward & Stanford (1983) used the serial discontinuity concept to suggest the differential effects of dam location along a river upon its lotic system. For example, direct allochthonous inputs (primarily leaf-litter) form an important part of the energy budget in forested headwater streams, and a headwater dam would greatly depress the ratio of coarse particulate to fine organic matter (Y_{neg}), severely impacting the invertebrate shredders. A dam on the lower reaches has little effect on the size-composition of the detritus (Y_0), so that the functional relationships within the receiving stream may not be greatly altered.

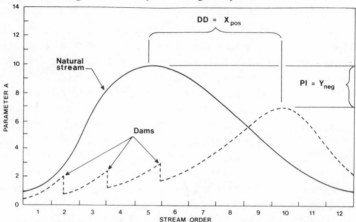

Fig. 34. Theoretical framework for conceptualizing the influence of impoundment on ecological parameters in a river system. (From Ward & Stanford, 1983. Reproduced by permission of Butterworths, Ltd.) *See text* for explanation.

The length of river affected by an impoundment will also depend upon its location, as well as on its capacity in relation to annual and seasonal inflows, operational procedures, and downstream uncontrolled-catchment contributions. For 15 impounded rivers in the UK, morphological changes were observed downstream from dams until the impounded catchment area had been reduced on average to 40% of the total catchment draining to the river (Petts, 1978). Reduced peak-discharges, with 1.5 and 2.44 years' recurrence-intervals, have been detected—until the contributing catchment-area is about 10 times that of the area draining to the reservoir (Gregory & Park, 1974). Some dams will affect only a few kilometres, others hundreds of kilometres, of the river below the impoundment, and the effects of yet others may extend to the estuaries and near-shore zone. Nevertheless, the river-system response may require a considerable period of time, and during the adjustment phase a variety of changes may occur along the channel.

The progressive adjustment of a single river-channel over time may be described in terms of a simple model (Fig. 35). In the short-term, discharges below the dam may not exceed the threshold for sediment transport, so that the flows will be accommodated within the existing channel-form through reach 1 (A and B). A dense periphyton may establish in 1A, and this may be supplemented by aquatic angiosperms in 1B. The invertebrate benthos will change in response to the altered chemical and thermal conditions, modified for a short distance below the dam by the release of high concentrations of plankton. The altered thermal regime below hypolimnial-release dams decreases species-diversity, but biomass can be particularly high as the amounts of primary production and detritus increase downstream. Sediment introduced by non-regulated tributaries (1 and 2) will be deposited in the main-stream, channel depth and channel capacity will be reduced, while the gravel substrate will be compacted, and pools filled, by the sedimentation of 'fines'. Neither periphyton nor benthic invertebrates may be able to establish themselves in such unstable channel conditions. However, downstream from the aggrading reaches, the channel will be stable (reach 2B and 3B), and a higher diversity, but lower biomass, of organisms will establish itself in comparison with reach 1. The effects of an impoundment decrease downstream as the relative contributions from the uncontrolled catchment become dominant; the modified thermal regime, in particular, may persist for only a short distance below the dam. Thus, channel changes are not observed in reach 4, and the lotic system, dominated by tributary inputs, may undergo no detectable change. Changes initiated at isolated locations can migrate progressively downstream from these loci; positive-feedback mechanisms will dominate in the first instance, but eventually negative-feedback mechanisms will develop to establish a new equilibrated condition.

As the time-scale increases, the probability of competent discharges also increases. Thus, within reach 1A, high releases of spillweir discharges will initiate channel degradation—increasing channel capacity—and the eroded material will be redistributed downstream, with the coarser fraction of the load deposited in reach 1B. Furthermore, as the reservoir matures, the quality of the releases will change, a trophic upsurge during the early years being followed by reduced nutrient loadings but often higher oxygen levels. As a result of these changes, the periphyton and benthic community may be disrupted. Sediment will continue to be deposited below tributary confluences, but with morphological change and the occasional high-dam release, the sediment will be distributed further downstream, initiating the process of aggradation in reach 2B and 3B, and again disrupting the benthic communities. However, in the long-term, a new quasi-equilibrium channel-form will establish itself, and the lotic biota will reflect this form as well as the altered warm-quality conditions and regulated flows. Throughout reach 1 the degraded channel will be characterized by a relatively coarse but stable substrate, and a dense periphyton may re-establish itself, with an invertebrate benthos dominated by a large standing-crop of a few species.

Downstream the water-quality and flow-variability characteristics will return to 'natural' conditions, and this will be paralleled by the channel's form. In detail, the lotic system within reach 2 will be controlled by the character of the substrate sediments: the persistence of a sand-bed can eliminate most forms of invertebrates and inhibit periphyton establishment, but reaches having clean, gravel substrates can develop a diverse invertebrate community and high levels of primary production. The third reach may have relatively high turbidities which would reduce primary production, so that the energy input will be dependent upon allochthonous organic matter from upstream, and the fauna may, by inoculation from tributary and downstream sources, re-establish a community comparable with that of the unregulated river.

The character of individual locations will change in an episodic manner, duly controlled by erosional dam-releases and/or depositional tributary-flows. Periods of temporary 'stability' will exist between short phases of change, but the duration of 'stable' periods can be highly variable and measured in terms of a few days to several years. Modifications will occur during the 'stable' periods, and these will relate to variations in water quality (particularly during reservoir maturation), to time-lags associated with response of biotic components (e.g. the time required for colonization), and to feedback effects between neighbouring locations. Nevertheless, some general patterns of change can be identified. In reach 1A, the high biomass, associated with a dense periphyton and stable substrate, may persist for several years before a sediment-free, high-magnitude, dam-release or spillweir flow initiates channel degradation. Periods of substrate instability will be associated with reduced periphyton cover and macroinvertebrate numbers. However, when once the channel morphology has established a new 'equilibrium', the stable, coarse-armour layer, flushed of 'fines', will, provided there is a moderate supply of nutrients, allow the re-establishment of the periphyton, and an enhanced macroinvertebrate community, although of relatively low diversity, will normally develop. At downstream locations, below tributary confluences, the periods of stability between episodes of change will be shorter than above—a reflection of the real-time interval between uncontrolled tributary inflows. Furthermore, the time-lag between dam closure and the initiation of channel change is shorter, for example, in reach 2A than in reach 1A, although downstream from the confluence (e.g. 2B) the time-lag may be considerable; certainly it is not uncommon for this time-lag to be measured in tens of years.

A particular problem in the assessment of the impact of impoundment upon downstream channels is the time required for the river system to achieve a new 'equilibrium' condition. Some components will change more readily than others, or will operate on different time-scales. Different external factors operate over a range of frequencies, placing complex stresses upon different causal links in the chains of interrelationships between system components. The minimum time required for a system to adjust to the imposed conditions is dependent upon those variables that require the longest time to achieve a new stable structure. Generally, the abiotic responses to disturbance are different, and usually slower, than biotic responses, and Neuhold (1981) suggested that the ecosystem recovery process will keep pace with the

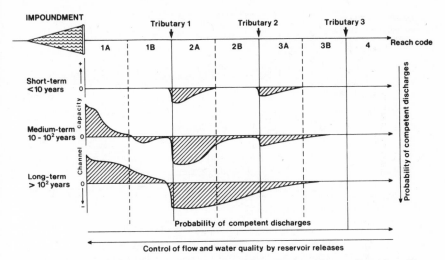

Fig. 35. A hypothetical model of river responses to impoundment (based on Petts, 1979). Reproduced by permission of Edward Arnold, Ltd.

physical recovery process. Thus, although apparently stable benthic communities have been observed only 5 years after dam closure, they will undergo a number of readjustments as the stochastic process of channel change progresses. Plant equilibrium on the Colorado River in Grand Canyon below Glen Canyon Dam had not been reached after 13 years (Turner & Karpiscak, 1980), and within this reach the bed-sediment may require between 10 and 1000 years to become stabilized (Laursen *et al.*, 1975). Channel changes below some British dams are continuing more than 50 years after dam completion (Petts, 1979). Moreover, the attainment of a new stable condition characteristically involves a sequence of alternating and opposing changes, which appear as fluctuations about a trend.

W. L. Graf (1977) demonstrated that the adjustment of many landforms to a Man-induced disturbance follows the negative expotential form of the rate-law, used to describe the decay of radioactive isotopes. Thus, after an initially rapid response, river channels will tend towards a condition of equilibrium at continuously decreasing rates of adjustment. Repeated observations of river channel change, however, have revealed that the response is certainly not continuous and regular over time, and Schumm (1977), for example, found that channel response in an experimental basin involved several alternating phases of erosion and deposition. Below dams, the alternation of erosional and depositional processes may result from: (1) variable sediment-discharges from tributaries; (2) main-stream erosion during high dam-releases; and (3) local sediment-transfers, and feedback effects, within the main channel. These interact to produce a typical complex response (Petts, 1980c, 1982). The response of channel morphology downstream from a major sediment-source, within a regulated river which still experiences destructive main-stream flows of moderate frequency, and with due consideration for the time-period required for vegetation establishment, is illustrated in Figure 36.

250

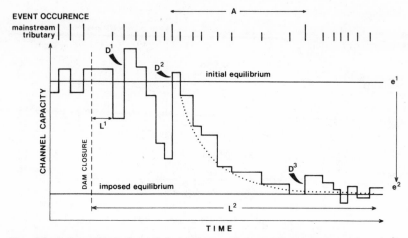

Fig. 36. A hypothetical model simulating the complex response of a single channel cross-section to flow regulation (based on Petts, 1980a). The relaxation path reflects the frequency of major reservoir releases (D^1, D^2, D^3) modified by tributary sediment injection, and vegetation established during period 'A'. Reproduced by permission of the Foundation for Environmental Conservation.

Referring to Fig. 36, the reaction time (L^1) reflects the time-period between dam closure and the first sediment-loaded tributary discharge, but the relaxation time (L^2) will be dependent upon the timing of peak tributary and mainstream flows in relation to the availability of conditions that are favourable to vegetation growth. Impoundment may create the potential for a reduction of channel capacity, but prior to vegetation establishment (phase 1), mainstream events (D^1 and D^2) will erode the deposits introduced by tributary flows, and maintain large channel capacities. However, the actual time-period between main-stream events, subsequently, is sufficient to permit the establishment of vegetation on the new deposits, and the effectiveness of subsequent main-stream events may be considerably reduced (D^3). When once a strong vegetation-cover has developed, relatively minor fluctuations of channel form will occur in response to varying discharge-sediment load combinations supplied by sub-basins. The readjustment of the channel dimensions will continue until adequate velocities are provided to maintain a stable 'quasi-equilibrium' form. For individual locations, the relaxation time may be relatively short; but because of the sequential nature of sediment transport, the complete readjustment of the channel form downstream from a dam may require a long period of time. Moreover, along a regulated river, the progressive redistribution of sediment, inter-adjustments between reaches, and interactions between the main river and tributary streams, will produce considerable variations in channel morphology over short time-scales.

The monitoring and surveillance of lotic systems below dams is problematical because of the difficulty of interpreting short-term changes with a view to anticipating long-term trends. Yet the need for monitoring and surveillance subsequent to dam completion stems from concern reflecting the, as yet, indeterminate nature of short-term environmental changes. For example, Buma & Day (1977) monitored channel-changes at eight locations over a 5 year period below a dam on Deer Creek, Ontario, Canada, following reservoir completion, but annual changes varied inconsistently, and considerable variations were observed between the locations during the 5 year period. A period of about 15 years may be required before dam-induced channel changes may be isolated (Petts & Pratts, 1983). Certainly the environmental scientist should seek to anticipate the consequences of human impacts over a 50, 100 or perhaps even 500 year time-scale as well as those of immediate significance (Frye, 1971).

PROSPECTS

'A rational man is forced to conclude that we simply do not know enough to be able to predict what would happen [as a result of river impoundment].
O'Connor et al. (1973 p. 772)

Traditionally, an assumption of high resilience has been implicit in many of Man's efforts to manage his environment, so that, no matter how large, or how permanent, the disturbance, the system should maintain its original stable condition. Encompassed within this view are the ideas that 'big' is beautiful, economically attractive, and environmentally allowable, and that a trial-and-error approach to resource development on any scale is acceptable (Holling, 1978). However, catchment ecosystems are fragile, and the river system—the lotic sub-system and the adjacent riparian and floodplain sub-systems—is particularly sensitive. The structure and function of river systems have more than one stable state, which is dependent upon the characteristics of the primary-control variables: discharge, sediment transport, and organic-matter supply. As long as these primary controls change within certain limits, which may be defined in terms of either magnitude or duration, stresses can be absorbed or accommodated; but if the limits are exceeded, then the system may be transferred into a totally different structure and mode of behaviour.

In the past, the most disturbing feature of impounded rivers was that the environmental changes resulting from their impoundment were not anticipated. Today, as a result of past experiences, we are able to make educated guesses as to the consequences of dam construction—some with, admittedly, higher degrees of likelihood than others—but for the most part our ability to predict them precisely before they occur is still poor. Matrices and checklists of potential impacts and issues from reservoir projects are available to facilitate trade-off analyses and decision-making between different choices, but the main need is for the acceptance and acknowledgement of uncertainty in environmental impact studies (Canter, 1983). River biophysical systems are unpredictable because a system's state at any one time may be at any point on a continuum of possible states. Many of these states have a low probability of occurrence, being related to the recurrence of an interaction between random processes: that is, the deterministic behaviour of a component sub-system is not simply related to a single control-factor, but is part of a larger system with which it interacts, so that the results produced may be indeterminate. Thus, the probability of selecting the correct alternative is small, as the behaviour of complex systems, when subjected to stress, is uncertain.

Pre-project studies have employed mathematical modelling procedures to simulate natural conditions, using traditional scientific descriptions of component processes. However, they contain many ill-defined parameters, and generally make many necessary, but often unjustifiable, assumptions—so that the reliance which can be placed upon the outcome of any single simulation is severely limited. Simulation models do, nevertheless, provide a procedure for the summary of causal relations and parameter values, and may be beneficially employed, quickly, to reveal inadequacies in current theory (Jeffers, 1972), and to identify critical components where research and new data are required (Mar, 1974). Simple simulation models can be valuable research tools when used to explore, or to generate, hypotheses (Stenseth, 1977); but their value as a management tool *per se* is limited, at least, to short-term projections. Simulation models can be useful only in a probabilistic context (Hornberger & Spear, 1981): that is, given that the model will contain inherent uncertainties in structure and parameter values, the only meaningful analysis must focus on the *likelihood* of various patterns of change. Mathematical models can play an important role in informing and assisting decision-makers, but quantitative methods are not essential to rational decisions, and fundamental problems exist that make a satisfactory quantitative decision-making method difficult to achieve (Hollick, 1981).

Despite the inadequacy of our ability to predict precisely what will happen to a river system subsequent to impoundment, an awareness of the uncertainty has provided a new stimulus for integrated environmental management within a long-term perspective. In the United States, the Task Committee on the Environmental Effects of Hydraulic Structures (T.C.E.E.H.S., 1978) recommended that site-specific matrices should be developed for the assessment of impacts of hydraulic structures, and that long-term effects, which may not be immediately apparent, must be considered. Indeed, although such matrices are liable to bias and require value-judgements, they have a general utility for planners and designers, at least in creating an awareness of the broader implications and possible directions of change. Post-construction studies have also been initiated to allow for the early detection of changes (e.g. Duthie & Ostrofsky, 1975). Such studies are particularly important in remote areas, where environmental considerations cannot often be easily accommodated, because of a lack of even basic data. Projects in remote areas of Canada, in particular, have suffered from a lack of environmental data, beyond simple descriptions, so that engineering plans proceeded, for economic reasons, without such data as a planning input (Berkes, 1981).

However, the most encouraging advances that are pertinent to the management of impounded rivers, have been the improvement in methodologies for determining in-stream flow needs, the incorporation of these needs into dam operating schedules, and the increasing adoption of multi-objective strategies based upon interdisciplinary communication and cooperation.

Determination of In-stream Flow Needs

Recognition of the needs of in-stream uses, including water-based recreation, fisheries, and aesthetic values, and of the conflicts between these and established uses, such as irrigation and hydroelectric power-production, during the early 1970s, led to the development of methodologies for formulating in-stream flow management procedures, and for assessing the in-stream effects of alternative development plans. The methodologies are designed to determine minimum and optimum flows, with due consideration to the carrying capacity for invertebrates and fish, which is related to the physical structure of the channel and to water quality. For example, as discharge increases, the usable area for invertebrate production rises until depth and/or velocity values exceed critical thresholds, and the weighted usable area is then reduced. However, management strategies must account for more than the flow-requirements of benthos and fish; flushing flows, for example, may be needed to maintain clean-gravel substrates. Thus, weighted usable-area determinations usually relate to annual flow-allotments, which represent the sum of short-time flow increments (often monthly) based upon special needs. Excessively high flows, however, may also be undesirable because in some instances they could transmit water of abnormal quality (confined to the tailwaters under low-flow conditions) for considerable distances downstream. Furthermore, whilst short durations of exposure to low-quality water may prove tolerable, long-duration exposure of the same magnitude can result in deleterious effects.

Solutions to technical problems of assessing in-stream flow-needs, in relation to other problems of communication, and the awareness of legal and social aspects of preserving in-stream values, were discussed at the Boise Symposium, Idaho, USA, in 1976 (Orsborn & Allman, 1976). In the same year, the 'Co-operative Instream Flow Service Group' (IFG) was formed to address and to interrelate four major components of in-stream flow activities (Wesche & Rechard, 1980): (1) to gain a basic understanding of possible changes in stream-channel morphology and hydraulics resulting from altered flows; (2) to develop improved methods to predict the effects that changes in channel configuration, flow regime, and other hydraulic factors, will have upon aquatic life and recreation uses; (3) to develop an awareness of management and operational processes; and (4) ditto of legal and institutional constraints. The need for including in-stream values as a goal in the preliminary planning stages was subsequently stressed by the 'Instream Flow Task Force', created in 1978 with specific regard for the general flow-characteristics needed to maintain the variety of in-stream uses—including fish and wildlife habitat, recreation, and aesthetics (Wesche & Rechard, 1980).

Numerous approaches have been developed. The simplest uses only mean annual flow; but more complex procedures allow due consideration of seasonally high-flows to flush fine sediments from pools and riffles, to recharge wetlands, or to facilitate fish-spawning migrations. Some approaches are designed to provide minimum-flow values that parallel the natural flow-regime during the yearly cycle. In 1980, eleven methodologies were predominantly in use in the western United States (Wesche & Rechard, 1980); some require no field data, but are based upon the manipulation, synthesis, and interpretation, of stream-flow records alone, whereas others utilize variable amounts of field data in establishing site-specific in-stream flow needs, using available fish habitat criteria to determine flow–habitat relationships.

An 'incremental methodology' for assessing relationships between stream-flow and fish-habitat structure has been developed by the IFG (Bovee, 1978a). The method uses water depth and velocity data from several cross-sections, each divided into between 9 and 20 subsections, to compute the water surface profile through a selected channel reach, at a range of discharges. Water widths, depths, and velocities, for the whole reach are then related to known optimum conditions for in-stream uses; for example, for a particular species and life-history phase, the weighted usable area is computed for specified discharges. Thus, it can be used both to formulate in-stream flow regulations, and to assess the effects of planned stream-flow changes and/or expected channel-morphology changes. Not least, the IFG 'incremental methodology' was designed as a communication link among fishery biologists, hydrologists, and hydraulic engineers.

Five basic steps are taken to formulate in-stream flow recommendations (Stalnaker, 1979): (1) field measurement of channel, physical, and chemical, characteristics; (2) hydraulic simulation of the spatial distribution of combinations of depths, velocities, substrates, and cover objects; (3) simulation of the temporal distribution of temperature and chemical constituents; (4) application of habitat-evaluation criteria for species and life-stages, for each flow regime and channel condition under investigation; and (5) display of the changing habitat usability over time. Bovee (1978b) has collated the flow requirements for 10 species of Salmonidae in different life-stages, using 'electivity curves' based upon weighted habitat criteria—water depth, mean velocity, temperature, and bed sediment. These have been used to evaluate the availability of physical microhabitats in streams under different conditions of discharge and channel configuration. However, several important variables were omitted, including the flow-sequence and provision of adequate cover. Binns & Eiserman (1979) developed a habitat-quality index based upon ten variables: critical-period stream-flows, annual stream-flow variation, maximum summer stream temperature, water velocity, cover, stream width, food abundance, food diversity, nitrate levels, and stream-bank stability. For 20 sites on streams in Wyoming, USA, a high correlation ($R = 0.95$) was obtained between the habitat-quality index and trout standing-crop.

River-system evaluation requires improved procedures for assessing the significance of physical and chemical parameters that are pertinent to specific needs. Attempts have been made to develop improved methodologies for quantifying substrate quality (Shirazi & Seim, 1981; Beschta, 1982), channel morphology (Petts, 1980b), and flow geometry (Mosley, 1982). The IFG is seeking to evaluate the statistical independence among variables that are used for the determination of in-stream flow recommendations (Wesche & Rechard, 1980). Many methodologies for formulating minimum-flow recommendations, for example, as developed by Gore (1978), based upon Bovee (1978a), fail to describe the dependence of flow velocity and depth in channel hydraulics. However, Gore & Judy (1981) included a function to describe this dependence in a new model for determining optimum conditions for stream macroinvertebrates, which stresses the prediction of density maxima and flow predictions for the maintenance of suitable lotic habitat.

Stalnaker (1979) has shown that prediction of the weighted usable area for specified discharges is an effective method of producing adequate flow-recommendations, and Gore & Judy (1981) concluded that the combination of flow requirements for spawning and for maintenance of fish-stocks, for aquatic macroinvertebrates, for flushing of accumulated detrital material, and for maintaining good riparian growth, should result in effective streamflow management. However, management strategies may be improved if the in-stream flow requirements are considered not in isolation, but in conjunction with catchment management programmes to 'control' tributary erosion and sediment yield, effluent sources, and floodplain development. The river is neither structurally, nor functionally, distinct from other components within the 'catchment ecosystem'. *A holistic philosophy dictates that radical progress in our understanding and management of individual ecosystems will ensue only by expanding the scales of observations to embrace such interrelated groupings.*

Rivers may be viewed as large-scale multi-faceted systems, which require that a multiplicity of models be developed, since a single all-encompassing model, however desirable as a conceptual goal, is not a practicable objective. By reforming questions and modelling objectives into an ordered structure, yet integrated into a synergistic whole, useful partial models may be constructed, validated, and employed (Halfon, 1979). Similarly, the application of multi-objective management programmes, formulated by the integration of discipline-specific aims—concerned with in-stream, floodplain/riparian, and catchment, management problems—and utilizing an holistic approach and long-term perspective, should provide for a new assessment of impacts within impounded rivers. For future projects such programmes would lead to an improved awareness of potential impacts and, therefore, to better project evaluations, and to improved post-project monitoring and surveillance schemes. For existing impounded rivers, they present a realistic framework for the assessment of restorative measures.

Trinity River Resurrected?

Trinity River is one of California's most famous fishing rivers, of importance both for angling and as a nursery for anadromous fishes. However, river impoundment led to a rapid rate of river-channel deterioration, such that 14 years after dam completion Bush (1976) suggested that Trinity River was a 'dying' stream. Environmental changes have involved habitat loss and aesthetic impacts that have economic implications through loss of scenic quality, loss of access, and a consequent loss of tourist trade. Concern was voiced by a broad spectrum of individuals, representing a wide range of expertise and interests—ranging from Federal Government agency personnel, to fishermen and citizens from many parts of California and other states.

In 1974 a Task Force was established to address the problems of river management and the actions needed to restore, and maintain, the fish and wildlife. The Main Report (Draft), published in 1979, suggested that the future of the Trinity River—and other impounded rivers—could, given certain economic concessions, be viewed with guarded optimism.

The Trinity River Basin

Originating in the Scott Mountains of northern California, the Trinity River (Fig. 37) flows for 210 km before joining the Klamath River 65 km from the Pacific Ocean. The 7637 km^2 drainage basin has a relief of 2135 m, and precipitation generally varies with altitude and distance inland from 750 to 1500 mm p.a., with a marked winter peak. The mountainous catchment supports a 90% coniferous forest-cover, most of which is commercially managed, but soil disturbance, through logging practices and road construction within areas of decomposed granitic bedrock, produces high sediment-yields approaching 400 m^3 per km^2 per yr. At Lewiston, river flow-variability was high, with a monthly range of 333 $\times$ 10^6 m^3 in May to 12 $\times$ 10^6 m^3 during September, and with instantaneous flows of above 500 m^3 per s and below 3.5 m^3 per s having a recurrence interval of 5 years. Sediment discharged from tributary sub-basins produced temporary accumulations within the channel, but these were regularly flushed by main-stream flows. Under historical conditions before river impoundment, at Lewiston, flows of sufficient magnitude to transport riffle gravels were exceeded between 10 and 40% of the time (D.W.R., pers. comm.). For 50 km below Lewiston, the channel migrated actively across a gravel-and-sand floodplain which supported only a sparse vegetation cover, and the channel-width ranged from 35 to 100 m. At Lewiston, the number of *Salmo gairdneri* (Steelhead Trout) counted annually between 1958 and 1964 averaged 3000, *Oncorhynchus kisutch* (Silver or Coho Salmon) averaged 230, and *O. tshawytscha* (King or Chinook Salmon) averaged over 6500; while more than 30 000 adults (in some years more than 75 000) spawned in the River downstream.

The Trinity River Project

Early planning investigations during the 1950s concluded that the fullest, and most economic, conservation and use of water resources of the Trinity River, for the broadest possible public benefit, would be obtained through the diversion of surplus runoff into the Central Valley of California (Sasaki, 1976). The United States Bureau of Reclamation completed construction of the Trinity River Project in the early 1960s, not only to provide a supply for irrigation, but also for flood-control, hydroelectric power essential for stimulating industrial growth, and potentially valuable new recreational

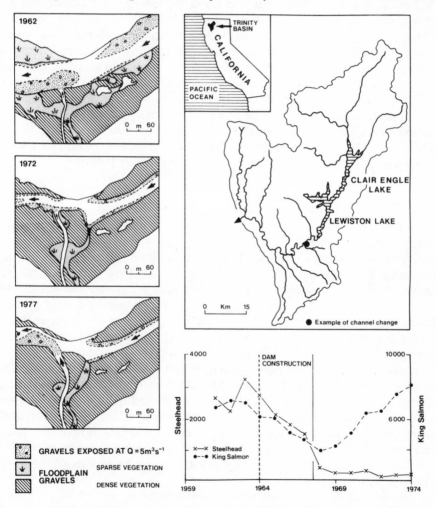

Fig. 37. The Trinity River, California: catchment map (upper right), an example of channel change (left), and 5-years' running means of Salmon and Steelhead Trout runs (lower right). (Data provided by the Department of Water Resources, Red Bluff, California, USA.)

opportunities. Trinity Dam impounds runoff from the 1885 km^2 headwater catchment, creating a 3100 × 10^6 m^3 reservoir, Clair Engle Lake, with a surface area of some 70 km^2 and inundating nearly 4% of the drainage basin. Water is released from Clair Engle Lake into Lewiston Reservoir for subsequent diversion, or for discharge into the Trinity River: 90% of the annual discharge at Lewiston is transferred to the Central Valley Project.

The dams isolate 100 km and 175 km of *O. tshawytscha* and *S. gairdneri* spawning, and nursery, habitats, respectively (Smith, 1976). Efforts were made to mitigate these impacts by providing a fish hatchery below the dams at Lewiston, with a capacity of about four million eggs. Moreover, the in-stream flow requirements of the river fishery had been considered, and 10% of the annual historic runoff, or about 150 × 10^6 m^3, was reserved for releases downstream. Flow regulation was implemented to provide higher discharges during the autumn, from a 'norm' of 4.25 m^3 per s to 7.10 m^3 per s, in order to facilitate natural spawning. With these measures, it was considered that the in-stream requirements of the Trinity River downstream would not be impaired, and that there would be no loss of fisheries (Bush, 1976). However, channel deterioration was noted within 1 year of dam completion (Serr, 1972), and fish and wildlife problems have occurred of a magnitude that were not foreseen in the early phases of the project (Sasaki, 1976).

Impacts

Directly, or indirectly, as a result of the Project construction and operation, major environmental changes occurred for nearly 65 km below the dams (Serr, 1972). The most significant impact upon lotic habitats has been the discharge regulation provided by Clair Engle Lake. Flow data (Table LIII) suggest that peak discharges have been significantly reduced: the magnitude of the post-dam 10-years' discharge, for example, is only one-third of that described in the pre-reservoir data. Headwater sediment sources have been isolated from the River below the dam, but channel degradation has been limited to only the first few kilometres downstream (Haley *et al.*, 1976). The frequency of bed-material transport has, in fact, been reduced (Table LIV), so that for 50 km below Lewiston Dam the duration of the minimum discharge competent to move gravels has been reduced from 40% of the time to less than 3%. Turbidity remains high for several months of the year, and storm events within tributary basins produce more sediment than the River can transport: the sediment-transport capacity has been reduced to only 8000 m^3—less than one-quarter of the load delivered annually by a single tributary (D.W.R., 1978).

Eight sites of sedimentation were observed within 4 years of dam completion (Ritter, 1968); also, tributary deltas have expanded, pools have been infilled with fine sediments, and the floodplain area has been enlarged. Channel widths have been reduced by up to 65% for 50 km below Lewiston, and major adjustments of channel morphology had occurred during the first

10 years after project completion. Vegetation became established rapidly on the floodplain and on new deposits. Between 1962 and 1972 the floodplain area increased by 14% within one 100 m reach at the expense of the river channel, and the vegetation cover increased from 58% to over 90% (Fig. 37, left). During the following 5 years, the floodplain area enlarged only slightly, but vegetation continued to form a 98% cover of non-river areas in 1977. Whilst the establishment of riparian vegetation adjacent to, and overhanging, the River provides shelter and contributes to the fish-food supply—both directly through the provision of terrestrial insects, and indirectly through the provision of organic matter—the negative impact of channel substrate deterioration has proved to be most significant.

Table LIII   Peak discharges below Trinity and Lewiston Reservoirs, Trinity River, North California. (Data provided by the Department of Water Resources, Red Bluff, California.)

Recurrent interval (yrs)	Peak discharge (m³ per s)		Ratio
	Pre-dam	Post-dam	
10	906.16	339.81	0.375
6	750.41	169.90	0.226
2.33	481.40	26.90	0.056
1.50	368.13	14.16	0.038

Table LIV   Channel adjustment below Trinity and Lewiston Dams, Trinity River, North California.

Channel length below dam (km)	Drainage area impounded (%)	Percentage of time minimum discharge required to move gravels is equalled or exceeded*		Channel width reduction (%)
		Pre-dam	Post-dam	
0—	99	50	0.05	0–30
4—	93	45	0.05	0–60
9—	88	45	3	50–60
20—	83	25	1	30–40
23—	78	40	3	30–55
25—	76	40	5	33–65
31—	70	35	5	No data
43—	64	30	4	20–38
53	51	60	40	No data

* Data provided by the Department of Water Resources, Red Bluff, California.

The accumulation of sand in riffles has had an adverse impact on both the food-producing, and spawning, capabilities of the Trinity River. Additionally, temperature alteration by the hypolimnial-release reservoir has had an important effect. Natural riffles receiving large quantities of sands during spring runoff from a tributary, lacked any organisms during the spring and summer months, although the subsequent partial flushing of the sediment allowed the more common species to return, albeit at a decreased-population level (Boles, 1979). Winter temperatures have been increased from 1°C to 5°C, and summer temperatures have been lowered from 26°C to 21°C. Reduced flows have resulted in earlier and more rapid warming of downstream areas than occurred under natural-flow conditions. These thermal changes have prevented colonization by all but a few organisms (Boles, 1979). Impoundments block recruitment of benthic invertebrates from upstream, and this also leads to a depauperate community-diversity. The combined effects of restricted recruitment, substrate deterioration, and thermal changes, have decreased macroinvertebrate species-diversity, density, and biomass, which are important for maintaining the fishery.

Riffle sedimentation also impacted upon the fish-stocks directly (Bush, 1976). Between 1963 and 1967, an estimated 28% of critical spawning-habitat was lost in the 26 km reach between Lewiston and Douglas City, the most heavily-used spawning portion of the River: 44 spawning riffles, providing 1935 m^2 of area in 1956, were reduced to only 10, covering less than 300 m^2, in 1970—a loss of 85%. Thus, although the number of *O. tshawytscha* returning to the hatchery at Lewiston has remained fairly constant since river impoundment, with an average annual run of 6200 (Fig. 36, lower right), the number of adult *O. tshawytscha* spawning in the Trinity River itself has declined markedly to only just over 4000 (Smith, 1976). The numbers of *O. kisutch* returning to the hatchery have increased eleven-fold during the post-construction period, but this is almost entirely a hatchery-generated and -sustained run, and runs of *S. gairdneri* have been reduced by 90% to a nominal average of just over 300 (D.W.R., 1978). Aesthetically, the vegetation and channel changes are, debatably, of benefit. Some judge the changes to be beneficial, in comparison with the former barren gravels. Others feel that the 'beach' shoreline was aesthetically superior. Certainly, the forested floodplain obstructs the view across the River, and hinders freedom of access and sport fishing.

Management Programme

Inspired in part by the concern of private citizens, the 'Trinity River Basin Fish and Wildlife Task Force', involving ten agencies and entities, was established in 1974. A *Multiobjective Water Resources Planning Approach* was adopted, within which environmental and social values are weighed equally with economic ones (Sasaki, 1976). Unforeseen degradation of lotic habitat within the Trinity River is a reflection of the failure to appreciate, and then to evaluate, the link between the River and its catchment, during project formulation. Flow regulation imposed by the construction of the Trinity and Lewiston Dams, has served to aggravate impacts that were induced by ineffective land-use management. Historically, and with increasing rapidity, resource-based activities within tributary sub-basins have contributed excessive quantities of sediment to the main-stream. Hydraulic mining and gravel abstraction, followed by uncontrolled housing development, poor road-construction practices, and inadequately regulated commercial deforestation, accelerated natural erosion-rates. The observed changes of the Trinity River are not manageable simply by flow manipulation through controlled reservoir releases; remedial action to compensate for, or eliminate, the deleterious effects, must include an extensive catchment-management programme.

The effective implementation of an in-stream management programme required the control of sediment delivery-rates from tributary sub-basins, through a watershed rehabilitation and protection programme, supplemented by the construction of debris dams (e.g. D.W.R., 1978). The scouring of filled pools, and the cleaning of riffles, by flow releases from the dam, were not possible without causing permanent damage to the riffles, or without causing flood damage to structures on the floodplain. Initially, the cleaning of pools had to be accomplished by using mechanical means. Boles (1980a) showed that riffle gravels from which sand deposits had been flushed, experienced a marked increase in benthic productivity, biomass, and diversity, and therefore food-producing capabilities to support a larger biomass of fish. However, some denudation of benthic organisms will occur immediately below the dam, due to the exclusion of those organisms which recolonize restored riffles solely from drift, and to the limitations imposed by the altered thermal regime. Nevertheless, the degradation of channel substrate and the loss of gravels below dams, due to uncontrolled spillweir discharge, can be mitigated by replacement with artificial riffles. At two restored riffles within 500 m of the dam, Boles (1979) reported that the abundant detrital and periphyton food supplies, and lack of competition and predation, allowed large populations of a few organisms to develop: mainly species of *Baetis, Simulium*, and Chironominae. Furthermore, restoration projects incorporating buffer areas of complex substrate, composed of gravel sizes up to those normally found in the Trinity River, favoured the diversification and stabilization of the benthos, and provided important rearing-areas for juvenile salmonids (Boles, 1980b).

In 1973, an improved flow-schedule had been proposed to increase the survival of out-migration juvenile salmonids; to increase the spawning area; to provide better attraction-flows; to increase the area for fish holding; and to decrease water temperatures for 50 km downstream. However, at the time, the costs of increasing the annual river-flow by about 240 m³ would have led to a loss of revenue from irrigation supply of over $2 million, and power production would have been reduced by about 50 000kW—equal to the demand of over 30 000 households (Sasaki, 1976). The Task Force (Draft) Report (Frederiksen, Kamine & Associates Inc., 1979) utilized the optimization of in-stream habitat requirements to propose the implementation of modified flow-release schedules with a minimum annual flow of either 265 hm³ or 234 hm³. In all, three options were proposed, each integrating a modified release schedule with the rehabilitation of the main-stream and its tributaries, and the use of hatchery facilities. The emphasis upon each aspect varied, however, to provide, below Lewiston, first, for 100% natural fish production, secondly, for 50% natural and 50% artificial, and thirdly, for 100% artificial. In each case, the flow schedule would provide minimum flows that would, with appropriate in-stream work, maintain habitats that would be favourable for all life-stages of natural and artificially produced salmonids. For the establishment of high natural production-levels, however, a period of up to 10 years may be required before sufficient habitat could be maintained continually. For each option, the costs of changing water-use, from agricultural irrigation-supply and hydroelectric power-production to river conservation, were weighed against the economic value of the proposed salmonid fishery. Nevertheless, each option would perpetuate anadromous species of historic, economic, and environmental, significance, and each encourages public involvement in river management.

The proposed (1979) 'Trinity River Basin Management Programme' had three primary objectives: (A) to establish and maintain (i) water releases from Lewiston Dam and (ii) channel characteristics that adequately maintain habitat for spawning, rearing, holding, and conveyance, of anadromous fish; (B) to establish and execute a conjunctive in-stream and hatchery management programme for in-stream fishes; and (C) to maintain the maximum diversity and population of wildlife species through consideration of the habitat (specifically, vegetation) requirements. However, a fundamental problem for the management of impounded rivers is that the perceived impacts stem from a value-scale that appears to be highly weighted and biased towards the benefit of the major water-users (Bush, 1976).

A value-system must accordingly be established that recognizes the need to place a new emphasis upon the maintenance of ecological and environmental standards within, and downstream from, water-source areas. Planners must consider the full range of impacts, and must view the river as a product of the whole 'catchment ecosystem'. The multi-objective framework adopted by the Trinity River Basin Fish and Wildlife Task Force exemplifies one direction towards achieving the successful management of impounded rivers. The unification of objectives, responsibilities, goals, and programmes, is essential; open communication and mutual goodwill are prerequisites; and compromise and, perhaps most importantly, innovation, are key concepts for the development of a successful environmental management strategy (Hall, 1976).

CONSERVATION OR BUST?

' . . . we should plan to preserve at least a few aquatic Gardens of Eden for our successors to enjoy. At present we are in real danger of altering everything just a bit, with little knowledge of the consequences, and one day we may discover that there is no wild river left.'

Hynes (1970b p. 423)

During the last decade, international awareness and activity, relating to environmental impacts, has grown rapidly. This growth reflects both an improved scientific understanding of the potential problems and an intensification of public emotion. However, many decision-makers still expect the environment to respond in a simple, linear manner to an applied stress, and only perceive the immediate short-term consequences. In reality, river systems consist of numerous interrelated hydrological, sedimentological, morphological, and biological, components. The response of a river system to an applied stress will inevitably be complex, as the components have to become adjusted towards a new steady-state condition. Responses will be episodic and controlled by thresholds, so that a considerable time-period may elapse before any semblance of equilibrium is achieved. Readjustment may involve relatively minor changes of individual components, but the cumulative long-term effects of those changes may totally alter the characteristics of a river. Indeed, the potential for 'destruction by insignificant increments' (Gamble, 1979) is apparent from the analysis of impounded rivers.

The advancement of scientific knowledge will not, of itself and alone, ensure improvement in environmental managment: a major gap will remain in the perception of decision-makers. It will only be with due recognition of the intricate composition of natural systems, and with a realization of the delicate balance in which the systems are maintained, that the widespread present-day indifference towards a long-term perspective will be remedied and, hopefully, reversed. Thus, a considerable effort needs to be maintained in stimulating environmental awareness, not least among the general public throughout the world, because positive public participation in the decision-making process could assure the avoidance or modification of schemes induced by short-term economies or for political gains.

The SCOPE Working Group on Man-made Lakes (SCOPE, 1972) made three recommendations involving public policy (p. 9):

(1) Assessment of alternatives—any decision to build a man-made lake should be based upon a comparison of its likely effects with those which would result from other technical or social actions to reach the same public goals.

(2) Canvass of impacts—any decision to build a man-made lake should include an attempt, however rudimentary, to canvass the full consequences of the dam and the new lake—consequences in the socio-cultural system—and in those of the Earth's crust, hydrology, and biological production in the water of the reservoir, upstream, downstream, and in the surrounding land.

(3) Consultation with the people affected—such assessment should treat the people directly affected by the reservoir as a dominant part of the ecosystem to be transformed, and should involve them appropriately in the decision-making process.

Dams are important now, and will be in the future, for water supply and power, and their utility is being enhanced by increasing recreational usage. However, river impoundment is not necessarily the most desirable alternative. Power may also be generated from fuel sources; flood losses may be reduced by improved land-management practices; and water supplies may be enhanced by the artificial recharge of aquifers, improved treatment, or recycling. It is erroneous to assume that the appraisal of alternatives is an easy task, but, equally, it should not be assumed that a dam—large or small—is necessarily desirable. The decision-making process involved in the appraisal of alternatives is based on the prior definition of social goals which differ among nations and change over time. A reservoir is often judged to be economically preferable to other alternatives, because of its multipurpose potential, and because it enhances the range of economic opportunity in the area affected. However, the maximum long-term resources potential will be realized by mastering Nature, not by force, but by understanding (Bronowski, 1973). Several hundred new large dams are being completed each year, and river regulation schemes are becoming more and more complex. Increasingly, long-distance water transfer from water-surplus to water-deficient areas is becoming a serious alternative in water planning for many countries (Biswas, 1983). The Ob Dam below the Irtysh confluence, USSR, for example, due for completion in 2010, is planned to divert impounded water south for 2500 km in order to augment flows within the Amu-Darya River, and to provide irrigation supplies within Soviet central Asia and Kazakhstan.

We may only guess at the ecological consequences of such major manipulations to the pattern of water-flow. Perhaps we have now reached a point in time when the *preservation* of a few river systems as 'Gardens of Eden' should be seriously considered—at least until we can gain a fuller understanding of the long-term environmental changes resulting from river impoundment (which requires acceptance of the often slow rate of environmental response). Certainly there is an increasing volume of evidence to show that even medium-term forecasts of water needs (i.e. of 10–20 years) often give exaggerated projections, because they fail to include improvements in the efficiency of water-use and other technological advances, which reduce the rate of increase in water-demand. The justification for continued dam construction must undergo detailed scrutiny in the light of the environmental changes which may result. Without doubt, further information is needed to ensure that river management is based upon the dispassionate, rational, evaluation of alternative project proposals and operational procedures.

Whether the changes to a river caused by an impoundment are desirable, or undesirable, is obviously a matter of subjective judgement. On the Vaal River, South Africa, the storage behind a dam of the heavy, natural, sandload which formerly reduced invertebrate and fish populations, has been seen as an improvement (Chutter, 1963). The discharge of cold water from a reservoir's hypolimnion during summer into a warm-water river (Tarzwell, 1939), and the elimination of predator and competitor fish species, thereby enhancing salmon and trout production, may be viewed as a beneficial impact in terms of economic fisheries. However, the elimination of 'less valuable' species could also be viewed as a major deleterious impact from the viewpoint of conserving naturally-stable ecosystems. The reduction in species richness and river diversity caused by impoundment, and the creation of less variable systems, may also be viewed as undesirable effects.

Traditionally, the goals of economic and social advance have been most promptly achieved by subduing Nature. Such an approach has generated the current protective response of society which, having perceived the magnitude of environmental destruction, seeks to buy time by environmental preservation and use-regulation. However, such a reactive response is commonly emotive rather than based upon any objective assessment. Reactive responses make the practice of environmental assessment arbitrary, inflexible, and unfocused, and are therefore intolerable (Holling, 1978). Rationality requires consideration of all alternative strategies, and the definition and evaluation of all the resultant consequences of each strategy. All too commonly, ecological factors have been considered merely as constraints on technical and economic rationality. Alternatively, environmental preservation is demanded without due consideration for socio-economic needs, or for the implications and costs of avoiding river impoundment. Ecological rationality requires a concern for the conservation and enhancement of the natural environment in relation to societal needs or desires, rather than preservation *per se*. A particular river-system may, arguably, be conserved even though it may have no present economic use because: (1) there is an unknown probability that an economic use of unknown value will be found for it in the future; and (2) because it has potential value for aesthetic, educational, or scientific purposes. Commonly, all such possible future environmental values are discounted from economic analysis. Decisions on inevitable changes are not simply a matter of social preference, but have a moral dimension, because not all of those who would be affected by the decision can contribute to it, or be meaningfully represented. Ecological decisions involve moral judgements regarding the rights of both future generations and species of fauna and flora (Hollick, 1981). Thus it would be irrational for a society to destroy ecosystems if it believed that future generations, or the involved species, had rights. However, consideration of general damage to ecosystems identified in environmental impact assessment (EIAs) prior to river impoundment, have carried little weight in courts, although a few selected topics of immediately demonstrable importance to local inhabitants have been effectively argued (Penn, 1975).

An approach to the uncertainty of impacts caused by river impoundment must be positive. Dams can realize considerable economic and socio-cultural benefits, but their construction must be based upon sufficient knowledge of the river and its catchment to allow for the due consideration of the long-term ecological consequences. Multi-objective river planning requires not only the examination of rivers as parts of the 'catchment ecosystem', but also the equal weighting of economic, social, and environmental, values. Without doubt, the advances made in the assessment of in-stream flow-needs and through the application of multidisciplinary approaches, suggests that the future may be viewed with guarded optimism. However, for a responsive attitude towards impacts to be of any real and lasting value, a more flexible approach to management must be introduced. Flexible operational procedures are needed during project planning—to allow for the re-allocation of water to meet unforeseen river protection and conservation requirements as they become known through post-project monitoring and surveillance.

Future management problems will be exacerbated as rivers experience increasing control on their flows by variable dam-releases. Severe ecological impacts may be anticipated if the determination of in-stream flow-needs during project planning fails to consider the maximum desirable flow, and acceptable range, rate, and frequency, of flow variation, as well as the minimum and optimum flows. Hydroelectric power regulation, with daily fluctuations of discharge from zero to 1000 m³ per s, is planned for some Swedish rivers (Henricson & Müller, 1979). For the Colorado River, USA, below Glen Canyon Dam, there has been a proposal to widen the range of daily-release variations, in order to enhance peak-power generation, from between 140 m³ per s and 600 m³ per s today, to between approximately 30 m³ per s and 1200 m³ per s in 1990 (Howard & Dolan, 1981).* The introduction of such extreme flow-patterns might virtually eliminate the fauna and flora for a considerable distance below the dams. Certainly they introduce a new scale of problem, which would have long-term ramifications throughout the whole stream system.

The optimal use of a particular environment in terms of resource exploitation, recreation, and amenity, requires action towards attaining the geomorphological, ecological, and aesthetic, harmony of the results of development. Sound management programmes require adequate data-bases with regard to sediment-source areas, production rates, and transport capacities; to channel morphology, substrate composition, and pool-riffle volumes; to the availability and dynamics of allochthonous organic-matter sources; to macroinvertebrate composition and biomass; to macrophyte and periphyton distribution; to water quality; and to floodplain, and even near-shore, dynamics and production.

Without such information, the development of sound methodologies, analyses, and evaluations, in dealing with the management of impounded rivers, is certainly problematic and, arguably, impossible. To date, broad and detailed monitoring programmes are the exception; yet data collected over several years are required in order to understand the dynamic and site-specific behaviour of lotic environments. An holistic approach within an interdisciplinary framework is, of itself and alone, not enough (although perhaps more than we can hope to see adopted!); environmental management must also evolve an anticipatory approach. 'Peaking' hydroelectric power-dams, especially, can completely—and rapidly—decimate the natural river-systems. Restoration may be impossible; protection may be the only solution.

* The American Wilderness Alliance has successfully led a national effort to stop the Bureau of Reclamation from adding new 'peaking power' generators to Glen Canyon Dam. *On the Wild Side*, American Wilderness Alliance, 42, June 1984.

267

REFERENCES

Abdurakhmanov, Yu. A. (1958). The effect of regulation of the flow of the Kura River on the behaviour and abundance of fishes in the region below the Mingechaur hydroelectric station. *Rybnoe Khoziaistvo*, **34** (12), 13–15. (Translated by W. E. Ricker, Fisheries Research Board of Canada, Translation Series No. 258, 1960.)

Abell, D. L. (1961). The role of drainage analysis in biological work on streams. *Verhandlungen Internationale Vereinigung für Theoretische und Angewandte Limnologie*, **24**, 533–7.

Abu-Zeid, M. (1983). The River Nile: Main water transfer projects in Egypt and impacts on Egyptian agriculture. Pp. 15–34 in *Long-distance Water Transfer* (Ed. A. K. Biswas). Tycooly International, Dun Laoghaire, Ireland: xvi + 417 pp., illustr.

Abul-Atta, A. A. (1978). *Egypt and the Nile After the Construction of the High Aswan Dam*. Ministry of Irrigation and Land Reclamation, Cairo. (Cited in Goudie, A. (1981) (q.v.).

Ackermann, W. C., White, G. F., & Worthington, E. B. (Ed.) (1973). *Man-Made Lakes: Their Problems and Environmental Effects*. Geophysical Monograph 17, American Geophysical Union, Washington, DC, USA: xv + 847 pp., illustr.

Ackers, P., & White, W. R. (1973). Sediment transport: New approach and analysis. *Journal of the Hydraulics Division, ASCE*, **99**, HY11, (Nov.), 2041–60.

Adams, B. L., Zangg, W. S., & McLain, L. R. (1975). Inhibition of saltwater survival and Na–K–ATPase elevation in Steelhead Trout (*Salmo gairdneri*) by moderate water temperatures. *Transactions of the American Fisheries Society*, **104**, 766–9.

Adeniji, H. A. (1973). Preliminary investigation into the composition and seasonal variation of the plankton in Lake Kainji, Nigeria. Pp. 617–9 in *Man-Made Lakes: Their Problems and Environmental Effects* (Ed. W. C. Ackermann, G. F. White, & E. B. Worthington). Geophysical Monograph 17, American Geophysical Union, Washington, DC, USA: xv + 847 pp., illustr.

Alabaster, D. S. (1970). River flow and upstream movement and catch of migratory salmonids. *Journal of Fisheries Biology*, **2**, 1–13.

Ambasht, R. S. (1971). Ecosystem study of a tropical pond in relation to primary production of different vegetational zones. *Hydrobiologia*, **12**, 57–61.

American Water Works Association (1971). Quality control in reservoirs. *Journal of the American Water Works Association*, **63**, 597–604.

Anderson, E. R., & Pritchard, D. W. (1951). *Physical Limnology of Lake Mead*. Report 258, US Navy Electronics Laboratory, San Diego, California, 43 pp.

Andrews, E. D. (1982). Bank stability and channel width adjustment, East Fork River, Wyoming. *Water Resources Research*, **18** (4), 1184–92.

Anon. (1983). Hydropower prospects for Spain. *Water, Power and Dam Construction*, **35** (2), 24–7.

Antonio, R. A. (1969). The Arkansas River Project civil engineering. *American Society of Civil Engineers*, **39** (12), 44–9.

Arai, T. (1973). Thermal structure of the artificial reservoir. Pp. 536–8 in *Man-Made Lakes: Their Problems and Environmental Effects* (Ed. W. C. Ackermann, G. F. White, & E. B. Worthington). Geophysical Monograph 17, American Geophysical Union, Washington, DC, USA: xx + 847 pp., illustr.

Armitage, P. D. (1976). A quantitative study of the invertebrate fauna of the River Tees below Cow Green Reservoir. *Freshwater Biology*, **6**, 229–40.

Armitage, P. D. (1977). Invertebrate drift in the regulated River Tees, and an unregulated tributary Maize Beck, below Cow Green Dam. *Freshwater Biology*, **7**, 167–83.

Armitage, P. D. (1978). Downstream changes in the composition, number and biomass of bottom fauna in the Tees below Cow Green Reservoir and an unregu-

268

lated tributary Maize Beck, in the first five years after impoundment. *Hydrobiologia*, **58**, 145–56.

Armitage, P. D. (1979). Stream regulation in Great Britain. Pp. 165–82 in *The Ecology of Regulated Streams* (Ed. J. V. Ward, & J. A. Stanford). Plenum Press, New York: xx + 398 pp., illustr.

Armitage, P. D., & Capper, M. H. (1976). The numbers, biomass and transport downstream of microcrustaceans and *Hydra* from Cow Green Reservoir (upper Teesdale). *Freshwater Biology*, **6**, 425–32.

Armitage, P. D., Machale, A. M., & Crisp, D. T. (1974). A survey of stream invertebrates in the Cow Green basin (upper Teesdale) before inundation. *Freshwater Biology*, **4**, 369–98.

Armstrong, L. (1973). Dam construction and the environment. Pp. 217–36 in *Transactions of the Eleventh International Congress of Large Dams, Madrid, Spain, Vol. 1*. International Commission on Large Dams, Paris: xxviii + 974 pp., illustr.

Arroyo, S. (1925). Channel improvements of Rio Grande below El Paso. *Engineering News Record*, **95**, 374–6.

Arumugam, P. T., & Furtado, J. I. (1980). Physico-chemistry, destratification and nutrient budget of a lowland eutrophicated Malaysian Reservoir and its limnological implications. *Hydrobiologia*, **70**, 11–24.

Attwell, R. I. G. (1970). Some effects of Lake Kariba on the ecology of a floodplain of the Mid-Zambezi Valley of Rhodesia. *Biological Conservation*, **2** (3), 189–96.

Baker, B. W., & Wright, G. L. (1978). The Murray Valley: its hydrologic regime and the effects of water development on the river. *Proceedings of the Royal Society of Victoria*, **90**, 103–10.

Balon, E. K. (1975). The eels of Lake Kariba: distribution, taxonomic status, age, growth and density. *Journal of Fish Biology*, **7**, 797–815.

Banks, J. W. (1969). A review of the literature on upstream migration of adult salmonids. *Journal of Fish Biology*, **1**, 85–136.

Banks, J. W., Holden, M. J., & McConnell, R. H. (1965). Fishery report. Pp. 21–42 in *The First Scientific Report of the Kainji Biological Research Team* (Ed. E. White). Unpublished report of the Ministry of Overseas Development, London: 89 pp., illustr.

Baranov, I. V. (1966). Biohydrochemical classification of the reservoirs in European USSR. Pp. 139–83 in *The Storage Lakes of the USSR and their Importance for Fishery* (Ed. P. V. Tyurin). Israel Program for Scientific Translations, Jerusalem: 244 pp., illustr.

Barila, T. Y., Williams, R. D., & Stauffer, J. R. (1981). The influence of stream order and selected stream-bed parameters on fish diversity in Raystown Branch, Susquehanna River Drainage, Pennsylvania. *Journal of Applied Ecology*, **18**, 125–31.

Barrett, G. W. (1981). Stress Ecology: an integrative approach. Pp. 3–12 in *Stress Effects on Natural Ecosystems* (Ed. G. W. Barrett, & R. Rosenberg). (Environmental Monographs & Symposia, Convener & Gen. Ed. N. Polunin). John Wiley & Sons, Chichester–New York–Brisbane–Toronto–Singapore: xviii + 305 pp., illustr.

Barrett, G. W., Dyne, G. M. van, & Odum, E. P. (1976). Stress ecology. *Bioscience*, **26**, 192–4.

Barrett, G. W., & Rosenberg, R. (Ed.) (1981). *Stress Effects on Natural Ecosystems* (Environmental Monographs & Symposia, Convener & Gen. Ed. Nicholas Polunin). John Wiley & Sons, Chichester–New York–Brisbane–Toronto–Singapore: xviii + 305 pp., illustr.

Barry, R. G. (1969). The world hydrological cycle. Pp. 11–29 in *Water, Earth and Man* (Ed. R. J. Chorley). Methuen & Co. Ltd, London: xix + 588 pp., illustr.

Bauer, F., & Burz, J. (1968). Dar Einfluss der Festoffführung alpiner Gewässer auf die Stauraumverlandung und Flussbetteintiefung. *Wasserwirtschaftliche Mitteilungen*, **4**, 114–21.

Baxter, R. M. (1977). Environmental effects of dams and impoundments. *Annual Review of Ecology and Systematics*, **8**, 255–83.

Baxter, R.M., & Glaude, P. (1980). Enviromental effects of dams and impoundments in Canada: Experience and prospects. *Canadian Bulletin of Fisheries and Aquatic Sciences*, **205**, vi + 34 pp., illustr.

Beadle, L. C. (1974). *The Inland Waters of Tropical Africa: An Introduction to Tropical Limnology*. Longman Inc., New York: 365 pp., illustr.

Beaumont, P. (1978). Man's impact on river systems: a world-wide view. *Area*, **10**, 38–41.

Beckett, D. C., & Miller, M. C. (1982). Macroinvertebrate colonization of multiplate samplers in the Ohio River: the effect of dams. *Canadian Bulletin of Fisheries and Aquatic Sciences*, **39** (12), 1622–7.

Beckinsale, R. P. (1969). River regimes. Pp. 455–71 in *Water, Earth and Man* (Ed. R. J. Chorley). Methuen & Co. Ltd., London: xix + 588 pp., illustr.

Beckinsale, R. P. (1972). Clearwater erosion: its geographical and international significance. *International Geography*, **2**, 1244–6.

Begg, G. W. (1973). The biological consequences of discharge above and below Kariba Dam. Pp. 421–30 in *Transactions of the Eleventh International Congress of Large Dams, Madrid, Spain*, Vol. 1. International Commission on Large Dams, Paris: xxviii + 974 pp., illustr.

Beiningen, K. T., & Ebel, W. J. (1970). Effect of John Day Dam on discharged nitrogen concentrations and salmon in the Columbia River. *Transactions of the American Fisheries Society*, **4**, 664–71.

Bell, H. S. (1942). *Stratified Flow in Reservoirs and its Use in Prevention of Silting.* US Department of Agriculture Miscellaneous Publication No. **491**, Sept.*

Bennett, G. W. (1970). *Management of Lakes and Ponds*. Van Nostrand Reinhold, New York: xv + 375 pp., illustr.

Bentley, W. W., & Raymond, H. L. (1976). Delayed migrations of yearling Chinook Salmon since completion of lower Monumental and Little Goose Dams on the Snake River. *Transactions of the American Fisheries Society*, **105**, 422–4.

Bergman, D. L., & Sullivan, C. W. (1963). Channel changes on Sandstone Creek near Cheyenne, Oklahoma. *US Geological Survey, Professional Paper* **475c**, 145–8.

Berkes, F. (1981). Some environmental and social impacts of the James Bay Hydro-electric Project, Canada. *Journal of Environmental Management*, **12**, 157–72.

Beschta, R. L. (1978). Long-term patterns of sediment production following road construction and logging in the Oregon Coast Range. *Water Resources Research*, **14** (6), 1011–6.

Beschta, R. L. (1982). Comment on 'Stream System Evaluation with Emphasis on Spawning Habitat for Salmonids' by Mostafa, A., Shirazi, & Wayne K. Seim. *Water Resources Research*, **18** (4), 1292–5.

Beschta, R. L., Jackson, W. L., & Knoop, K. D. (1981). Sediment transport during a controlled reservoir release. *Water Resources Bulletin*, **17** (4), 635–41.

Bettess, R., & White, W. R. (1981). Mathematical simulation of sediment movement in streams. *Proceedings of the Institution of Civil Engineers*, **71** (2), 879–92.

Binns, N. A., & Eiserman, F. M. (1979). Quantification of fluvial trout habitat in Wyoming. *Transactions of the American Fisheries Society*, **108** (3), 215–28.

Bishop, J. E., & Hynes, H. B. N. (1969). Downstream drift of the invertebrate fauna in a stream ecosystem. *Archiv für Hydrobiologie*, **66** (1), 56–90.

Bishop, K. A., & Bell, J. D. (1978). Observations on the fish fauna below Tallowa Dam (Shoalhaven R., New South Wales) during river flow stoppages. *Australian Journal of Marine and Freshwater Research*, **29**, 543–9.

Biswas, A. K. (1978). *United Nations Water Conference: Summary and Main Documents.* (Water Development, Supply and Management, Vol. 2.) Pergamon Press, Oxford, England, UK: xvii + 217 pp.

270

Biswas, A. K. (Ed.). (1983). *Long-distance Water Transfer: A Chinese Case Study and International Experiences*. (Water Resources Series, Volume 3.) Tycooly International, Dun Laoghaire, Ireland: xvi + 417 pp., illustr.

Black, P. E. (1970). The watershed in principle. *Water Resources Bulletin*, 6 (2), 153–62.

Blake, R. F. (1977). The effect of the impoundment of Lake Kainji, Nigeria, on the indigenous species of mormyrid fishes. *Freshwater Biology*, 7 (1), 37–42.

Blench, T. (1972). Morphometric changes. Pp. 287–308 in *River Ecology and Man* (Ed. R. T. Oglesby, C. A. Carlson, & J. A. McCann). Academic Press, New York, USA: xvii + 465 pp., illustr.

Blezzard, N., Crann, H. H., Iremonger, D. J., & Jackson, E. (1971). Conservation of the environment by river regulation. *Yearbook of the Association of River Authorities*, 1971, pp. 70–115.

Blum, J. L. (1960). Algae population in flowing waters. *Special Publication of the Pymatuning Laboratory for Field Biology, University of Pittsburgh*, USA, 2, 11–22.

Boehmer, R. J. (1973). Ages, lengths, and weights of paddlefish caught in Gavins Point Dam tailwaters, Nebraska. *Proceedings of the South Dakota Academy of Science*, 52, 104–46.

Boesch, D. F., Diaz, R. J., & Virnstein, R. W. (1976). Effects of Tropical Storm Agnes on soft-bottom macro-benthic communities of the James and York Estuaries and the Lower Chesapeake Bay. *Chesapeake Scientist*, 17, 246–59.

Boesch, D. F., & Rosenberg, R. (1981). Response to stress in marine benthic communities. Pp. 179–200 in *Stress Effects on Natural Ecosystems* (Ed. G. W. Barrett, & R. Rosenberg). (Environmental Monographs & Symposia, Convener & Gen. Ed. N. Polunin). John Wiley & Sons Ltd, Chichester–New York–Brisbane–Toronto–Singapore: xvii + 305 pp., illustr.

Boles, G. L. (1979). *Macroinvertebrate colonization of Replacement Substrate below a Hypolimnial Release Reservoir*. California Department of Water Resources Report, Red Bluff, California: 46 pp.

Boles, G. L. (1980a). *Differential Selection of Smolting Steelhead Trout* (Salmo gairdneri) *at Trinity River Hatchery, California*. California Department of Water Resources. Northern District, Information Report 77–2: iii + 13 pp.

Boles, G. L. (1980b). *Macroinvertebrate Abundance and Diversity as influenced by Substrate Size in the Trinity River*. California Department of Water Resources, Northern District, Information Report, January: 9 pp.

Bombówna, M., Bucka, H., & Huk, V. (1978). Impoundments and their influence on the rivers studied by bioassays. *Verhandlungen Internationale Vereinigung für Theoretische und Angewandte Limnologie*, 20, 1629–33.

Bondurant, D. C., & Livesey, R. H. (1973). Reservoir sedimentation studies. Pp. 364–7 in *Man-Made Lakes: Their Problems and Environmental Effects* (Ed. W. C. Ackermann, G. F. White, & E. B. Worthington). Geophysical Monograph 17, American Geophysical Union, Washington, DC: xv + 847 pp., illustr.

Bonetto, A. A., Yuan, E. C., Pignalberi, C., & Oliveros, O. (1971). Informaciones complementaries sobre migrociones de peces contenidas en las cuencas temporarieas de su valle de inundación. *Physis* (Buenos Aires), 29, 213–23.

Boon, P. J. (1979). Adaptive strategies of *Amphipsyche* larvae (Trichoptera: Hydropsychidae) downstream of a tropical impoundment. Pp. 237–56 in *The Ecology of Regulated Streams* (Ed. J. V. Ward, & J. A. Stanford). Plenum Press, New York: xx + 398 pp., illustr.

Borland, W. L. (1951). Unpublished data in the files of the Bureau of Reclamation, Denver, Colorado. Described in Brune, G. M. (1953) (q.v.).

Borland, W. M., & Miller, C. R. (1960). Sediment problems of the Lower Colorado River. *Proceedings of the American Society of Civil Engineers, Journal of the Hydraulics Division*, HY4, 61–88.

Bormann, F. H., & Likens, G. E. (1969). The watershed–ecosystem concept and studies of nutrient cycles. Pp. 49–76 in *The Ecosystem Concept in Natural Resources*

Management (Ed. G. M. van Dyne). Academic Press, New York & London: xii + 383 pp., illustr.

Bovee, K. D. (1978*a*). The incremental method of assessing habitat potential for coolwater species, with management implications. *American Fisheries Society Special Publication* 11, 340–6.

Bovee, K. D. (1978*b*). *Probability-of-use criteria for the Family Salmonidae*. Information Paper No. 4, Fish & Wildlife Service, Co-operative Instream Flow Service Group, Fort Collins, Colorado, USA: 80 pp.

Bowmaker, A. P. (1976). The physico-chemical limnology of Mwenda River Mouth, Lake Kariba. *Archiv für Hydrobiologie*, 77, 66–108.

Brádka, J. (1966). Bilance sedimentu v údolni nádrźi. *Vodni Hospodârstvi*, 16, 1–14.

Brádka, J., & Rehacková, V. (1964). Mass destruction of fish in the Slapy Reservoir in winter 1962–63. *Vodni Hospodârstvi*, 14, 451–2.

Branson, B. A. (1974). The American paddlefish: origins of distress. *National Parks Conservation Magazine*, 48 (1), 21–3.

Bray, D. I., & Kellerhals, R. (1979). Some Canadian examples of the response of rivers to man-made changes. Pp. 351–72 in *Adjustments of the Fluvial System* (Ed. D. D. Rhodes, & G. P. Williams). Kendall/Hunt, Dubuque, Iowa: ix + 372 pp., illustr.

Brett, J. R. (1957). Salmon research and hydro-electric power development. *Bulletin of the Fisheries Research Board, Canada*, 114, 1–26.

Briggs, J. C. (1948). The quantitative effects of a dam upon the bottom fauna of a small California stream. *Transactions of the American Fisheries Society*, 78, 70–81.

Bronfman, A. M. (1977). The Azov Sea water economy and ecological problems: investigation and possible solutions. Pp. 39–58 in *Environmental Effects of Complex River Development* (Ed. G. F. White). Westview, Boulder, Colorado: xvi + 416 pp., illustr.

Bronowski, J. (1973). *The Ascent of Man*. British Broadcasting Corporation, London, England, UK: 448 pp., illustr.

Brook, A. J., & Rzóska, J. (1954). The influence of the Gebel Aulyia Dam on the development of Nile plankton. *Journal of Animal Ecology*, 23, 101–15.

Brook, A. J., & Woodward, W. B. (1956). Some observations on the effects of water inflow and outflow on the plankton of small lakes. *Journal of Animal Ecology*, 25 (1), 22–5.

Brooker, M. P. (1981). The impact of impoundments on the downstream fisheries and general ecology of rivers. *Advances in Applied Biology*, VI, 91–152.

Brooker, M. P., & Hemsworth, R. J. (1978). The effect of the release of an artificial discharge of water on invertebrate drift in the R. Wye, Wales. *Hydrobiologia*, 59 (3), 155–63.

Brooks, N. H., & Koh, R. C. Y. (1969). Selective withdrawal from density stratified reservoirs. *Journal of the Hydraulics Division, ASCE*, 95, HY4, 1369–400.

Brown, A. W. A., & Deom, J. D. (1973). Summary: Health aspects of man-made lakes. Pp. 755–64 in *Man-Made Lakes: Their Problems and Environmental Effects* (Ed. W. C. Ackermann, G. F. White, & E. B. Worthington). Geophysical Monograph 17, American Geophysical Union, Washington, DC: xv + 847 pp., illustr.

Brown, C. B. (1944). Sedimentation in reservoirs. *Transactions of the American Society of Civil Engineers*, 109, 1085.

Brown, C. B. (1950). Sediment transportation. Chapt. XII in *Engineering Hydraulics* (Ed. H. Rouse). John Wiley & Son, New York: 797 pp., illustr.

Bruk, S., Miloradov, V., & Milišić, V. (1981). Discussion of 'Sedimentation in Iron Gates Reservoir on the Danube', by Vlad Focsa. *Journal of the Hydraulics Division, ASCE*, 107, HY12, 1746–9.

Brune, G. M. (1953). Trap efficiency of reservoirs. *Transactions of the American Geophysical Union*, 34 (3), 407–18.

Buma, P. G., & Day, J. C. (1977). Channel morphology below reservoir storage projects. *Environmental Conservation*, 4 (4), 279–84.

Burns, F. L. (1977). Localised destratification of large reservoirs to control discharge temperatures. *Progress in Water Technology*, **9**, 59–63.

Burton, G. W., & Odum, E. P. (1945). The distribution of stream fish in the vicinity of Mountain Lake, Virginia. *Ecology*, **26**, 182–94.

Bush, A. (1976). Is the Trinity River dying? Pp. 112–22 in *Instream Flow Needs* Vol. 11 (Ed. J. F. Orsborn, & C. H. Allman). American Fisheries Society, Bethesda, Maryland: vi + 657 pp., illustr.

Butcher, A. D. (1967). A changing aquatic fauna in a changing environment. *IUCN Publication* (new series), **9**, 197–219.

Butorin, N. V., Vendrov, S. L., Dyakonov, K. N., Reteyum, A. Yu, & Romanenko, V. I. (1973). Effect of the Rybinsk Reservoir on the surrounding area. Pp. 242–50 in *Man-Made Lakes: Their Problems and Environmental Effects* (Ed. W. C. Ackermann, G. F. White, & E. B. Worthington). Geophysical Monograph 17, American Geophysical Union, Washington, DC: ix + 847 pp., illustr.

Buttling, S., & Shaw, T. L. (1973). Predicting the rate and patterns of storage loss in reservoirs. Pp. 565–80 in *Transactions of the Eleventh International Congress of Large Dams, Madrid, Spain*. Vol. 1. International Commission on Large Dams, Paris: xxviii + 974 pp., illustr.

Cadwallader, P. L. (1977). J. O. Langtry's 1949–50 Murray River investigation. *Fisheries and Wildlife Papers, Victoria*, **13**, 70 pp.

Cadwallader, P. L. (1978). Some causes of the decline in range and abundance of native fish in the Murray–Darling river system. *Proceedings of the Royal Society of Victoria*, **90**, 211–24.

Cairns, J. (1974). Indicator Species *vs.* the Concept of Community Structure as an Index of Pollution. *Water Resources Bulletin*, **10**, 338–47.

Cairns, J., & Dickson, K. L. (1977). Recovery of streams from spills of hazardous materials. Pp. 24–42 in *Recovery and Restoration of Damaged Ecosystems* (Ed. J. Cairns, K. L. Dickson, & E. E. Herricks). University of Virginia Press, Charlottesville, Virginia: x + 531 pp., illustr.

Cairns, J., Dickson, K. L., & Herricks, E. E. (Ed.) (1977). *Recovery and Restoration of Damaged Ecosystems*, University of Virginia Press, Charlottesville, Virginia: x + 531 pp., illustr.

Cambray, J. A., & Jubb, R. A. (1977). Dispersal of fishes via the Orange–Fish tunnel, S. A. *Journal of the Limnological Society of South Africa*, **3** (1), 33–5.

Canter, L. W. (1983). Impact studies of dams and reservoirs. *Water Power and Dam Construction*, **35** (7), 18–23.

Carpenter, K. E. (1928). *Life in Inland Waters*. Sidgwick & Jackson, London: 267 pp., illustr.

Chandler, D. C. (1937). Fate of tropical lake plankton in streams. *Ecological Monographs*, **7**, 455–79.

Chandler, D. C. (1939). Plankton entering the Huron River from Portage and Base-Line Lake, Michigan. *Transactions of the American Microscopy Society*, **58**, 24–41.

Changming, L., & Dakang, Z. (1983). Impact of south-to-north water transfer upon the natural environment. Pp. 169–80 in *Long-distance Water Transfer* (Ed. A. K. Biswas). Tycooly International, Dun Laoghaire, Ireland: xvi + 471 pp., illustr.

Chaston, I. (1968). Endogenous activity as a factor in invertebrate drift. *Archiv für Hydrobiologie*, **64**, 324–34.

Cheret, I. (1973). General report of the consequences on the environment of building dams. Pp. 1–104 in *Transactions of the Eleventh International Congress of Large Dams, Madrid, Spain*, Vol. IV. International Commission on Large Dams, Paris: ix + 1072 pp., illustr.

Chikova, V. M. (1974). Species and age composition of fishes in the lower reach (downstream) of the V. I. Lenin Hydroelectric Station. Pp. 185–92 in *Biological*

273

and *Hydrological Factors of Local Movements of Fish in Reservoirs* (Ed. B. S. Kuzin). Amerind Publishing Co., New Delhi: iv + 389 pp., illustr.

Chorley, R. J. (1969). The drainage basin as the fundamental geomorphic unit. Pp. 77–100 in *Water, Earth and Man* (Ed. R. J. Chorley). Methuen & Co., London: xix + 588 pp., illustr.

Chow, V. T. (Ed.) (1964). *Handbook of Applied Hydrology: A Compendium of Water Resources Technology.* McGraw-Hill, New York: xiv + 29 sec., pp. var., illustr.

Churchill, M. A. (1947). Effect of density currents upon raw water quality. *Journal of the American Water Works Association,* **39**, 357–60.

Churchill, M. A. (1957). Effects of storage impoundments on water quality. *Journal of the Sanitary Engineering Division, ASCE,* **83**, SA1, Feb., 48 pp.

Churchill, M. A., & Nicholas, W. R. (1967). Effects of impoundments on water quality. *Journal of the Sanitary Engineering Division, ASCE,* **93**, SA6, Dec., pp. 73–90.

Chutter, F. M. (1963). Hydrobiological studies on the Vaal River in the Vereeniging area, Part 1: Introduction, water chemistry and biological studies on the fauna of habitats other than muddy bottom sediments. *Hydrobiologia,* **21**, 1–65.

Chutter, F. M. (1968). On the ecology of the fauna of stones in the current in a South African river supporting a very large *Simulium* (Diptera) population. *Journal of Applied Ecology,* **5**, 531–61.

Clark, M. J. (1978). Geomorphology in coastal-zone management. *Geography,* **63**, 273–82.

Coad, B. W. (1980). Environmental change and the freshwater fishes of Iran. *Biological Conservation,* **19** (1), 51–80.

Collet, K. O. (1978). The present salinity position in the River Murray Basin. *Proceedings of the Royal Society of Victoria,* **90**, 111–23.

Collins, G. B. (1976). Effects of dams on Pacific Salmon and Steelhead Trout. *Marine Fisheries Review,* **38** (1), 39–46.

Copeland, B. J. (1970). Estuarine classification and response to disturbances. *Transactions of the American Fisheries Society,* **99**, 826–35.

Cornwallis, L. (1968). Some notes on the wetlands of the Niriz basin in S. W. Iran. (Proceedings of the Technical Meeting on Wetland Conservation, Ankara–Bursa–Istanbul, 9–16 October 1967.) *Publications of the International Union for Conservation of Nature and Natural Resources,* Vol. **12**, pp. 152–60.

Coskun, E., Claborn, B. J., & Moore, W. L. (1969). Application of continuous accounting techniques to evaluate the effects of small structures on Mukewater Creek, Texas. Pp. 79–99 in *Effects of Watershed Changes on Streamflow* (Ed. W. L. Moore, & C. W. Morgan). University of Texas Press, Austin, Texas: xi + 289 pp., illustr.

Costa, J. E. (1974). Response and recovery of a piedmont watershed from tropical storm Agnes, June 1972. *Water Resources Research,* **10**(1), 106–12.

Coutant, C. C. (1962). The effect of a heated water effluent upon the macroinvertebrate riffle fauna of the Delaware River. *Proceedings of the Pennsylvania Academy of Science,* **36**, 38–71.

Coutant, C. C. (1963). Stream plankton above and below Green Lake Reservoir. *Proceedings of the Pennsylvania Academy of Science,* **37**, 122–6.

Cowell, B. C. (1967). The Copepoda and Cladocera of a Missouri River reservoir: a comparison of sampling in the reservoir and the discharge. *Limnology and Oceanography,* **12**, 125–36.

Cramer, F. K., & Oligher, R. C. (1964). Passing fish through hydraulic turbines. *Transactions of the American Fisheries Society,* **93**, 243–59.

Crisp, D. T. (1977). Some physical and chemical effects of the Cow Green (upper Teesdale) impoundment. *Freshwater Biology,* **7**, 109–20.

274

Croome, R. L., Tyler, P. A., Walker, K. F., & Williams, W. D. (1976). A limnological survey of the River Murray in the Albury–Wodonga area. *Search*, **7**(1), 14–7.
Culp, J. M., & Davies, R. W. (1982). Analysis of longitudinal zonation and the river continuum concept in the Oldman–South Saskatchewan River system. *Canadian Journal of Fisheries and Aquatic Sciences*, **39**, 1258–66.
Cummins, K. W. (1975). The ecology of running waters: theory and practice. Pp. 277–93 in *Proceedings of the Sandusky River Basin Symposium*. International Joint Committee on the Great Lakes, Heidelberg College, Tiffin, Ohio: 346 pp.
Cummins, K. W., Klug, J. J., Wetzel, R. G., Peterson, R. C., Suberkroff, K. F., Manny, B. A., Wuycheck, J. C., & Howard, F. D. (1972). Organic enrichment with leaf leachate in experimental lotic ecosystems. *BioScience*, **22**, 719–22.
Cummins, W. A., & Potter, H. R. (1972). Rate of erosion in the catchment of Cropston Reservoir, Charnwood Forest, Leicestershire. *Mercian Geologist*, **6**, 149–57.
Curry, R. R. (1972). Rivers—a geomorphic and chemical overview. Pp. 9–32 in *River Ecology and Man* (Ed. R. T. Oglesby, C. A. Carlson, & J. A. McCann). Academic Press, New York: xvii + 465 pp., illustr.
Cushing, C. E. (1963). Filter-feeding insect distribution and planktonic food in the Montreal River. *Transactions of the American Fisheries Society*, **92**, 216–9.
Cyberski, J. (1973). Accumulation of debris in water storage reservoirs of Central Europe. Pp. 359–63 in *Man-Made Lakes: Their Problems and Environmental Effects* (Ed. W. C. Ackermann, G. F. White, & E. B. Worthington). Geophysical Monograph 17, American Geophysical Union, Washington, DC, USA: xx + 847 pp., illustr.
Daget, J. (1960). Effets du barrage de Markala sur les migrations des poissons dans le Moyen-Niger. Pp. 352–6 in *Proceedings of the Seventh Technical Meeting, IUCN/ FAO, Athens, Sept. 1958, Vol. 4*, IUCN, Brussels: 4 vols, pp. var., illustr.
Davidson, F. A., Vaughn, E., Hutchinson, S. J., & Pritchard, A. L. (1943). Factors influencing the upstream migration of the Pink Salmon (*Oncorhynchus gorbuscha*). *Ecology*, **24** (2), 149–68.
Davies, B. R. (1975). Cabora Bassa hazards. *Nature* (London), **254**, 477–8.
Davies, B. R. (1979). Stream regulation in Africa: A review. Pp. 113–42 in *The Ecology of Regulated Streams* (Ed. J. V. Ward, & J. A. Stanford). Plenum Press, New York, USA: xi + 398 pp., illustr.
Davies, B. R., Hall, A., & Jackson, P. B. N. (1975). Some ecological aspects of the Cabora Bassa Dam. *Biological Conservation*, **8**, 189–201.
Dawley, E. M., Schiewe, M., & Monk, B. (1976). Effects of long-term exposure to supersaturation of dissolved atmospheric gases on juvenile Chinook Salmon and Steelhead Trout in deep and shallow test-tanks. Pp. 1–10 in *Gas Bubble Disease* (Ed. D. H. Fickeisen, & N. J. Schneider). National Technological Information Service, US Department of Commerce, Springfield, Virginia: vii + 132 pp., illustr.
Décamps, H., & Casanova-Batut, T. (1978). Les matières en suspension et la turbidité de l'eau dans la rivière Lot. *Annales de Limnologie*, **14**, 59–84.
Décamps, H., Capblancq, J., Casanova, H., & Tourenq, J. M. (1979). Hydrobiology of some regulated rivers in the South-West of France. Pp. 273–88 in *The Ecology of Regulated Streams* (Ed. J. V. Ward, & J. A. Stanford). Plenum Press, New York, USA: xi + 398 pp., illustr.
Decoursey, D. G. (1975). Implications of flood-water retarding schemes. *Transactions of the American Society of Agricultural Engineers*, **18**, 897–904.
Dendy, F. E., Champion, W. A., & Wilson, R. B. (1973). Reservoir sedimentation surveys in the United States. Pp. 349–58 in *Man-Made Lakes: Their Problems and Environmental Effects*. (Ed. W. C. Ackermann, G. F. White, & E. B. Worthington), Geophysical Monograph 17, American Geophysical Union, Washington, DC, USA: xv + 847 pp., illustr.

Dendy, J. S., & Stroud, R. H. (1949). The dominating influence of Fontana Reservoir on temperature and dissolved oxygen in the Little Tennessee River and its impoundments. *Journal of the Tennessee Academy of Science.* **24**, 41–51.

Denisova, A. I. (1978). Long-term changes in runoff of nutrients and organic matter due to control of flow of the Dnieper. *Hydrobiological Journal,* **14** (2), 60–8.

Department of Water Resources—*see D.W.R.*

Derby, R. L. (1956). Chlorination of deep reservoirs for taste and odor control. *Journal of the American Water Works Association,* **48** (7), 775–80.

Diacon, A., Constantinescu, C., Dragu, S., & Mihai, R. (1973). Modifications physiques dans le milieu environmant, determinées par des barrages et des usines, en Roumanie. Pp. 445–62 in *Transactions of the Eleventh International Congress of Large Dams, Madrid, Spain,* Vol 1. International Commission on Large Dams, Paris, France: xxviii + 974 pp., illustr.

Dickson, L. W. (1975). Hydroelectric development of the Nelson River system in Northern Manitoba. *Journal of the Fisheries Research Board of Canada,* **32** (1), 10–16.

Din, S. H. Sharaf (1977). Effects of the Aswan High Dam on the Nile flood on the estuarine and coastal circulation pattern along the Mediterranean Egyptian coast. *Limnology and Oceanography,* **22** (2), 194–207.

Dittmar, H. (1955). Ein Sauerlandbach Untersuchungen an einem Viesen-Mittelgebirgsbach. *Archiv für Hydrobiologie,* **50**, 305–552.

Dolan, R., Howard, A., & Gallenson, A. (1974). Man's impact on the Colorado River in the Grand Canyon. *American Scientist,* **62**, 392–401.

Dominy, C. L. (1973). Recent changes in Atlantic Salmon (*Salmo salar*) in the light of environmental changes in the Saint John River, New Brunswick, Canada. Biological Conservation, **5** (2), 105–13.

Dorst, J. (1970). *Before Nature Dies.* Collins, London: 352 pp., illustr.

Douglas, I. (1967). Man, vegetation and the sediment yield of rivers. *Nature* (London), **215**, 925–8.

Douglas, I. (1968). The effects of precipitation chemistry and catchment area lithology on the quality of river water in selected catchments in Eastern Australia. *Earth Science Journal,* **2**, 126–44.

Douglas, I. (1972). The geographical interpretation of river water quality data. *Progress in Geography,* **4**, 1–61.

Dudley, R. G. (1974). Growth of *Tilapia* of the Kafue floodplain, Zambezi: Predicted effects of the Kafue Gorge Dam. *Transactions of the American Fisheries Society,* **103**, 281–91.

Dunne, T. (1978). Field studies of hillslope flow processes. Pp. 227–89 in *Hillslope Hydrology* (Ed. M. J. Kirkby). John Wiley & Sons, Chichester: xvi + 389 pp., illustr.

Duthie, H. C., & Ostrofsky, M. L. (1975). Environmental impact of the Churchill Falls (Labrador) Hydroelectric Project: a preliminary assessment. *Journal of the Fisheries Research Board of Canada,* **32**, 117–25.

D. W. R. (Department of Water Resources) (1978). *Grass Valley Creek: Sediment Control Study.* For the Trinity River Basin Fish and Wildlife Task Force, California Department of Water Resources, Northern District, Red Bluff, California: vii + 73 pp. + App., illustr.

Ebel, W. J. (1969). Supersaturation of nitrogen in the Columbia River and its affect on Salmon and Steelhead Trout. *Fisheries Bulletin,* **68** (1), 1–11.

Ebel, W. J., & Koski, C. H. (1968). Physical and chemical limnology of Brownlee Reservoir. *Fisheries Bulletin,* **67**, 295–335.

Ebel, W. J., & Raymond, H. L. (1976). Effect of atmospheric gas saturation on Salmon and Steelhead Trout on the Snake and Columbia Rivers. *Marine Fisheries Review,* **38**(7), 1–14.

Edwards, R. J. (1978). The effect of hypolimnion releases on fish distribution and species diversity. *Transactions of the American Fisheries Society,* **107**, 71–7.

Edwards, R. W., & Crisp, D. T. (1982). Ecological implications of river regulation in the UK. Pp. 843–65 in *Rivers: Fluvial Processes, Engineering & Management* (Ed. R. D. Hey, J. C. Bathurst, & J. Thorne). John Wiley & Sons, Chichester: 892 pp., illustr.

Efford, I. E. (1975). Environmental impact assessment and hydroelectric projects, hindsight and foresight in Canada. *Journal of the Fisheries Research Board of Canada,* **32**, 98–100 + appendix.

Egglishaw, H. J., & Mackay, D. W. (1967). A survey of the bottom fauna of streams in the Scottish Highlands, 3: Seasonal changes in the fauna of three streams. *Hydrobiologia,* **30**, 305–34.

Einstein, H. A. (1961). Needs in sedimentation. *Proceedings of the American Society of Civil Engineers, Journal of the Hydraulics Divison,* **HY2**, 1–8.

Einstein, H. A., & Chien, N. (1958). Discussion of 'Mechanics of streams with movable beds of fine sand' by N. H. Brooks. *Transactions of the American Society of Civil Engineers,* **123**, 526–94.

Elder, R. A., & Wunderlich, W. O. (1968). Evaluation of Fontana Reservoir Field Measurements. Pp. 221–86 in *Proceedings of the Speciality Conference on Current Research into the Effects of Reservoirs on Water Quality* (Ed. R. A. Elder, P. A. Krenkel, & E. L. Thackston). Technical Report 17, Department of Environment & Water Resources Engineering, Vanderbilt University, Nashville, Tennessee: vii + 390 pp. illustr.

Eliseev, A. L., & Chikova, W. M. (1974). Conditions of fish production in the lower reach (downstream) of the V. I. Lenin Volga hydro-electric station. Pp. 193–200 in *Biological and Hydrological Factors of Local Movements of Fish in Reservoirs* (Ed. B. S. Kuzin). Amerind Publishing Co., New Delhi: iv + 389 pp., illustr.

Elliot, J. M. (1967). The life-histories and drifting of the Plecoptera and Ephemeroptera in a Dartmoor stream. *Journal of Animal Ecology,* **38** (1), 19–32.

Elliot, R. A., & Engstrom, L. R. (1959). Controlling floods on the Tennessee. *Civil Engineering,* **29**, 60–3.

El-Shamy, F. M. (1977). Environmental impacts of hydro-electric power plants. *Journal of the Hydraulics Division, Proceedings of the American Society of Civil Engineers,* **HY9**, 1007–20.

El-Zarka, S. El-Din (1973). Kainji Lake, Nigeria. Pp. 197–219 in *Man-Made Lakes: Their Problems and Environmental Effects* (Ed. W. C. Ackermann, G. F. White, & E. R. Worthington). Geophysical Monograph 17, American Geophysical Union, Washington, DC: xv + 847 pp., illustr.

Entz, B. (1976). Lake Nasser and Lake Nubia. Pp. 271–98 in *The Nile: Biology of an Ancient River* (Ed. J. Rzóska). (Monographiae Biologicae, 29.) Dr W. Junk Publishing Co., The Hague: xix + 417 pp., illustr.

Esch, G. W., & McFarlane, R. W. (1976). *Thermal Ecology II.* ERDA Symposium Series 40, Conference–750425, National Technical Information Service, Springfield, Virginia, USA: 323 pp., illustr.

Eschmeyer, R. W., & Smith, C. G. (1943). Fish spawning below Norris Dam. *Journal of the Tennessee Academy of Science,* **18**, 4–5

Eustis, R. B., & Hillen, R. H. (1954). Stream sediment removal by controlled reservoir releases. *Progressive Fish Culturist,* **16**, 30–5.

Ewer, D. W. (1966). Biological investigations on the Volta Lake, May 1964 to May 1965. Pp. 21–30 in *Man-Made Lakes.* Synoposia of the Institute of Biology No. 15 (Ed. R. H. Lowe-McConnell). Academic Press, London: xiii + 218 pp.

FAO—*see* Food and Agriculture Organization.

Fedorev, B. G. (1969). Erosion below hydro-electric dams. *Trudy Tsnievt,* No. 58, pp. 1*.

Fels, E., & Keller, R. (1973). World register on man-made lakes. Pp. 43–9 in *Man-Made Lakes: Their Problems and Environmental Effects* (Ed. W. C. Ackermann, G. F. White, & E. B. Worthington). Geophysical Monograph 17 American Geophysical Union, Washington, DC: xv + 847 pp., illustr.

Fenglan, Y., & Wenkai, W. (1983). Analysis of storage for the regulation of surface water in the Huang-Huai-Hai Plain for the south-to-north water transfer. Pp. 289–98 in *Long-distance Water Transfer* (Ed. A. K. Biswas). Tycooly International, Dun Laoghaire, Ireland: xvi + 417 pp., illustr.

Fiala, L. (1966). Akinetic spaces in water supply reservoirs. *Verhandlungen Internationale Vereinigung für Theoretische und Angewandte Limnologie*, **16**, 685–92.

Filatova, T. N., & Kafejarv, T. O. (1973). Some results of 20 year network observations on currents on Tschudsko-Pskovskoe Lake, USSR. Pp. 316–5 in *Man-Made Lakes: Their Problems and Environmental Effects* (Ed. W. C. Ackermann, G. F. White, & E. B. Worthington). Geophysical Monograph 17, American Geophysical Union, Washington, DC: xv + 847 pp., illustr.

Fisher, S. G., & LaVoy, A. (1972). Differences in littoral fauna due to fluctuating water levels below a hydroelectric dam. *Journal of the Fisheries Research Board of Canada*, **29**, 1472–6.

Fisher, S. G., & Likens, G. E. (1973). Energy flow in Bear Brook, New Hampshire: an integrative approach to ecosystem metabolism. *Ecological Monographs*, **42**, 421–39.

Fleming, G. (1969). Design curves for suspended load estimation. *Proceedings of the Institute of Civil Engineers*, **43**, 1–9.

Food and Agriculture Organization (cited as FAO) (1968). *First and Second Group Fellowship Study-tours on Inland Fisheries Research, Management and Fish Culture in the USSR*. TA 2547, Food and Agriculture Organization, Rome: 183 pp.

Ford, M. E. (1963). Air injection for control of reservoir limnology. *Journal of the American Water Works Association*, **55**, 267–74.

Foulger, T., & Petts, (1984). Water-quality implications of artificial flow fluctuations in regulated rivers. *The Science of the Total Environment*, (in press).

Fraley, J. J. (1979). Effects of elevated stream temperatures below a shallow reservoir on a cold-water macroinvertebrate fauna. Pp. 257–72 in *The Ecology of Regulated Streams* (Ed. J. V. Ward, & J. A. Stanford). Plenum Press, New York: xi + 398 pp. illustr.

Fraser, J. C. (1972). Regulated discharge and the stream environment. Pp. 26–85 in *River Ecology and Man* (Ed. R. T. Oglesby, C. A., Carlson, & J. A. McCann). Academic Press, New York: xvii + 465 pp., illustr.

Frederiksen, Kamine & Associates Inc. (1979). *Proposed Trinity River Basin Fish and Wildlife Management Program*. Main Report (draft) for the US Bureau of Reclamation and Trinity River Basin Fish and Wildlife Task Force. Frederiksen, Kamine and Associates Inc., Sacramento, California: xxi + 344 pp., illustr. + 3 app.

Frey, D. G. (1967). Reservoir research—objectives and practices with an example from the Soviet Union. Pp. 26–36 in *Reservoir Fishery Resources Symposium*. American Fisheries Society, Washington, DC: viii + 569 pp., illustr.

Frickel, D. G. (1972). *Hydrology and the Effects of Conservation Structures, Willow Creek Basin, Valley Country, Montana, 1954–68*. US Geological Survey Water Supply Paper, 1532–G, 35pp., illustr.

Frith, H. J. (1977). *Waterfowl in Australia*. Reed, Sydney, Australia: xii + 328 pp., illustr.

Froehlich, C. G., Arcifa-Zago, M. S., & Carvalho, M. A. J. De (1978). Temperature and oxygen stratification in Americana Reservoir, State of São Paulo, Brazil. *Verhandlungen Internationale Vereinigung für Theoretische und Angewandte Limnologie*, **20**, 1710–9.

278

Fruh, E. G., & Clay, H. M. (1973). Selective withdrawal as a water quality management tool for southwestern impoundments. Pp. 335–41 in *Man-Made Lakes: Their Problems and Environmental Effects* (Ed. W. C. Ackermann, G. F. White, & E. R. Worthington). Geophysical Monograph 17, American Geophysical Union, Washington, DC: xv + 847 pp., illustr.

Frye, J. C. (1971). A geologist views the environment. *Illinois State Geological Survey. Environmental Geology Notes*, **42**, 9.

Furness, H. D. (1978). *Ecological Studies on the Pongola River Floodplain*. Working Document IV, Workshop on Man and the Pongolo Floodplain. C.I.S.R., Pietermaritzburg, South Africa, No. 14/106/7C.*

Gamble, D. J. (1979). Destruction of insignificant increments: Arctic offshore development the circumpolar challenge. *Northern Perspectives*, **7** (6), 1–4.

Ganapati, S. V. (1973). Man-made lakes in South India. Pp. 65–73 in *Man-made Lakes: Their Problems and Environmental Effects* (Ed. W. C. Ackermann, G. F. White, & E. B. Worthington). Geophysical Monograph 17, American Geophysical Union, Washington, DC: xv + 847 pp., illustr.

Garton, J. E., Rice, C. E., & Steichen, J. M. (1976). Modification of reservoir water quality by artificial destratification. *Annals of the Oklahoma Academy of Science*, **5**, 47–56.

Geen, G. H. (1974). Effects of hydroelectric development in Western Canada on aquatic ecosystems. *Journal of the Fisheries Research Board of Canada*, **31**, 913–27.

Geen, G. H. (1975). Ecological consequences of the proposed Moran Dam on the Fraser River. *Journal of the Fisheries Reseach Board of Canada*, **32**, 126–35.

Gerasimov, I. P., & Gindin, A. M. (1977). The problem of transferring runoff from northern and Siberian rivers to the arid regions of the European USSR, Soviet Central Asia, and Kazakhstan. Pp. 59–70 in *Environmental Effects of Complex River Development* (Ed. G. F. White). Westview, Boulder, Colorado: xvi + 417 pp., illustr.

Gibbs, R. (1970). Mechanisms controlling world water chemistry. *Science*, **170**, 1088–90.

Gilbert, C. R., & Sauer, S. P. (1970). *Hydrologic Effects of Floodwater Retarding Structures, Garza–Little Elm Reservoirs, Texas*. US Geological Survey Water Supply Paper 1984: 95 pp., illustr.

Gill, D. (1971). Damming the Mackenzie: A theoretical assessment of the long-term influence of river impoundment on the ecology of the Mackenzie River Delta. Pp. 204–22 in *Proceedings of the Peace–Athabaska Delta Symposium*. Water Resources Centre, University of Alberta, Edmonton, Alberta: xvi + 359 pp., illustr.

Gill, M. A. (1968). River-bed degradation below dams. *Proceedings of the American Society of Civil Engineers, Journal of the Hydraulics Division*, **HY2**, 593–5.

Gilvear, D. J. & Petts, G. E. (1985). Turbidity and suspended solids variations downstream of a regulating reservoir. *Earth Surface Processes and Landforms*, (in press).

Gnilka, A. (1975). Some chemical and physical aspects of Center Hill Reservoir. *Journal of the Tennessee Academy of Science*, **50** (1), 7–11.

Godoy, M. P. de (1975). *Piexes do Brasil Suborden Characoides Bacia do Rio Mogi Guassu*. Piracicaba, Editoria Franciscam, São Paulo, Brazil: 4 vols. pp. var.

Golterman, H. L. (1975). *Physiological Limnology*. Elsevier Scientific Publishing Co., New York: 489 pp.

Goodman, D. (1975). The theory of diversity–stability relationships in ecology. *Quarterly Review of Biology*, **50**, 237–66.

Gore, J. A. (1977). Reservoir manipulations and benthic macroinvertebrates in a prairie river. *Hydrobiologia*, **55**, 113–23.

Gore, J. A. (1978). Technique for predicting instream flow requirements of benthic macroinvertebrates. *Freshwater Biology,* **8**, 141–51.

Gore, J. A. (1980). Ordinational analysis of benthic communities upstream and downstream of a prairie storage reservoir. *Hydrobiologia,* **69**, 33–44.

Gore, J. A., & Judy, R. D. (1981). Predictive models of benthic macroinvertebrate density for use in instream flow studies and regulated flow management. *Canadian Journal of Fisheries and Aquatic Science,* **38**, 1363–70.

Gottschalk, L. C. (1948). Analysis and use of reservoir sedimentation data. Pp. 131–41 in *Proceedings of the Federal Inter-Agency Sedimentation Conference,* Washington, USA, January 1948.*

Gottschalk, L. C. (1962). Effects of watershed protection measures on reduction of erosion and sediment damage in the United States. *International Association for Scientific and Hydrological Publications,* **59**, 426–47.

Gottschalk. L. C. (1964). Reservoir sedimentation. Pp. 17–5 & 17–6 in *Handbook of Applied Hydrology* (Ed. V. T. Chow). McGraw-Hill, New York: xiv + 29 sec, pp. var., illustr.

Goudie, A. (1981). *The Human Impact: Man's Role in Environmental Change.* Basil Blackwell, Oxford, England, UK: x + 316 pp., illustr.

Gould, H. R. (1951). Some quantitative aspects of Lake Mead turbidity currents. *Society of Economic Palaeontologists and Mineralogists Special Publication,* No. 2, pp. 34–52.

Gould, H. R. (1960). Turbidity currents: Comprehensive survey of sedimentation in Lake Mead, 1948–49. *US Geological Survey,* **295**, 201–7.

Gower, A. M. (Ed.) (1980). *Water Quality in Catchment Ecosystems.* John Wiley & Sons, Chichester: xii + 335 pp., illustr.

Graf, W. L. (1983). The hydraulics of reservoir sedimentation. *Water Power and Dam Construction,* **35** (4), 45–52.

Graf, W. L. (1977). The rate law in fluvial geomorphology. *American Journal of Science,* **272**(2), 178–91.

Graf, W. L. (1979). The development of montane arroyos and gullies. *Earth Surface Proceses,* **4**, 1–14.

Graf, W. L. (1980). The effect of dam closure on downstream rapids. *Water Resources Research,* **16**(1), 129–36.

Greenhalgh, G. (1980). *The Necessity for Nuclear Power.* Graham & Trotman, London: xi + 250 pp., illustr.

Gregory, K. J. (1976a). Drainage basin adjustments and Man. *Geographia Polonica,* **34**, 155–73.

Gregory, K. J. (1976b). Changing drainage basins. *Geographical Journal,* **142**, 238–47.

Gregory, K. J. (Ed.) (1977). *River Channel Changes.* John Wiley & Sons, Chichester: xi + 448 pp., illustr.

Gregory, K. J., & Ovenden, J. C. (1979). Drainage network volumes and precipitation in Britain. *Transactions of the Institute of British Geographers,* **4**(1), 1–11.

Gregory, K. J., & Park, C. C. (1974). Adjustment of river channel capacity downstream from a reservoir. *Water Resources Research,* **10**(4), 870–3.

Gregory, K. J., & Park, C. C. (1976). Stream channel morphology in northwest Yorkshire. *Revue de Geomorphologie Dynamique,* **xxv**(2), 63–72.

Grimard, Y., & Jones, H. G. (1982). Trophic upsurge in new reservoirs: a model for total phospherous concentrations. *Canadian Journal of Fisheries and Aquatic Science,* **39**, 1473–83.

Grimshaw, D. L., & Lewin, J. (1980a). Reservoir effects on sediment yield. *Journal of Hydrology,* **47**, 163–71.

Grimshaw, D. L., & Lewin, J. (1980b). Source identification for suspended sediments. *Journal of Hydrology,* **47**, 151–63.

280

Grizzle, J. M. (1981). Effects of hypolimnetic discharge on fish health below a reservoir. *Transactions of the American Fisheries Society*, 110, 29–43.

Grover, N. C., & Howard, C. S. (1938). The passage of turbid water through Lake Mead. *Transactions of the American Society of Civil Engineers*, 103, 720–*

Guy, P. R. (1981). River bank erosion in the mid-Zambezi Valley, downstream of Lake Kariba. *Biological Conservation*, 20, 199–212.

Gvelesiani, L. G., & Shmalkmzel, N. P. (1971). Studies of storage work silting of H. E. P. plants on mountain rivers and silt deposition fighting. *International Association of Hydraulic Research, 14th Congress*, Vol. 5, pp. 17–20.

Hack, J. T. (1957). *Studies of Longitudinal Stream Profiles in Virginia and Maryland*. US Geological Survey Professional Paper, 294B, 53 pp.

Hadley, R. F., & Schumm, S. A. (1961). Sediment sources and drainage basin characteristics in upper Cheyenne River basin. *US Geological Survey Water Supply Paper*, 1531-B, 137–96.

Hales, Z. L., Shindala, A., & Denson, K. H. (1970). Riverbed degradation prediction. *Water Resources Research*, 6, 549–56.

Haley, R., Burton, T., & Stone, T. (1976). *Potential Effects of Sediment Control Operations and Structures on Grass Valley Creek and Trinity River Fish and Wildlife*. California Department of Fish and Game Report, Red Bluff, California: 44 pp.

Halfon, E. (1979). Preview: theory in ecosystem analysis. Pp. 1–13 in *Theoretical Systems Ecology* (Ed. E. Halfon). Academic Press, London: xvi + 516 pp.

Hall, A., Valente, I., & Davies, B. R. (1977). The Zambezi River in Mozambique: the physico-chemical status of the Middle and Lower Zambezi prior to the closure of the Cabora Bassa Dam. *Freshwater Biology*, 7, 187–206.

Hall, A. E. (1976). The Forest Service Role in the Trinity River Basin Fish and Wildlife Restoration Program: functioning of a public land management in an interagency framework. Pp. 123–8 in *Instream Flow Needs* Volume 11 (Ed. J. F. Orsborn, & C. H. Allman). American Fisheries Society, Bethesda, Maryland: vi + 657 pp., illustr.

Hall, G. E. (1951). Fish population of the stilling basin below Wisher Dam. *Proceedings of the Oklahoma Academy of Science*, 30 (1949), 59–62.

Hall, G. E. (Ed.) (1971). *Reservoir Fisheries and Limnology*. Special Publication No. 8, American Fisheries Society, Washington, DC, USA: x + 511 pp., illustr.

Hall, J. R., & Pople, W. (1968). Recent vegetational changes in the lower Volta River. *Ghana Journal of Science*, 8, 24–9.

Hamilton, R., & Buell, J. W. (1976). *Effects of Modified Hydrology on Campbell River Salmonids*. Environment Canada, Fisheries and Marine Services Technical Report. Series PAC/T–76–20, 156 pp.

Hammad, H. Y. (1972). Riverbed degradation after closure of dams. *American Society of Civil Engineers, Journal of Hydraulics Division*, 98, 591–607.

Hammerton, D. (1972). The Nile River—a case history. Pp. 171–214 in *River Ecology and Man* (Ed. R. T. Oglesby, C. A. Carlson, & J. A. McCann). Academic Press, New York, USA: xvii + 465 pp., illustr.

Hamvas, F. (1963). Measurement of bed changes in the vicinity of river barrages. *International Association of Scientific & Hydrological Publications*, 62, 430–7.

Hannan, H. H. (1979). Chemical modifications in reservoir-regulated streams. Pp. 75–94 in *The Ecology of Regulated Streams* (Ed. J. V. Ward, & J. A. Stanford). Plenum Press, New York, USA: xi + 398 pp., illustr.

Hannan, H. H., & Broz, L. (1976). The influence of a deep-storage and an underground reservoir on the physico-chemical limnology of a permanent central Texan river. *Hydrobiologia*, 51, 43–63.

Hannan, H. H., & Dorris, T. C. (1970). Succession of a macrophyte community in a constant-temperature river. *Limnology and Oceanography*, 15, 442–53.

Hannan, H. H., Fuchs, I. R., & Whitenberg, D. C. (1979). Spatial and temporal patterns of temperature, alkalinity, dissolved O_2, and conductivity in an oligo-mesotrophic deep-storage reservoir in central Texas. *Hydrobiologia*, **66**, 209–21.

Hannan, H. H., Young, W. C., & Mayhew, J. J. (1973). Nitrogen and phosphorous in a stretch of the Guadalupe River with five mainstream impoundments. *Hydrobiologia*, **43**, 419–41.

Harrel, R. C., Davis, B. J., & Dorris, T. C. (1967). Stream order and species diversity of fishes in an intermittent Oklahoma stream. *American Midland Naturalist*, **78**, 428–36.

Harrel, R. C., & Dorris, T. C. (1968). Stream order, morphometry, physico-chemical conditions, and community structure of benthic macroinvertebrates in an intermittent stream system. *The American Midland Naturalist*, **80**(1), 220–51.

Harrison, A. S. (1950). Report on special investigation of bed sediment segregation in a degrading bed. *California Institute of Engineering Research Series*, **33**, 1.*

Hartman, M. O., Nicks, A. D., Rhoades, E. D., Schoof, R. R., & Allen, P. B. (1969). Field experiment on Washita River. Pp. 136–49 in *Effects of Watershed Changes on Streamflow* (Ed. W. L. Moore, & C. W. Morgan). University of Texas Press, Austin, Texas, USA: xi + 289 pp., illustr.

Hartman, R. T., & Himes, C. L. (1961). Phytoplankton from Pymatuning Reservoir in downstream areas of the Shenango River. *Ecology*, **42**, 180–3.

Harvey, A. M. (1969). Channel capacity and the adjustment of streams to hydrologic regime. *Journal of Hydrology*, **8**, 82–98.

Haslam, S. M. (1978). *River Plants*. Cambridge University Press, London, England, UK: xii + 396 pp., illustr.

Hathaway, G. A. (1948). Observations on channel changes degradation and scour below dams. Pp. 267–307 in *Report of the Second Meeting of the International Association of Hydraulic Research*. Stockholm, Sweden, 7–9 June 1948.*

Hauer, F. R., & Stanford, J. A. (1982). Ecological responses of hydropsychid caddisflies to stream regulation. *Canadian Journal of Fisheries and Aquatic Sciences*, **39**(9), 1235–42.

Haut, M. (1954). Biologie, profils en long et en travers des eaux courantes. *Bulletin, francaise Pisciculture*, **175**, 41–53.

Hawkes, H. A. (1975). River zonation and classification. Pp. 313–74 in *River Ecology* (Ed. B. A. Whitton). Blackwell Scientific Publications, Oxford: x + 725 pp., illustr.

Hawkins, C. A., & Sedell, J. R. (1981). Longitudinal and seasonal changes in functional organizations of macroinvertebrate communities in four Oregon streams. *Ecology*, **62**, 387–97.

Hayes, F. R. (1953). Artificial freshets and other factors controlling the ascent and population of Atlantic Salmon in the La Have River, Nova Scotia. *Bulletin of the Fisheries Research Board of Canada*, **99**, 47 pp.

Hazan, A. (1914). Storage to be provided in impounding reservoirs for municipal water supply. *Transactions of the American Society of Civil Engineers*, **77**, 1539–40.

Heath, W. A., (1961). Compressed air revives polluted Swedish lakes. *Water and Sewage Works*, **108**, 200.

HEC. (1977). *HEC–6. Scour and Deposition in Rivers and Reservoirs. User's Manual*. US Army Corps of Engineers, Hydrologic Engineering Center, Davis, California, USA: 149 pp.

Heeg, J., Breen, C. M., Colvin, P. H., Furness, H. D., & Musil, C. F. (1978). On the dissolved solids of the Pongola flood-plain pans. *Journal of the Limnological Society of South Africa*, **4** (1), 59–64.

Heinemann, H. G., Holt, R. F., & Rausch, D. L. (1973). Sediment and nutrient research on selected corn-belt reservoirs. Pp. 381–6 in *Man-Made Lakes: Their Problems and Environmental Effects* (Ed. W. C. Ackermann, G. F. White, & E. B. Worthington). Geophysical Monograph 17, American Geophysical Union, Washington, DC, USA: xv + 847 pp., illustr.

Henricson, J., & Müller, K. (1979) Stream regulation in Sweden with some examples from Central Europe. Pp. 183–200 in *The Ecology of Regulated Streams* (Ed. J. V. Ward, & J. A. Stanford). Plenum Press, New York, USA: xi + 398 pp., illustr.

Hergenrader, G. L., (1980). Eutrophication of the Salt Valley Reservoir, 1968–73, 1: The effects of eutrophication of standing crop and composition of phytoplankton. *Hydrobiologia,* **71**, 61–82.

Hergenrader, G. L., & Hammer, M. J. (1973). Eutrophication of small reservoirs on the Great Plains. Pp. 560–6 in *Man-Made Lakes: Their Problems and Environmental Effects* (Ed. W. C. Ackermann, G. F. White, & E. B. Worthington). Geophysical Monograph 17, American Geophysical Union, Washington, DC: xv + 847 pp., illustr.

Hewlett, J. D., & Hibbert, A. R. (1967). Factors affecting the response of small watersheds to precipitation in humid areas. Pp. 275–90 in *International Symposium on Forest Hydrology* (Ed. W. E. Sopper, & H. W. Lull). Pergamon Press, Oxford, England, UK: vii + 561 pp.

Highler, L. W. G. (1975). Reaction of some caddis larvae (Trichoptera) to different types of substrate in an experimental stream. *Freshwater Biology,* **5**, 151–7.

Hilsenhoff, W. L. (1971). Changes in the downstream insect and amphipod fauna caused by an impoundment with a hypolimnion drain. *Annals of the Entomological Society of American,* **64**, 743–6.

Hocutt, C. H., & Stauffer, J. R. (1975). Influence of gradient on the distribution of fishes in Conomingo Creek, Maryland and Pennsylvania. *Chesapeake Science,* **16**, 143–7.

Hoffman, C. E., & Kilambi, R. V. (1970). *Environmental Changes Produced by Cold-water Outlets from Three Arkansas Reservoirs.* Publication No. 5, Water Research Center, University of Arkansas, Fayetteville, Arkansas, USA.*

Holden, P. B. (1979). Ecology of riverine fishes in regulated stream systems with emphasis on the Colorado River. Pp. 57–74 in *The Ecology of Regulated Streams* (Ed. J. V. Ward, & J. A. Stanford). Plenum Press, New York, USA: xi + 398 pp., illustr.

Holden, P. B., & Crist, L. W. (1979). *Documentation of Changes in the Macroinvertebrate and Fish Populations in the Green River due to Inlet Modification of Flaming Gorge Dam.* BIO/WEST, Logan, Utah, USA, PR–16–2.*

Holden, P. B., & Stalnaker, C. B. (1975). Distribution and abundance of mainstream fishes of the middle and upper Colorado River basins, 1967–1973. *Transactions of the American Fisheries Society,* **104** (2), 217–31.

Hollick, M. (1981). The role of quantitative decision-making methods in environmental impact assessment. *Journals of Environmental Management,* **12**, 65–78.

Holling, C.S. (1973). Resilence and stability of ecological systems. *Annual Review of Ecological Systematics,* **4**, 1–23.

Holling, C. S. (Ed.) (1978). *Adaptive Environmental Assessment and Management.* (International series on Applied Systems Analysis 3.) John Wiley & Sons, Chichester: xviii + 377 pp., illustr.

Hollis, G. E. (1979). *The Impact of Man Upon the Hydrological Cycle.* Geobooks, Norwich, England, UK: ix + 278 pp., illustr.

Holmes, N. T. H., & Whitton, B. A. (1977). The macrophytic vegetation of the River Tees in 1975: observed and predicted changes. *Freshwater Biology,* **7**, 43–60.

Holmes, N. T. H., & Whitton, B. A. (1981). Phytobenthos of the River Tees and its tributaries. *Freshwater Biology,* **11**, 139–68.

Hooke, J. M. (1980). Magnitude and distribution of rates of river bank erosion. *Earth Surface Processes,* **5** (2), 143–58.

Hornberger, G. M., & Spear, R. C. (1981). An approach to the preliminary analysis of environmental systems. *Journal of Environmental Management,* **12** (1), 7–18.

Horton, R. E. (1945). Erosional development of streams and their drainage basins: hydrophysical approach to quantitative morphology. *Bulletin of the Geological Society of America,* **263**, 303–12.

Horwitz, R. J. (1978). Temporal variability patterns and the distributional patterns of stream fishes. *Ecological Monographs*, **48**, 307–21.

Howard, A. D., & Dolan, R. (1981). Geomorphology of the Colorado River in the Grand Canyon. *Journal of Geology*, **89**, 269–98.

Hrbáček, J. (1969). Water passage and the distribution of plankton organisms in Slapy Reservoir. Pp. 144–54 in *Man-Made Lakes: The Accra Symposium* (Ed. L. E. Obeng). Ghana Universities Press, Accra, Ghana: 398 pp.

Hubbs, C. (1972). Some thermal consequences of environmental manipulations of water. *Biological Conservation*, **4**, 185–8.

Hubbs, C., & Pigg, J. (1976). The effects of impoundments on threatened fishes of Oklahoma. *Annals of the Oklahoma Academy of Science*, **5**, 133–77.

Hudson, H. E., Brown, C. B., Shaw, H. B., & Longwell, J. S. (1949). Effect of land use on reservoir situation. *Journal of the American Water Works Association*, October, pp. 913–32.

Huet, M. (1959). Profiles and biology of West European streams as related to fish management. *Transactions of the American Fisheries Society*, **88**, 155–63.

Huggins, A. F., & Griek, M. R. (1974). River regulation and peak discharge. *Proceedings of the American Society of Civil Engineers, Journal of the Hydraulics Division*, **100**, 901–18.

Hussainy, S. U., & Abdulappa, M. K. (1973). A limnological reconnaissance of Lake Gonewada, Nagpur, India. Pp. 500–6 in *Man-Made Lakes: Their Problems and Environmental Effects* (Ed. W. C. Ackermann, G. F. White, & E. B. Worthington). Geophysical Monograph 17, American Geophysical Union, Washington, DC, USA: xv + 847 pp., illustr.

Hydraulic Engineering Center—*see* HEC.

Hynes, H. B. N. (1955). Distribution of some freshwater Amphipoda in Britain. *Verbandlungen Internationale Vereinigung für Theoretische und Angewandte Limnologie*, **12**, 620–8.

Hynes, H. B. N. (1970a). The ecology of flowing waters in relation to management. *Journal of the Water Pollution Control Federation*, **42** (3), 418–24.

Hynes, H. B. N. (1970b). *The Ecology of Running Waters*. Liverpool University Press, Liverpool, England, UK: xxiv + 555 pp., illustr.

Hynes, H. B. N. (1975). The stream and its valley. *Verhandlungen Internationale Vereinigung für Theoretische und Angewandte Limnologie*, **19**, 1–15.

Hynes, H. B. N., Williams, D. D., & Williams, N. E. (1976). Distribution of benthos within the substratum of a Welsh mountain stream. *Oikos*, **27**, 307–10.

I.C.E. (1933). *Floods in Relation to Reservoir Practice*. The Institute of Civil Engineers, London, England, UK: 67 pp.

I.C.E. (1975). *Reservoir Flood Standards: Discussion Paper*. The Institute of Civil Engineers, London, England, UK: 5 pp.

Ichim, I., & Radoane, M. (1980). On the anthropogenic influence time in morphogenesis, with special regard to the problem of channel dynamics. *Revue Roumanie de Geographie*, **24**, 35–40.

Illies, J. (1952). Die Mölle: Faunistischökologische Untersuchungen an eimen Forellenbach in Lipper Bergland. *Archiv für Hydrobiologie*, **46**, 424–612.

Illies, J. (1961). Versuch einer allgemein biozönotischen Gleiderung der Fliessgewässer. *Internationalen Revue der gesamten Hydrobiologie*, **46**, 205–13.

Illies, J., & Botosaneau, L. (1963). Problèmes et méthodes de la classification et de la zonation écologique des eaux courantes, considerées surtout du point de vue faunistique. *Mitteilungen Internationale Vereinigung Theoretische und Angewandte Limnologie*, **12**, 1–57.

Imevbore, A. M. A. (1967). Hydrology and plankton of Eleiyele Reservoir, Ibadan, Nigeria. *Hydrobiologia*, **30**, 154–76.

Imevbore, A. M. A. (1970). The chemistry of the River Niger in the Kainji Reservoir area. *Archiv für Hydrobiologie*, **67** (3), 412–31.

Ingols, R. S. (1957). Pollutional effects of hydraulic power generation. *Sewage and Industrial Wastes*, **29** (3), 292–6.

Ingols, R. S. (1959). Effect of impoundment on downstream water quality, Catawba River, S.C. *Journal of the American Water Works Association*, **51**, 42–6.

Institute of Civil Engineers, London—*see* I.C.E.

International Commission on Large Dams (1973). *World Register of Dams*. International Commission on Large Dams, Paris: xii + 998 pp.

Irwin, W. H., Symons, J. M., & Robeck, G. G. (1966). Impoundment destratification by mechanical pumping. *Proceedings of the American Society of Civil Engineers, Journal of the Sanitary Engineering Division*, **92** (6), 21–40.

Isom, B. G. (1971). Effects of storage and mainstream reservoirs on benthic macroinvertebrates in the Tennesee Valley. Pp. 179–91 in *Reservoir Fisheries and Limnology* (Ed. G. E. Hall). Special Publication No. 8, American Fisheries Society, Washington, DC, USA: x + 511 pp., illustr.

Jaag, O., & Ambühl, H. (1964). The effect of the current on the composition of biocoenoses in flowing-water streams. *Advances in Water Pollution Research*, **1**, 31–49.

Jackson, P. B. N. (1966). The establishment of fisheries in man-made lakes in the tropics. Pp. 53–69 in *Man-Made Lakes* (Ed. R. H. Lowe-McConnell). Symposia of the Institute of Biology, 15, Academic Press, London, England, UK: xiii + 218 pp., illustr.

Jackson, P. B. N. (1975). Fish. Pp. 259–76 in *Man-Made Lakes and Human Health* (Ed. N. F. Stanley, & M. P. Alpers). Academic Press, London, England, UK: xvi + 495 pp., illustr.

Jackson, P. B. N., & Davies, B. R. (1976). Cabora River in its first year: some ecological aspects and comparisons. *Rhodesian Science News*, **10** (5), 128–33.

Jackson, P. B. N., & Rogers, K. H. (1976). Cabora Basin fish populations before and during the first filling phase. *Zoologica Africana*, **11** (2), 373–97.

Jain, S. K., Naegamřala, J. P., & Sahasrabudhe, S. R. (1973). Impact of Damodani Valley Reservoirs on the environmental status of the region. Pp. 783–803 in *Transactions of the Eleventh International Congress on Large Dams, Madrid, Spain*. Vol. 1. International Commission of Large Dams, Paris: xxviii + 974 pp., illustr.

Jansen, P.Ph., Bendegom, L. Van, Berg, J. Van Den, Vries, M. de, & Zanen, A. (1979). *Principles of River Engineering: The Non-Tidal Alluvial River*. Pitman, London, England, UK: xv + 509 pp., illustr.

Jaske, J. T., & Goebel, J. B. (1967). Effects of dam construction on the temperature of the Columbia River. *Journal of the American Water Works Association*, **59**, 935–42.

Jassby, A. D. (1980). The environmental effects of hydroelectric power development. Pp. 32–43 in *Energy and the Fate of Ecosystems*. Supporting Paper 8 for the Study of Nuclear & Alternative Energy Systems (National Research Council Committee on Nuclear & Alternative Energy Systems). National Academy Press, Washington, DC, USA: xvii + 397 pp.

Jeffers, J. N. R. (1972). The challenge of modern mathematics to the ecologist. Pp. 1–11 in *Mathematical Models in Ecology* (Ed. J. N. R. Jeffers). Blackwell Scientific Publications, Oxford: v + 398 pp., illustr.

Joglekar, D. V., & Wadekar, G. T. (1951). Effects of weirs and dams on the regime of rivers. *International Association of Hydraulic Research, 4th Meeting, Bombay Symposium*, App. 16, pp. 349–53.

Johnson, N. M., Likens, G. E., Bormann, F. H., Fisher, D. W., & Pierce, R. S. (1969). A working model for the variation of stream water chemistry at the Hubbard Brook Experimental Forest, New Hampshire. *Water Resources Research*, **5**, 1353–63.

Jovanovic, D. (1973). Some effects on the environment of the building of dams in East Africa. Pp. 69–86 in *Transactions of the Eleventh International Congress of*

285

Large Dams, Madrid, Spain, Vol. 1. Interntional Commission on Large Dams, Paris, France: xxviii + 974 pp., illustr.

Jubb, R. A. (1960). Some Lake Kariba fish problems. *Piscator*, **47**, 112–9.

Jubb, R. A. (1972). The J. G. Strydom Dam: Pongolo River: northern Zululand. The importance of floodplain pans below it. *Piscator*, **86**, 104–109.

Jubb, R. A. (1976). Unintentional introduction of fishes via hydro-electric power stations and centrifugal pumps. *Journal of the Limnological Society of South Africa*, **2**, 29–30.

Kalleberg, H. (1958). Observations in a stream tank of territoriality and competition in juvenile salmon and trout. *Institute of Freshwater Research, Drottningholm, Sweden*, Report No. 39, pp. 55–98.

Kellerhals, R. (1971). Factors controlling the level of Lake Athabasca. Pp. 57–105 in *Proceedings of the Peace–Athabaska Delta Symposium*. Water Resources Centre, University of Alberta, Edmonton, Alberta, Canada: xvi + 359 pp., illustr.

Kellerhals, R. (1982). Effect of river regulation on channel stability. Pp. 685–716 in *Gravel-Bed Rivers* (Ed. R. D. Hey, J. C. Bathurst, & C. R. Thorne). John Wiley & Sons, Chichester, England, UK: xv + 875 pp., illustr.

Kellerhals, R., & Gill, D. (1973). Observations and potential downstream effects of large storage projects in Northern Canada. Pp. 731–54 in *Transactions of the Eleventh International Congress of Large Dams, Madrid, Spain*, Vol. 1. International Commission on Large Dams, Paris: xxviii + 979 pp., illustr.

Kenmuir, D. H. S. (1975). Sardines in Cabora Bassa Lake? *New Scientist*, **13**, 379–80.

Kenmuir, D. H. S. (1976). Fish spawning under artificial flood conditions on the Mana floodplain, Zambezi River. *Kariba Studies*, **6**, 86–97.

Kennard, M. F. (1972). Examples of the internal condition of some old earth dams. *Journal of the Institute of Water Engineers*, **26**, 135–47.

Kerr, M. A. (1968). Annual Report, Zambezi Valley Research Station. (Described in Attwell, R. I. G. (1970) (q.v.).

Kimsey, J. B. (1957). Fisheries problems in impounded waters of California and the lower Colorado River. *Transactions of the American Fisheries Society*, **87**, 39–57.

Kinawy, I. Z., Wafa, T. A., Labib, A. H., & Shenouda, W. E. (1973). Effects of sedimentation in the High Aswan Dam Reservoir. Pp. 879–98 in *Transactions of the Eleventh International Congress of Large Dams, Madrid, Spain*, Vol. 1. International Commission on Large Dams, Paris, France: xviii + 974 pp., illustr.

King, N. J. (1961). An example of channel aggradation induced by flood control. *US Geological Survey Professional Paper*, 424B, pp. 29–32.

King, R. D., & Tyler, P. A. (1982). Downstream effects of the Gordon River Power Development, South-West Tasmania. *Australia Journal of Marine and Freshwater Research*, **33**, 431–42.

Kittrell, F. W., & Quinn, J. J. (1949). Multi-puprose reservoirs and downstream water supplies. *Engineering News Record*, May 26th, pp. 174–6.

Knappen, T. T., Stratton, J. H., & Davis, C. V. (1952). River regulation by reservoirs. Pp. 1–21 in *Handbook of Applied Hydraulics*, 3rd edition (Ed. C. V. Davis, & K. E. Sorensen). McGraw-Hill, New York: xi + 42 sec. pp. var., illustr.

Kofoid, C. A. (1903). The plankton of the Illinois River, 1894–1899, Part 1. *Bulletin of the Illinois State Laboratory of Natural History*, **6**, 95–629.

Kofoid, C. A. (1908). The plankton of the Illinois River, 1894–1899, Part 2. *Bulletin of the Illinois State Laboratory of Natural History*, **8**, 1–355.

Koh, R. C. (1964). *Viscous Stratified Flow Towards a Line Sink*. Report of the W. M. Keck Laboratory of Hydraulics and Water Resources, California Institute of Technology, Pasadena, California, USA: KH-R-6.*

Komura, S., & Simons, D. B. (1967). River bed degradation below dams. *Journal of the Hydraulics Division of the American Society of Civil Engineers*, **93**, 1–14.

Koryak, M. (1976). The influence of mainstream navigation dams on water quality

286

and fisheries in the upper Ohio River basin. Pp. 158–73 in *Instream Flow Needs*, Vol. II (Ed. J. F. Orsborn, & C. H. Allman). American Fisheries Society, Bethesda, Maryland, USA: 657 pp., illustr.

Kroger, R. L. (1973). Biological effects of fluctuating water levels in the Snake River, Grand Teton National Park, Wyoming. *The American Midland Naturalist*, **89**, 478–81.

Lack, T. J. (1971). Quantitative studies on the phytoplankton of the River Thames and Kennet at Reading. *Freshwater Biology*, **1**, 213–24.

Lake, J. S. (1975). Fish of the Murray River. Pp. 213–24 in *The Book of the Murray* (Ed. G. C. Lawrence, & G. K. Smith). Rigby, Adelaide, Australia: 264 pp., illustr.

Lane, E. W. (1934). Retrogression of levels in riverbeds below dams. *Engineering News Record*, **112**, p. 838.

Langbein, W. B. & Leopold, L. B. (1964). Quasi-equilibrium states in channel morphology. *American Journal of Science*, **262**, 782–94.

Langbein, W. B. & Schumm, S. A. (1958). Yield of sediment in relation to mean annual precipitation. *Transactions of the American Geophysical Union*, **39**, 1076–84.

Laursen, E. M., Ince, S., & Pollack, J. (1975). On sediment transport through the Grand Canyon. Pp. 4–76 to 4–87 in *Proceedings of the Third Federal Interagency Sedimentation Conference, Denver, Colorado*. SED-COM-03, Water Resources Council, Washington, DC, USA: xiv + 7 sec., pp. var., illustr.

Lauterbach, D., & Leder, A. (1969). The influence of reservoir storage on statistical peak flows. Pp. 821–6 in *Floods and their Computation*, Vol. 2. Publication No. 85 of International Association of Scientific Hydrology, Gentbrugge, Belgium: 396 pp., illustr.

Lavis, M. E., & Smith, K. (1971). Reservoir storage and the thermal regime of rivers, with special reference to the River Lune, Yorkshire. *Science of the Total Environment*, **1**, 81–90.

Lawson, J. M. (1925). Effect of Rio Grande storage on river erosion and deposition. *Engineering News Record*, Sept. 3rd, pp. 327–34.

Lawson, L. L., & Rushforth, S. R. (1975). The diatom flora of the Provo River, Utah, USA. *Bibliography of Phycology*, **17**, 149 pp.

Ledger, D. C. (1964). Some hydrological characteristics of West African rivers. *Transactions of the Institute of British Geographers*, **35**, 73–90.

Leentvaar, P. (1966). The Brokopondo Research Project, Surinam. Pp. 33–41 in *Man-Made Lakes* (Ed. R. H. Lowe-McConnell). Symposia of the Institute of Biology, No. 15, Academic Press, London: xviii + 218 pp.

Lehmann, C. (1927). Uber den Einfluss der Talspeuen auf die unterhalb liegende Bachand Flussfischerei. *Zeitschrift für Fischerei und deren Hilfswissenschaften*, **25**, 467–76.

Lehmkuhl, D. M. (1972). Change in thermal regime as a cause of reduction of benthic fauna downstream of a reservoir. *Journal of the Fisheries Research Board of Canada*, **29**, 1329–32.

Lelek, A., & El-Zarka, S. (1973). Ecological comparison of the pre-impoundment and post-impoundment fish fauna of the River Niger and Kainji Lake, Nigeria. Pp. 655–60 in *Man-Made Lakes: Their Problems and Environmental Effects* (Ed. W. C. Ackermann, G. F. White, & E. B. Worthington). Geophysical Monograph 17, American Geophysical Union, Washington, DC: xv + 847 pp., illustr.

Leopold, L. B., & Maddock, T. (1953). *The Hydraulic Geometry of Stream Channels and Some Physiographic Implications*. US Geological Survey Professional Paper, No. 252, 56 pp.

Leopold, L. B. & Maddock, T. (1954). *The Flood Control Controversy*. Ronald, New York: 278 pp.

Leopold, L. B., Wolman, M. G., & Miller, J. P. (1964). *Fluvial Processes in Geomorphology*. Freeman, San Francisco: 522 pp.

Lewis, D. S. C. (1974). The effects of the formation of Lake Kainji (Nigeria) upon the indigenous fish population. *Hydrobiologia*, **45** (2–3), 281–301.

Liepolt, R. (1972). Uses of the Danube River. Pp. 233–49 in *River Ecology and Man* (Ed. R. T. Oglesby, C. A. Carlson, & J. A. McCann). Academic Press, New York: xvii + 465 pp., illustr.

Lillehammer, A., & Saltveit, S. J. (1979). Stream regulation in Norway. Pp. 201–14 in *The Ecology of Regulated Streams* (Ed. J. V. Ward, & J. A. Stanford). Plenum Press, New York: xi + 398 pp., illustr.

Lind, O. T. (1971). Organic matter budget of a central Texas reservoir. Pp. 193–202 in *Reservoir Fisheries and Limnology* (Ed. G. E. Hall). Special Publication No. 8, American Fisheries Society, Washington, DC: x + 511 pp., illustr.

Lindroth, A. (1957). Abiogenic gas supersaturation of river water. *Archiv für Hydrobiologie*, **53**, 589–97.

Livesey, R. H. (1963). Channel armouring below Fort Randall Dam. *US Department of Agriculture Miscellaneous Publication*, No. 970, pp. 461–70.

Livingstone, D. A. (1963). *Chemical Composition of Rivers and Lakes*. US Geological Survey Professional Paper, 440G, 64 pp.

Logan, P., & Brooker, M. P. (1983). The macroinvertebrate fauna of riffles and pools. *Water Research*, **17** (3), 262–70.

Long, C. S., & Krema, R. F. (1969). Research on a system for bypassing juvenile salmon and trout around low-head dams. *Commercial Fishing Review*, **31** (6), 27–9.

Long, G. E. (1974). Model stability, resilience and management of an aquatic community. *Oecologia*, **17**, 65–85.

Lotspeich, F. B. (1980). Watersheds as the basic ecosystem: this conceptual framework provides a basis for a natural classification system. *Water Resources Bulletin*, **16** (4), 581–6.

Love, S. K. (1961). Relationships of impoundment and water quality. *Journal of the American Water Works Association*, **53**, 559–68.

Lowe, R. L. (1979). Phytobenthic ecology and regulated streams. Pp. 25–34 in *The Ecology of Regulated Streams* (Ed. J. V. Ward, & J. A. Stanford). Plenum Press, New York: xi + 398 pp. illustr.

Lowe-McConnell, R. H. (Ed.) (1976). *Man-Made Lakes*. (Symposia of the Institute of Biology No. 15.) Academic Press, London & New York: xii + 218 pp., illustr.

Lowenthal, D. (Ed.) (1965). *Man and Nature* by G. P. Marsh. Balknap Press of Harvard University Press, Cambridge, Mass.: xxix + 472 pp.

Lowrance, W. W. (1976). *Of Acceptable Risk: Science and the Determination of Safety*. William Kaufmann, Los Altos, California: x + 180 pp.

Lugo, A. E. (1978). Stress and ecosystems. Pp. 61–101 in *Energy and Environmental Stress in Aquatic Ecosystems* (Ed. J. H. Thorp, & J. W. Gibbons). DOE Symposium Series (Conf.-771114), Oak Ridge, Tennessee: xxii + 854 pp., illustr.

Lvovitch, M. I. (1958). Streamflow formation factors. *International Association of Scientific and Hydrological Publications*, **45**, 122–32.

Lvovitch, M. I. (1973). The global water balance. *US International Hydrological Decade Bulletin*, **23**, 28–42.

Macan, T. T. (1963). *Freshwater Ecology*. Longmans, London, England, UK & John Wiley, New York, USA: 338 pp., illustr.

McClure, R. G., & Stewart, K. W. (1976). Life cycle and production of the mayfly *Choroterpes* (*Neochoroterpes*) *mexicanus* Allen (Ephemeroptera: Leptophlebiidae). *Annals of the Entomological Society of America*, **69**, 134–44.

MacCurdy, E. (1954). *The Notebooks of Leonardo Da Vinci*. The Reprint Society, London: 3 vols. pp. var., illustr.

MacDonald, J. R., & Hyatt, R. A. (1973). Supersaturation of nitrogen in water during passage through HEP turbines at Mactaquac Dam. *Journal of the Fisheries Research Board of Canada*, **30**, 139–224.

McIntire, C. D. (1966). Some effects of current velocity on the periphyton communities in laboratory streams. *Hydrobiologia*, **27**, 559–70.

McIntire, C. D., & Colby, J. A. (1978). A hierarchical model of lotic ecosystems. *Ecological Monographs*, **48**, 167–90.

McIntosh, B. M., Gear, J. H. S., & Pritchford, R. J. (1973). The consequences on the environment of building dams: biological effects with special reference to medical aspects. Pp. 289–304 in *Transactions of the Eleventh International Congress on Large Dams, Madrid, Spain*. Vol. 1. International Commission of Large Dams, Paris, France: xxviii + 974 pp., illustr.

Mackie, G. L. Rooke, J. B., Roff, J. C., & Gerrath, J. F. (1983). Effects of changes in discharge level on temperature and oxygen regimes in a new reservoir and downstream. *Hydrobiologia*, **101**, 179–88.

MacKinnon, D., & Brett, J. R. (1964). Some observations on the movement of Pacific Salmon fry through a small impounded water basin. *Journal of the Fisheries Research Board of Canada*, **12** (3), 362–8.

MacKinnon, D., Edgeworth, L., & McLaren, R. E. (1961). An assessment of Jones Creek spawning channel. *Canadian Fish Culturist*, **30**, 25–31.

Makkaveyev, N. I. (1970). The impact of large water engineering projects on geomorphic processes in stream valleys. *Geomorphologia*, **2**, 28–34.

Malhotra, S. L. (1951). Effects of barrages and weirs on the regime of rivers. *International Association of Hydraulic Research 4th Meeting, Bombay Symposium*, pp. 335–47.

Mannion, A. M. (1982). Diatoms: their use in physical geography. *Progress in Physical Geography*, **6** (2), 233–60.

Mar, B. W. (1974). Problems encountered in multidsciplinary resources and simulation models development. *Journal of Environmental Management*, **2**, 83–100.

Marsh, G. P. (1864). *Man and Nature* (Ed. D. Lowenthal [1965]). Balknap Press of Harvard University Press, Cambridge, Mass., USA: xxix + 472 pp.

Marshall, B. E., & Falconer, A. C. (1973). Physico-chemical aspects of Lake McIlwaine (Rhodesia), a eutrophic tropical impoundment. *Hydrobiologia*, **42**, 15–62.

Martin, D. B., & Arneson, R. D. (1978). Comparative limnology of a deep-discharge reservoir and a surface-discharge lake on the Madison River, Montana. *Freshwater Biology*, **8**, 33–42.

Maystrenko, Y. G., & Denisova, A. I. (1972). Method of forecasting the content of organic and biogenic substances in the water of existing and planned reservoirs. *Soviet Hydrology Selected Papers*, **6**, 515–40.

Meier, R. L. (1972). Communication stress. *Annual Review of Ecological Systematics*, **3**, 289–314.

Merkley, W. B. (1978). Impact of Red Rock Reservoir on the Des Moines River. Pp. 62–76 in *Current Perspectives on River–Reservoir Ecosystems* (Ed. J. Cairns, jr, E. F. Benfield, & J. R. Webster). North American Benthological Society, Springfield, Ill., USA: v + 85 pp., illustr.

Mermel, T. W. (1976). International activity in dam construction. *Water, Power and Dam Construction*, **28**(4), 66–9.

Mermel, T. W. (1981). Major dams of the world. *Water, Power and Dam Construction*, **33**(5), 55–64.

Mermel, T. W. (1982) Major dams of the world. *Water, Power and Dam Construction*, **34**(5), 93–104.

Mermel, T. W. (1983). Major dams of the World—1983. *Water, Power and Dam Construction*, **35**(8), 43.

Micklin, P. P. (1979). Disciplinary plans for USSR rivers. *Geographical Magazine*, **51**(10), 701–6

Milhous, R. T. (1982). Effect of sediment transport and flow regulation on the ecology of gravel-bed rivers. Pp. 819–42 in *Gravel-bed Rivers* (Ed. R. D. Hey, J. C. Bathurst, & C. R. Thorne). John Wiley and Sons, Chichester, England, UK: xv + 875 pp., illustr.

Miller, C. R. (1962). Discussion of 'The Process of Channel Degradation' by E. R.

Tinney. *Journal of Geophysical Research,* **67**(4), 1481–3.

Minckley, W. L. (1964). Upstream movements of *Gammarus* in Doe Run, Kentucky. *Ecology,* **45**, 185–97.

Minckley, W. L., & Deacon, J. E. (1968). Southwestern fishes and the enigma of 'endangered species'. *Science,* **159**, 1424–32.

Minshall, G. W. (1978). Autotrophy in stream ecosystems. *BioScience,* **28**, 767–71.

Minshall, G. W., & Minshall, J. N. (1977). Microdistribution of benthic invertebrates in a Rocky Mountain (USA) stream. *Hydrobiologia,* **55**, 231–49.

Minshall, G. W., & Winger, P. V. (1968). Effect of reduction in streamflow on invertebrate drift. *Ecology,* **49**, 580–2.

Mitchell, D. S. (1978). *Aquatic Weeds in Australian Inland Waters.* Australian Government Publishing Service, Canberra: xxii + 189 pp., illustr.

Mitchell, D. S., & Marshall, B. E. (1974). Hydrobiological observations on three Rhodesian reservoirs. *Freshwater Biology,* **4**, 61–74.

Moffett, J. W. (1949). The first four years of King Salmon maintenance below Shasta Dam, Sacramento River, California. *California Fish and Game,* **35**, 77–102.

Moore, C. M. (1969). Effects of small structures on peak flow. Pp. 101–17 in *Effects of Watershed Changes on Streamflow* (Ed. W. L. Moore, & C. W. Morgan). University of Texas Press, Austin, Texas, USA: xi + 289 pp., illustr.

Mosley, M. P. (1982). Analysis of the effect of channel morphology and instream uses in a braided river, Ohan River, New Zealand. *Water Resources Research,* **18**(4), 800–12.

Motwani, N. P., & Kanwai, Y. (1970). Fish and fisheries of the coffer-dammed right channel of the River Niger at Kainji. Pp. 27–48 in *Kainji Lake Studies, 1: (Ecology)* (Ed. S.A. Visser). University of Ibadan Press, Ibadan, Nigeria.*

Mullan, J. W., Starostka, V. J., Stone, J. L., Wiley, R. W., & Wiltzius, W. J. (1976). Factors affecting Upper Colorado River Reservoir tailwater trout fisheries. Pp. 405–23 in *Instream Flow Needs,* Vol. II (Ed. J. F. Orsborn, & C. E. Allman). American Fisheries Society, Bethesda, Maryland, USA: 657 pp., illustr.

Müller, K. (1962). Limnologisch–Fischereibiologische Untersuchungen in regulierten Gewässern Schwedish Lapplands. *Oikos,* **13**, 125–34.

Mundie, J. H. (1974). Optimization of the salmonid nursery stream. *Journal of the Fisheries Research Board of Canada,* **31**, 1827–37.

Mundie, J. H. (1979). The regulated stream and salmon management. Pp. 307–20 in *The Ecology of Regulated Streams* (Ed. J. V. Ward, & J. A. Stanford). Plenum Press, New York, USA: xi + 398 pp., illustr.

Natural Environment Research Council—*see* N.E.R.C.

Nebeker, A. V. (1971). Effect of high winter water temperatures on adult emergence of aquatic insects. *Water Research,* **5**, 777–83.

Neel, J. K. (1963). Impact of reservoirs. Pp. 575–93 in *Limnology in North America* (Ed. D. G. Frey). University of Wisconsin Press, Madison, Wisconsin, USA: xvii + 734 pp. illustr.

Neel, J. K. (1967). Reservoir eutrophication and dystrophication following impoundment. Pp. 322–32 in *Reservoir Fishery Resources Symposium.* American Fisheries Society, Washington, DC, USA: viii + 569 pp., illustr.

N.E.R.C. (1976). *Report of the Working Party on Hydrology.* Unpublished. Natural Environment Research Council, London, England, UK: 10 pp.

Neu, H. J. A. (1975). Runoff regulation and its effects on the ocean environment. *Canadian Journal of Civil Engineering,* **2**, 583–91.

Neuhold, J. H. (1981). Strategy of stream ecosystem recovery. Pp. 261–5 in *Stress Effects on Natural Ecosystems* (Ed. G. W. Barrett, & R. Rosenberg). (Environmental Monographs and Symposia, Convenor & Gen. Ed. N. Polunin) John Wiley & Sons, Chichester–New York–Brisbane–Toronto–Singapore: xvii + 305 pp., illustr.

Newson, M. D. (1980*a*). The erosion of drainage ditches and its effect on bed-load yields in mid-Wales: reconnaissance case studies. *Earth Surface Processes,* **5** (3), 275–90.

290

Newson, M. D. (1980*b*). The geomorphological effectiveness of floods—contribution stimulated by two recent events in mid-Wales. *Earth Surface Processes,* **5**, 1–16.

Nichols, P. R. (1968). *Passage Conditions and Counts of Fish at the Snake Island Fishway, Little Falls Dam, Potomac River, Maryland, 1960–63.* Fisheries and Wildlife Service Special Report No. 565.*

Nilsson, B. (1976). The influence of Man's activities in rivers on sediment transport. *Nordic Hydrology,* **7**, 145–60.

Nisbet, M. (1961). Un exemple de pollution de rivière par vidage d'une retenue hydro-electrique. *Verhandlungen Internationale Vereinigung für Theoretische und Angewandte Limnologie,* **14**, 678–80.

Nishizawa, T., & Yamabe, K. (1970). Changes in downstream temperature caused by the construction of reservoirs. *Tokyo University Scientific Report, Section C,* **10**, 27–42.

Northrop, W. L. (1965). Republican River channel deterioration. *US Department of Agriculture Miscellaneous Publication* No. 970, pp. 409–24.

N.W.W.A. (1972). *The Effect on Fish Migration and Behaviour of Water Bank Releases from Stocks Reservoir.* North West Water Authority Report, Preston, UK: 15 pp.

Obeng, L. E. (Ed.) (1969). *Man-Made Lakes: The Accra Symposium.* Ghana University Press, Accra, Ghana: 398 pp., illustr.

Obeng, L. E. (1981). Man's impact on tropical rivers. Pp. 265–88 in *Perspectives in Running Water Ecology* (Ed. M. A. Lock, & D. D. Williams). Plenum Press, New York, USA: x + 430 pp., illustr.

O'Connor, A. J., Ganong, G. H. D., & Gordon, A. Y., (1973). The Macraquac Development, effects on environment. Pp. 755–72 in *Transactions of the Eleventh International Congress on Large Dams, Madrid, Spain,* Vol. 1. International Commission of Large Dams, Paris, France: xxviii + 974 pp., illustr.

Odum, E. P. (1969). The strategy of ecosystem development. *Science,* **164**, 262–70.

Odum, E. P. (1981). The effects of stress on the trajectory of ecological succession, Pp. 43–8 in *Stress Effects on Natural Ecosystems* (Ed. G. W. Barrett, & H. Rosenberg). (Environmental Monographs & Symposia, Convener & Gen. Ed. N. Polunin) John Wiley & Sons, Chichester–New York–Brisbane–Toronto–Singapore: xiii + 305 pp., illustr.

Odum, E. P., Finn, J. T., & Franz, E. H. (1979). Perturbation theory and the subsidy-stream gradient. *BioScience,* **27**, 349–52.

Oglesby, R. T., Carlson, C. A., & McCann, J. A. (Ed.) (1972). *River Ecology and Man.* Academic Press, New York: xvii + 465 pp., illustr.

Ohle, W. (1937). Kalksystematik unserer Binnengewässer und Kalkgehalt Rügener Bache. *Geologie Meere und Binnengewässer,* **1**, 291–316.

Orsborn, J. F., & Allman, C. H. (1976). *Instream Flow Needs,* Vol. II. (Proceedings of Boise Symposium, Idaho, May 1976). American Fisheries Society, Bethesda, Maryland: vi + 657 pp., illustr.

Ortal, R., & Por, F. D. (1978). Effect of hydrological changes on aquatic communities in the Lower Jordan River. *Verhandlungen Internationale Vereinigung für Theoretische und Angewandte Limnologie,* **20**, 1543–51.

Ostrofsky, M. L. (1978). Trophic changes in reservoirs: an hypothesis using phosphorus budget models. *Internationale Revue der Gesantem Hydrobiologie,* **64**(4), 481–99.

Oswood, M. W. (1979). Abundance patterns of filter-feeding caddis fly (Trichoptera, Hydropsychidae) and seston in a Montana (USA) lake outlet. *Hydrobiologia,* **63**, 177–83.

Oyebande, L. (1981). Sediment transport and river basin management in Nigeria. Pp. 201–26 in *Tropical Agricultural Hydrology: Watershed Management and Land Use* (Ed. R. Lal, & E. W. Russell). John Wiley & Sons, Chichester: xiv + 483 pp., illustr.

Pardé, M. (1955). *Fleuves et Rivières,* 3rd edition. Colin, Paris: 224 pp.

Park, C. C. (1977). World-wide variations in hydraulic geometry exponents of stream channels: an analysis and some observations. *Journal of Hydrology*, **33**, 133–46.

Park, C. C. (1978). Allometric analysis and stream channel morphometry. *Geographical Analysis*, **10** (3), 211–28.

Park, C. C. (1980). *Ecology and Environmental Management*. Dawson, Folkestone, UK: 272 pp., illustr.

Park, D. L., & Farr, W. E. (1972). Collection of juvenile Salmon and Steelhead Trout passing through orifices in gatewalls of turbine intakes at Ice Harbor Dam. *Transactions of the American Fisheries Society*, **101** (2), 381–4.

Parker, F. A., & Krenkel, P. A. (1969). *Thermal Pollution—State of the Art*. National Centre for Research and Training in the Hydrologic and Hydraulic Aspects of Water Pollution Control, Vanderbilt University, Nashville, Tennesse, USA: Report No. 3.*

Parsons, J. W. (1957). The trout fishing of the tailwater below Dale Hollow Reservoir. *Transactions of the American Fisheries Society*, **85**, 75–92.

Pasch, R. W., Hackney, P. A., & Holbrook, J. A. II (1980). Ecology of paddlefish in Old Hickory Reservoir, Tennessee, with emphasis on first-year life history. *Transactions of the American Fisheries Society*, **109**(2), 157–67.

Patrick, D. M., Smith, L. M., & Whitton, C. B. (1982). Methods for studying accelerated fluvial change. Pp. 783–815 in *Gravel-bed Rivers* (Ed. R. D. Hey, J. C. Bathurst, & C. R. Thorne). John Wiley & Sons, Chichester: xv + 875 pp., illustr.

Patrick, R. (1970). Benthic stream communities. *American Scientist*, **58**, 546–9.

Pearson, W. D., & Franklin, D. R. (1968). Some factors affecting drift rates of *Baetis* and Simuliidae in a large river. *Ecology*, **49**, 75–81.

Pearson, W. D., Kramer, R. H., & Franklin, D. R. (1968). Macroinvertebrates in the Green River below Flaming Gorge Dam, 1964–65 & 1967. *Proceedings of the Utah Academy of Science, Arts and Letters*, **45**, 148–67.

Pechlander, R. (1964). Plankton production in natural lakes and hydroelectric basins in the alpine region of the Austrian Alps. *Verhandlungen Internationale Vereinigung für Theoretische und Angewandte Limnologie*, **15**, 375–83.

Pemberton, E. L. (1975). Channel changes in the Colorado River below Glen Canyon Dam. Ch. 5, pp. 61–73, in *Proceedings of the Third Federal Interagency Sedimentation Conference, Denver, Colorado, SED-COM-03*. Water Resources Council, Washington, DC: xiv + 7 sec., pp. var., illustr.

Peňáz, M., Kubícek, F., Marvan, P., & Zelinka, M. (1968). Influence of the Vír River Valley Reservoir on the hydrobiological and ichthyological conditions in the River Svratka. *Acta Scientiarum naturalium Academiae Scientiarum bohemoslovacae-Brno*, **2**, 1–60.

Penn, A. F. (1975). Development of James Bay: the role of environmental impact assessment in determining the legal right to an interlocutory injunction. *Journal of the Fisheries Research Board of Canada*, **32**, 136–60.

Pennak, R. W., (1971). Towards a classification of lotic habitats. *Hydrobiologia*, **38**(2), 321–34.

Pennak, R. W., & Gerpen, E. D. Van (1947). Bottom fauna production and physical nature of the substate in a northern Colorado trout stream. *Ecology*, **8**, 42–8.

Petr, T. (1978). Tropical man-made lakes—their ecological impacts. *Archiv für Hydrobiologie*, **8**(3), 368–85.

Petts, G. E. (1977). Channel response to flow regulation: the case of the River Derwent, Derbyshire. Pp. 145–64 in *River Channel Changes* (Ed. K. J. Gregory). Wiley–Interscience, Chichester: xiv + 448 pp., illustr.

Petts, G. E. (1978). *The Adjustment of River Channel Capacity Downstream from Reservoirs in Great Britain*. Unpublished PhD thesis, University of Southampton, Southampton, UK: v + 279 pp., illustr.

Petts, G. E. (1979). Complex response of river channel morphology subsequent to reservoir construction. *Progress in Physical Geography*, **3** (3), 329–62.

292

Petts, G. E. (1980a). Long-term consequences of upstream impoundment. *Environmental Conservation,* **7** (4), 325–32.

Petts, G. E. (1980b). Morphological changes of river channels consequent upon headwater impoundment. *Journal of the Institution of Water Engineers and Scientists,* **34** (4), 374–82.

Petts, G. E. (1980c). Implications of the fluvial process-channel morphology interaction below British reservoirs for stream habitat. *The Science of the Total Environment,* **16**, 149–63.

Petts, G. E. (1982). Channel changes within regulated rivers. Pp. 117–42 in *Papers in Earth Studies* (Ed. B. H. Adlam, C. R. Fenn, & L. Morris). Geobooks, Norwich, UK: iii + 204 pp., illustr.

Petts, G. E. (1984). Sedimentation within a regulated river: Afon Rheidol, Wales. *Earth Surface Processes and Landforms,* **9** (2), 125–134.

Petts, G. E., & Greenwood, M. (1981). Habitat changes below Dartmoor reservoirs. *Report and Transactions of the Devonshire Association for the Advancement of Science, Literature and Arts,* **113**, 13–27.

Petts, G. E., & Greenwood, M. (1985). Channel changes and invertebrate faunas below Nant-y-Môch Dam, River Rheidol, Wales, UK. *Hydrobiologia,* (in press).

Petts, G. E., & Lewin, J. (1979). Physical effects of reservoirs on river systems. Pp. 79–92 in *Man's Impact on the Hydrological Cycle in the United Kingdom* (Ed. G. E. Hollis). Geobooks, Norwich, UK: ix + 278 pp., illustr.

Petts, G. E., & Pratts, J. D. (1983). Channel changes resulting from low-flow regulation on a lowland river, England. *Catena,* **10** (V2), 77–85.

Pfitzer, D. W. (1954). Investigation of waters below storage reservoirs in Tennessee. *Transactions of the North American Wildlife Conference,* **19**, 271–82.

Pfitzer, D. W. (1962). Investigations of water below a large storage reservoir in Tennessee. Appendixes A & B in *Dingel-Johnson Rep., Proj, F-I-B,* Tennessee Game and Fish Commission, Nashville, Tennessee, USA: 233 pp.

Pfitzer, D. W. (1967). Evaluations of tailwater fishery resources resulting from high dams. Pp. 477–88 in *Reservoir Fishery Resources Symposium.* American Fisheries Society, Washington, DC, USA: viii + 569 pp., illustr.

Pflieger, W. L. (1975). *The Fishes of Missouri.* Missouri Department of Conservation, Jefferson City, Missouri, USA: 343 pp.

Pickup, G. (1980). Hydrologic and sediment modelling studies in the environmental impact assessment of a major tropical dam project. *Earth Surface Processes,* **5**(1), 61–75.

Pickup, G., & Reiger, W. A. (1979). A conceptual model of the relationship between channel characteristics and discharge. *Earth Surface Processes,* **4**, 37–42.

Poole, W. C., & Stewart, K. W.(1976). The vertical distribution of macrobenthos within the substratum of the Brazos River, Texas. *Hydrobiologia,* **50**, 151–60.

Powell, G. C. (1958). Evaluation of the effects of a power dam water release pattern upon the downstream fishery. *Colorado Cooperative Fisheries Unit Quarterly Report,* **4**, 31–7.

Pratts, J. D. (1983). *Channel Morphology and Bed-material Interactions within a Regulated River.* Paper presented at the 5th BGRG Postgraduate Symposium, Huddersfield, UK, May 1983.

Pritchard, A. L. (1936). Factors influencing the upstream spawning migration of the Pink Salmon (*Oncorhynchus gorbuscha* Walbaum). *Journal of the Biological Board of Canada,* **2**(4), 383–9.

Prowse, G. A., & Talling, J. H. (1958). The seasonal growth and succession of plankton Algae in the White Nile. *Limnology and Oceanography,* **3**, 222–37.

Purcell, L. T. (1939). The ageing of reservoir waters. *Journal of the American Water Works Association,* **31** (10), 1755.

Pyefinch, K. A. (1966). Hydro-electric schemes in Scotland: Biological problems and effects on salmonid fisheries. Pp. 139–48 in *Man-Made Lakes* (Ed. R. H. Lowe-McConnell). (Symposia of the Institute of Biology No. 15.) Academic Press, London & New York: xiii + 218 pp., illustr.

Rabeni, C. F., & Minshall, G. W. (1977). Factors affecting microdistribution of stream benthic insects. *Oikos, 29*, 33–43.

Radford, D. S. (1972). Some effects of hydroelectric power installations on aquatic invertebrates and fish in the Kananaskis River system. *Alberta Conservation,* Summer, pp. 19–21.

Radford, D. S., & Hartland-Rowe, R. (1971). A preliminary investigation of bottom fauna and invertebrate drift in an unregulated and a regulated stream in Alberta. *Journal of Applied Ecology, 8*, 883–903.

Ras, K. L., & Palta, B. R. (1973). Great man-made lakes of Bhakra, India. Pp. 170–85 in *Man-Made Lakes: Their Problems and Environmental Effects* (Ed. W. C. Ackermann, G. F. White, & E. B. Worthington). Geophysical Monograph 17, American Geophysical Union, Washington, DC: xv + 847 pp., illustr.

Rasid, H. (1979). The effects of regime regulation by the Gardiner Dam on downstream geomorphic processes in the South Saskatchewan River. *Canadian Geographer, 23*, 140–58.

Rausch, D. L., & Heinemann, H. G. (1975). Controlling trap efficiency. *Transactions of the American Society of Agricultural Engineers, 18* (6), 1105–13.

Raymond, H. L. (1968). Migration rates of yearling Chinook Salmon in relation to flows and impoundments in the Columbia and Snake Rivers. *Transactions of the American Fisheries Society, 97* (4), 356–79.

Raymond, H. L. (1969). Effect of the John Day Reservoir on the migration rate of juvenile Chinook Salmon in the Columbia River. *Transactions of the American Fisheries Society, 98* (3), 513–4.

Raymond, H. L. (1979). Effects of dams and impoundments on migration of juvenile Chinook Salmon and Steelhead Trout from the Snake River, 1966–75. *Transactions of the American Fisheries Society, 108* (6), 509–29.

Rees, W. A. (1978). The ecology of the Kafue Lechwe: as affected by the Kafue Gorge hydroelectric scheme. *Journal of Applied Ecology, 15*, 205–17.

Reif, C. B. (1939). The effect of stream conditions on lake plankton. *Transactions of the American Microscopy Society, 58*, 398–403.

Reynolds, F. R. (1976). Tagging important in River Murray fish study. *Australian Fisheries, 35*, 4–6.

Richards, K. S. (1980). A note on changes of channel geometry at tributary junctions. *Water Resources Research, 16*, 241–4.

Richards, K. S., & Wood, R. (1977). Urbanization, water redistribution and their effects on channel processes. Pp. 369–88 in *River Channel Changes* (Ed. K. J. Gregory). John Wiley & Sons, Chichester: xi + 448 pp., illustr.

Riddick, T. M. (1957). Forced circulation of reservoir waters. *Water Sewage Works, 104* (6), 231–7.

Ridley, J. E., & Steel, J. A. (1975). Ecological aspects of river impoundments. Pp. 565–87 in *River Ecology* (Ed. B. A. Whitton). Blackwell Scientific Publications, Oxford: x + 725 pp., illustr.

Rigler, F. H. (1964). The phosphorus fractions and the turnover time of inorganic phosphorus in different types of lakes. *Limnology and Oceanography, 9*, 511–8.

Ritter, J. R. (1968). *Changes in the Channel Morphology of Trinity River and Eight Tributaries, 1961–65.* US Geological Survey, Open File Report, August 1968.*

Robinson, W. L. (1978). The Columbia: a river system under siege. *Oregon Wildlife, 33* (6), 3–7.

Rodhe, W. (1964). Effects of impoundment on water chemistry and plankton in Lake Ransaren (Swedish Lappland). *Verhandlungen Internationale Vereinigung für Theoretische und Angewandte Limnologie, 15*, 437–43.

294

Roehl, J. W., & Holeman, J. N. (1973). Sediment studies pertaining to small reservoir design. Pp. 376–80 in *Man-Made Lakes: Their Problems and Environmental Effects* (Ed. W. C. Ackermann, G. F. White, & E. B. Worthington). Geophysical Monograph 17, American Geophysical Union, Washington, DC: xv + 847 pp., illustr.

Rogers, H. H., Raynes, J. J., Posey, F. H., & Ruland, W. E. (1973). Lake destratification by underwater air diffusion. Pp. 572–7 in *Man-Made Lakes: Their Problems and Environmental Effects* (Ed. W. C. Ackermann, G. F. White, & E. B. Worthington). Geophysical Monograph 17, American Geophysical Union, Washington, DC: xv + 847 pp., illustr.

Ross, L. E., & Rushforth, S. R. (1980). The effects of a new reservoir on the attached diatom communities in Huntington Creek, Utah, USA. *Hydrobiologia*, **68** (2), 157–65.

Ruane, R. J., Vigander, S., & Nicholas, W. R. (1977). Aeration of hydro-releases at Fort Patrick Henry Dam. *Proceedings of the American Society of Civil Engineers, Journal of the Hydraulics Division,* **HY10**, 1135–45.

Ruffner, G. A., & Carothers, S. W. (1975). Recent notes on the distribution of some mammals of the Grand Canyon region. *Plateau*, **47** (4), 154–60.

Rutter, E. J., & Engstrom, L. R. (1964). Reservoir regulation. Sec. 25–III in *Handbook of Applied Hydrology* (Ed. V. T. Chow). McGraw-Hill, New York, USA: xiv + 29 secs, pp. var., illustr.

Ruttner, F. (1963). *Fundamentals of Limnology* (3rd edition). University of Toronto Press, Toronto, Ontario, Canada: xv + 295 pp., illustr.

Rzóska, J. (Ed.) (1976). *The Nile: Biology of an Ancient River.* (Monographies Biologicae 29.) W. Junk Publishers, The Hague: xix + 417 pp., illustr.

Rzóska, J., Brook, A. J., & Prowse, G. A. (1955). Seasonal plankton development in the White and Blue Nile near Khartoum. *Verhandlungen Internationale Vereinigung für Theoretische und Angewandte Limnologie*, **12**, 327–34.

Sabol, K. J. (Ed.) (1974). *National Conference on Flood Plain Management* (League City, Texas). National Association of Conservation Districts, Texas, USA: 261 pp.

Sandison, E. E., & Hill, M. B. (1966). The distribution of *Balanus pallidus stutsburi* Darwin, *Gryphaea gasar* (Adanson) Pautzenberg, *Mercierella enigmatica* Fauvel and *Hydroides uncinata* (Philippi) in relation to salinity in Lagos Harbour and adjacent creeks. *Journal of Animal Ecology*, **35**, 235–50.

Sasaki, E. N. (1976). Water resources development and management—Trinity River, California. Pp. 129–38 in *Instream Flow Needs*, Vol. II (Ed. J. F. Orsborn, & C. H. Allman). American Fisheries Society, Bethesda, Maryland, USA: vi + 657 pp., illustr.

Sauer, S. P., & Masch, F. D. (1969). Effects of small structures on water yield in Texas. Pp. 118–35 in *Effects of Watershed Changes on Streamflow* (Ed. W. L. Moore, & C. W. Morgan). University of Texas Press. Austin, USA & London, England, UK : xi + 289 pp., illustr.

Saunders, J. W., & Smith, M. W. (1962). Physical alteration of stream habitat to improve brook trout production. *Transactions of the American Fisheries Society*, **91**, 185–8.

Savory, C. R. (1961). Report on the status of game in the Eastern Zambezi Valley. Described in Attwell, R.I.G. (1970) (q.v.).

Schlosser, I. J. (1982). Fish community structures and function along two habitat gradients in a headwater stream. *Ecological Monographs*, **52** (4), 395–414.

Schmitz, W. R., & Hasler, A. D. (1958). Artificially induced circulation of lakes by means of compressed air. *Science*, **128** (3331), 1088–9.

Schoeneman, D. E. (1959). Survival of downstream migrant salmon passing Alder Dam in an open flume. *Washington State Department of Fisheries, Fisheries Research Papers*, **2** (2), 31–7.

Schoeneman, D. E., & Junge, C. D. (1961). An evaluation of the mortalities to downstream migrant salmon at McNary Dam. *Transactions of the American Fisheries Society*, **90**, 58–72.

Schreiber, J. D., & Rausch, D. L. (1979). Suspended sediment—phosphorus relationships for the inflow and outflow of a flood detention reservoir. *Journal of Environmental Quality*, **8** (4), 510–4.

Schumm, S. A. (1963). *The Disparity Between Rates of Denudation and Orogeny*. US Geological Survey Professional Paper, 454H, 13 pp.

Schumm, S. A. (1969). River metamorphosis. *Proceedings of the American Society of Civil Engineers, Journal of the Hydraulics Division*, **HY1**, 255–73.

Schumm, S. A. (1973). Geomorphic thresholds and complex response of drainage basins. Pp. 299–310 in *Fluvial Geomorphology* (Ed. M. E. Morrisawa). Allen & Unwin, London, England, UK: xi + 314 pp., illustr.

Schumm, S. A. (1977). *The Fluvial System*. Wiley-Interscience, New York, USA: xvii + 338 pp., illustr.

Schumm, S. A., & Hadley, R. F. (1967). Arroyos and the semi-arid cycle of erosion. *American Journal of Science*, **255**, 161–74.

Schwartz, H. I. (1969). Hydrologic aspects of limnology in South Africa (1969). *Hydrobiologia*, **34**, 14–28.

SCOPE (1972). *Man-made Lakes as Modified Ecosystems*. (Report 2 of the Scientific Committee on Problems of the Environment (SCOPE).) International Council of Scientific Unions, Paris, France: 76 pp., illustr.

Scott, G. R. (1973). *Scour and Fill in Tujunga Wash—a Fanhead Valley in South California*. US Geological Survey Professional Paper, 732B: 29 pp., illustr.

Scullion, J., Parish, C. A., Morgan, N., & Edwards, R. W. (1982). Comparison of benthic macroinvertebrate fauna and substratum coomposition in riffles and pools in the impounded River Elan and the unregulated River Wye, mid-Wales. *Freshwater Biology*, **12**, 579–95.

Seavy, L. M. (1948). *Sedimentation survey of Arrowrock Reservoir, Boise, Idaho*. US Bureau of Reclamation, Denver, Colorado. Reported in Brune, G.M. (1953) (q.v.).

Selye, H. (1974). *Stress Without Distress*. The New American Library of Canada Ltd., Scarborough, Ontario, Canada: x + 193 pp., illustr.

Senuya, C. (1972). Metalimnia layer in Lake Kinneret, Israel. *Hydrobiologia*, **40** (3), 355–9.

Senuya, C., Senuya, S., & Berman, T. (1969). Preliminary observations of the hydro-mechanics, nutrient cycles and eutrophication status of Lake Kinneret (Lake Tiberias). *Verhandlungen Internationale Vereinigung für Theoretische und Angewandte Limnologie*, **17**, 342–51.

Serr, E. E. (1972). Unusual sediment problems in N. Coastal California. Chapt. 17 (13 pp.) in *Sedimentation* (Ed. H. W. Shen). Einstein Symposium Volume, Water Resources Publication, Fort Collins, Colorado, USA: 27 chapters, pp. var., 5 app., illustr.

Shalash, S. (1983*a*). Degradation of the River Nile Pt. 1. *Water Power and Dam Construction*, **35** (7), 37–43.

Shalash, S. (1983*b*). Degradation of the River Nile Pt. 2. *Water Power and Dam Construction*, **35** (8), 56–8.

Sharonov, I. V. (1963). Habitat conditions and the behaviour of fish in the tailwater of the Volga Hydroelectric Power Station. *Trudy Instituta Biologii Vnutrennikh Vod.*, No. **6** (9), 195–200. (Translated from the Russian, US Department of Commerce, Springfield, Illinois, TT68–50389.)

Shelford, V. E. (1911). Ecological succession, 1: Stream fishes and the method of physiographic analysis. *Biological Bulletin*, **21**, 9–35.

296

Sherman, B. J., & Phinney, H. K. (1971). Benthic algal communities of the Metolius River. *Journal of Phycology*, **7**, 269–73.

Shiel, R. J. (1976). Association of Entomostraca with weedbed habitats in a billabong of the Goulburn River, Victoria. *Australian Journal of Marine and Freshwater Research*, **27**, 533–49.

Shiel, R. J. (1978). Zooplankton communities of the Murray–Darling system. *Proceedings of the Royal Society of Victoria*, **90**, 193–202.

Shikhshabekov, M. M. (1971). Resorption of the gonads in some semi-diadromous fishes of the Arakum Lakes (Dagestan, USSR) as a result of the regulation of discharge. *Journal of Ichthyology*, **11** (3), 427–31.

Shirazi, M. A., & Seim, W. K. (1981). Stream system evaluation with emphasis on spawning habitat for salmonids. *Water Resources Research*, **17** (3), 592–94.

Shulits, S. (1934). Experience with bed degradation below dams on European rivers. *Engineering News Record*, June, pp. 838–9.

Simmons, G. M., jr., & Voshell, J. R., jr (1978). Pre- and post-impoundment benthic macroinvertebrate communities of the North Anna River. Pp. 45–61 in *Current Perspectives on River–Reservoir Ecosystems* (Ed. J. Cairns, jr, E. F. Benfield, & J. R. Webster). North American Benthological Society, Springfield, Illinois, USA: v + 85 pp., illustr.

Simons, D. B., & Li, Ruh-ming. (1982). Bank erosion in regulated rivers. Pp. 717–54 in *Gravel-Bed Rivers* (Ed. R. D. Hey, J. C. Bathurst, & C. R. Thorne). John Wiley & Sons, Chichester, England, UK: xv + 875 pp., illustr.

Skulberg, C. M., & Kotai, J. (1978). Miljøfaktorer og algeutvikling i strømmende vann: Noen observasjoner av innvirkningene av vassdrogsregulerunger på begroingsforhold i Glåma i Ølsterdalen. *Norsk Institutt for Vannforskning, Årbok*, 1977, pp. 63–73.

Smith, F. E. (1976). Water development impact on fish resources and associated values of the Trinity River, California. Pp. 98–111 in *Instream Flow Needs, Vol. 1.* (Ed. J. F. Orsborn, & C. H. Allman). American Fisheries Society, Bethesda, Maryland: 657 pp., illustr.

Smith, K. (1972). *Water in Britain*. Macmillan, London: xiii + 241 pp., illustr.

Smith, N. (1971). *A History of Dams*. Peter Davies, London: xiv + 279 pp., illustr.

Sokolov, A. A., Rantz, S. E., & Roche, M. (1976). *Floodflow Computation Methods Compiled from World Literature*. UNESCO, Paris: 294 pp.

Soltero, R. A., Gasperino, A. F., & Graham, W. G. (1974*a*). Chemical and physical characteristics of a eutrophic reservoir and its tributaries: Long Lake, Washington. *Water Research*, **8**, 419–33.

Soltero, R. A., Wright, J. C., & Horpestad, A. A. (1973). Effects of impoundment on the water quality of the Bighorn River. *Water Research*, **7**, 343–54.

Soltero, R. A., Wright J. C., & Horpestad, A. A. (1974*b*). The physical limnology of Bighorn Lake, Yellowtail Dam, Montana: internal density currents. *Northwest Scientist*, **48**, 107–24.

Spence, J. A., & Hynes, H. B. N. (1971). Differences in fish populations upstream and downstream of a mainstream impoundment. *Journal of the Fisheries Research Board of Canada*, **28**, 35–43, 45–6.

Sreenivasan, A. (1964). A hydrological study of a tropical impoundment, Bhavanisagar Reservoir, Madras State, India, for the years 1955–61. *Hydrobiologia*, **24**, 514–39.

Sreenivasan, A. (1977). Fisheries of the Stanley Reservoir (Mettur Dam) and three other reservoirs of Tamilnadu, India: a case history. *Proceedings of the IPFC Symposium on the Development and Utilization of Inland Fishery Resources, Colombo, Sri Lanka, 27–29 October 1976* (Ed. I. Dunn). FAO Regional Office for Asia and the Far East, Bangkok, Thailand: 17 pp.

Stalnaker, C. B. (1979). The use of habitat structure preferenda for establishing flow

297

regimes necessary for maintenance of fish habitat. Pp. 301–37 in *The Ecology of Regulated Rivers* (Ed. J. V. Ward, & J. A. Stanford). Plenum Press, New York, USA: xi + 398 pp., illustr.

Stanford, J. V., & Ward, J. V. (1979). Stream regulation in North America. Pp. 215–36 in *The Ecology of Regulated Streams* (Ed. J. V. Ward, & J. A. Stanford). Plenum Press, New York, USA: xi + 398 pp., illustr.

Stanford, J. V., & Ward, J. V. (1981). Preliminary interpretations of the distribution of Hydropsychidae in a regulated river. Pp. 323–8 in *Proceedings of the Third International Symposium on Trichoptera* (Ed. G. P. Moretti). W. Junk Publishers, The Hague: xxi + 472 pp., illustr.

Stanley, J. W. (1951). Retrogression of the Lower Colorado River after 1935. *Transactions of the American Society of Civil Engineers*, 116, 943–57.

Stenseth, N. C. (1977). Modelling the population dynamics of voles: models as research tools. *Oikos*, 29, 449–56.

Stober, Q. J. (1964). Some limnological effects of Tiber Reservoir on the Marias River, Montana. *Proceedings of the Montana Academy of Science*, 23, 111–37.

Stoddart, D. R. (1967). Growth and structure in geography. *Transaction of the Institute of British Geographers*, 41, 1–19.

Strahler, A. N. (1957). Quantitative analysis of watershed geomorphology. *Transactions of the American Geophysical Union*, 38, 913–20.

Strahler, A. N. (1975). *Physical Geography*, 4th Edition. John Wiley & Sons, New York: 643 pp., illustr.

Strakhov, N. M. (1967). *Principles of Lithogenesis*, I (Trans. J. P. Fitzsimmons, S. I. Tomkieff, & J. E. Hemingway). Consultants Bureau, New York, USA: 245 pp.

Straškraba, M. (1973). Limnological basis for modelling reservoir ecosystems. Pp. 517–35 in *Man-Made Lakes: Their Problems and Environmental Effects* (Ed. W. C. Ackermann, G. F. White, & E. B. Worthington). Geophysical Monograph 17, American Geophysical Union, Washington, DC, USA: xv + 847 pp., illustr.

Stromquist, L. (1981). Recent studies on soil erosion, sediment transport and reservoir sedimentation in semi-arid central Tanzania. Pp 180–200 in *Tropical Agricultural Hydrology: Watershed Management and Land Use* (Ed. R. Lal, & E. W. Russell). John Wiley & Sons, Chichester, England, UK: xiv + 482 pp., illustr.

Stroud, R. H., & Martin, R. G. (1973). Influence of reservoir discharge location on the water quality, biology, and sport, fisheries of reservoirs and tailwaters. Pp. 540–58 in *Man-Made Lakes: Their Problems and Environmental Effects* (Ed. W. C. Ackermann, G. F. White, & E. B. Worthington). Geophysical Monograph 17, American Geophysical Union, Washington, DC, USA: xv + 847 pp., illustr.

Swale, E. M. F. (1964). A study of the phytoplankton of a calcareous river. *Journal of Ecology*, 52, 433–46.

Symons, J. M., Weibel, S. R., & Robeck, G. G. (1965). Impoundment influences on water quality. *Journal of the American Water Works Association*, 57 (1), 51–75.

Szupryczyński, J. (1976). The effect of the reservoir near Wloclawek on the geographical environment. *Geographia Polonica*, 33 (1), 135–41.

Talling, J. F. (1969). The incidence of vertical mixing, and some biological and chemical consequences of tropical African lakes. *Verhandlungen Internationale Vereinigung für Theoretische und Angewandte Limnologie*, 17, 998–1012.

Talling, J. F. (1976). Chapters on water characteristics and phytoplankton. Pp. 357–84 and 385–402 in *The Nile: Biology of an Ancient River* (Ed. J. Rzóska). Monographiae Biologicae 29, W. Junk Publishers, The Hague: xix + 417 pp., illustr.

Talling, J. F., & Rzóska, J. (1967). The development of plankton in relation to hydrological regime in the Blue Nile. *Journal of Ecology*, 55, 637–62.

Tarzwell, C. M. (1939). Changing the Clinch River into a trout stream. *Transactions of the American Fisheries Society*, 68, 228–33.

T.C.E.E.H.S. (1978). Environmental effects of hydraulic structures. Task Committee on the Environmental Effects of Hydraulic Structures, Hydraulics Division, American Society of Civil Engineers. *Proceedings of the American Society of Civil Engineers*, **HY2**, pp. 203–21.

Task Committee on the Environmental Effects of Hydraulic Structures—*see* T.C.E.E.H.S.

Therenin, M., Labonde, J., Gras, M., Leynaud, M., & Mangerel, M. (1973). Modification de la qualité de l'eau dans les lacs artificiels de barrages: Le cas du barrage Seine. Pp. 29–54 in *Transactions of the Eleventh International Congress of Large Dams, Madrid, Spain*, Vol. 1. International Commission on Large Dams, Paris, France: xxviii + 974 pp., illustr.

Thomas, W. L. (Ed.) (1956). *Man's Role in Changing the Face of the Earth*. University of Chicago Press, Chicago: xxxvii + 1193 pp., illustr.

Thompson, D. H., & Hunt, F. D. (1930). The fishes of Champaign County: a study of the distribution and abundance of fishes in small streams. *Illinois Natural History Survey Bulletin*, **19**, 1–110.

Tinley, K. L. (1975). Marromeu wrecked by the big dam. *African Wildlife*, **29** (2), 22–5.

Tinney, E. R. (1962). The process of channel degradation. *Journal of Geophysical Research*, **67** (4), 1475–8.

Townsend, G. H. (1975). Impact of the Bennett Dam on the Peace–Athabaska delta. *Journal of Fisheries Research Board of Canada*, **32**, 171–6.

Tramer, E. J., & Rogers, P. M. (1973). Diversity and longitudinal zonation in fish populations of two streams entering a metropolitan area. *American Midland Naturalist*, **90**, 366–74.

Trautman, M. B., & Gartman, D. K. (1974). Re-evaluation of the effects of man-made modification on Gordon Creek between 1887 & 1973 and especially as regards fish fauna. *Ohio Journal of Science*, **74** (3), 162–73.

Trefethen, P. (1972). Man's impact on the Columbia River. Pp. 77–98 in *River Ecology and Man* (Ed. R. T. Oglesby, C. A. Carlson, & J. A. McCann). Academic Press, New York, USA: xvii + 465 pp., illustr.

Trotzky, H. M., & Gregory, R. W. (1974). The effects of water flow manipulation below a hydroelectric power dam on the bottom fauna of the upper Kennebec River, Maine. *Transactions of the American Fisheries Society*, **103**, 318–24.

Truesdale, G., & Taylor, G. (1978). Quality implications in reservoirs filled from surface water sources. *Progress in Water Technology*, **10**, 289–300.

Turner, D. I. (1971). Dams and ecology. *Civil Engineering (ASCE)*, **41**, 76–80.

Turner, R. M., & Karpiscak, M. M. (1980). *Recent Vegetation Changes Along the Colorado River Between Glen Canyon Dam and Lake Mead, Arizona*. US Geological Survey Professional Paper, No. 1132, 125 pp.

Tyler, P. A., & Buckney, R. T. (1974). Stratification and biogenic meromixis in Tasmanian reservoirs. *Australian Journal of Marine and Freshwater Research*, **25**, 299–313.

Tyler, P. A., & Marshall, K. C. (1967). Hyphomicrobia—a significant factor in manganese problems. *Journal of the American Water Works Association*, **59**, 1043–8.

Tyurin, P. V. (Ed.) (1966). *The Storage Lakes of the USSR and their Importance for Fishery*. Israel Program for Scientific Translations, Jerusalem: 244 pp., illustr.

Ulfstrand, S. (1968). Benthic animal communities in Lapland streams. *Oikos, Supplement*, **10**, 120 pp.

US Army Corps of Engineers—*see* U.S.A.C.E.

U.S.A.C.E. (1950). *Second Interim Report on Sedimentation in Conches Reservoir South Canadian River Watershed, Albuquerque District*. US Army Corps of Engineers, Albuquerque, New Mexico. Reported in Brune, G. M. (1953). (q.v.)

Vanicek, C. D., Kramer, R. H., & Franklin, D. R. (1970). Distribution of Green

River fishes in Utah and Colorado following closure of Flaming Gorge Dam. *Southwestern Naturalist*, **14**, 297–315.

Vannote, R. L., Minshall, G. W., Cummins, K. W., Sedell, J. R., & Cushing, C. E. (1980). The river continuum concept. *Canadian Journal of Fisheries and Aquatic Sciences*, **37**, 130–7.

Vannote, R. L., & Sweeney, B. W. (1980). Geographical analysis of thermal equilibria; conceptual model for evaluating the effect of natural and modified thermal regimes on aquatic insect communities. *American Naturalist*, **115**, 667–95.

Vaux, W. G. (1968). Intragravel flow and interchange of water in a streambed. *US Fisheries and Wildlife Service, Fisheries Bulletin*, **66**, 479–89.

Vincent, E. R. (1977). *Madison River Temperature Study*. (Job Program Report, Federal Aid Project F-G-R-25, Job No. 11(a).) Montana Department of Fish and Game, USA: 10 pp.*

Viner, A. B. (1969). Observation of the hydrobiology of the Volta Lake, April 1965–April 1966. Pp. 133–43 in *Man-Made Lakes: The Accra Symposium* (Ed. L. E. Obeng). Ghana University Press, Accra, Ghana: 398 pp., illustr.

Vitousek, P. M. (1977). The regulation of element concentration in mountain streams in the northeastern United States. *Ecological Monographs*, **47**, 65–87.

Vitousek, P. M., & Reiners, W. A. (1975). Ecosystem succession and nutrient retention: a hypothesis. *BioScience*, **25**, 376–81.

Vladykov, V. D. (1964). *Inland Fisheries Resources of Iran, Especially of the Caspian Sea with Special Reference to Sturgeon*. FAO Report No. 1818: 51 pp.*

Voeïkov, A. I. (1901). De l'influence de l'homme sur la Verre. *Annales de Géographie*, **x**, 97–114, 193–215.

Vollenweider, R. A. (1968). *Scientific Fundamentals of the Autotrophication of Lakes and Flowing Waters with Particular References to Nitrogen and Phosphorus as Factors in Eutrophication*. Technical Report OAS/CS1/68.27, OECD, Paris, France: 159 pp.

Wafa, T. A., & Labib, A. H. (1973). Seepage losses from Lake Nasser. Pp. 287–91 in *Man-Made Lakes: Their Problems and Environmental Effects* (Ed. W. C. Ackermann, G. F. White, & E. B. Worthington). Geophysical Monograph 17, American Geophysical Union, Washington, DC, USA: xv + 847 pp., illustr.

Walburg, C. H., Kaiser, G. L., & Hudson, P. L. (1971). Lewis and Clark Lake tailwater biota and some relations of the tailwater and reservoir fish populations. Pp. 449–67 in *Reservoir Fisheries and Limnology* (Ed. G. E. Hall). Special Publication No. 8, American Fisheries Society, Washington, DC, USA: x + 511 pp., illustr.

Walker, K. F. (1979). Regulated streams in Australia: the Murray–Darling River system. Pp. 143–64 in *The Ecology of Regulated Streams* (Ed. J. V. Ward, & J. A. Stanford). Plenum Press, New York: xi + 398 pp., illustr.

Walker, K. F., Hillman, T. J., & Williams, W. D. (1979). The effects of impoundment on rivers: an Australian case study. *Verhandlungen Internationale Vereinigung für Theoretische und Angewandte Limnologie*, **20**, 1695–701.

Wallace, J. B., & Merritt, R. W. (1980). Filter feeding ecology of aquatic insects. *Annual Review of Entomology*, **25**, 103–32.

Walling, D. E. (1981). Yellow River which never runs clear. *The Geographical Magazine*, **LIII** (9), 568–76.

Walling, D. E., & Webb, B. W. (1975). Spatial analysis of river water quality: a survey of the River Exe. *Transactions of the Institute of British Geographers*, **65**, 155–90.

Ward, J. V. (1974). A temperature-stressed ecosystem below a hypolimnial release mountain reservoir. *Archiv für Hydrobiologie*, **74**, 247–75.

Ward, J. V. (1975). Downstream fate of zooplankton from a hypolimnial release mountain reservoir. *Verhandlungen Internationale Vereinigung für Theoretische und Angewandte Limnologie*, **19**, 1798–804.

300

Ward, J. V. (1976a). Comparative limnology of differentially regulated sections of a Colorado mountain river. *Archiv für Hydrobiologie,* **78**, 319–42.

Ward, J. V. (1976b). Effects of flow patterns below large dams on stream benthos. Pp. 235–52 in *Instream Flow Needs,* Vol. II (Ed. J. F. Orsborn, & C. H. Allman). American Fisheries Society, Bethesda, Maryland, USA: 657 pp., illustr.

Ward, J. V. (1976c). Effects of thermal constancy and seasonal temperature displacement on community structure of stream macroinvertebrates. Pp. 302–7 in *Thermal Ecology II* (Ed. G. W. Esch, & R. W. McFarlane). (ERDA Symposium Series 40 [Conf.–750425].) National Technical Information Series, Springfield, Illinois, USA: 323 pp., illustr.

Ward, J. V., & Short, R. A. (1978). Macroinvertebrate community structure of four special lotic habitats in Colorado, USA. *Verhandlungen Internationale Vereinigung für Theoretische und Angewandte Limnologie,* **20**, 1382–7.

Ward, J. V., & Stanford, J. A. (Ed). (1979a). *The Ecology of Regulated Streams.* Plenum Press, New York, USA: xi + 398 pp., illustr.

Ward, J. V., & Stanford, J. A. (1979b). Ecological factors controlling stream zoobenthos. Pp. 35–56 in *The Ecology of Regulated Streams* (Ed. J. V. Ward, & J. A. Stanford). Plenum Press, New York: xi + 398 pp., illustr.

Ward, J. V., & Stanford, J. A. (1983). The serial discontinuity concept of lotic ecosystems. Pp. 29–42 in *Dynamics of Lotic Ecosystems* (Ed. T. D. Fontaine, & S. M. Bartell). Ann Arbor Science, Ann Arbor, Michigan, USA: xii + 284 pp., illustr.

Warner, R. F. (1981). The impacts of dams and weirs on the Hawkesbury–Nepean River System, New South Wales, Australia. Paper presented to the *2nd International Conference on Fluvial Sediments,* University of Keele, UK, 21–25 September 1981.

Waters, T. F. (1962). Diurnal periodicity in the drift of stream invertebrates. *Ecology,* **43**, 316–20.

Waters, T. F. (1964). Recolonization of denuded stream-bottom areas by drift. *Transactions of the American Fisheries Society,* **93**, 311–5.

Waters, T. F. (1965). Interpretation of invertebrate drift in streams. *Ecology,* **46**, 327–34.

Waters, T. F. (1966). Production rate, population density, and drift of a stream invertebrate. *Ecology,* **47**, 595–605.

Webb, B. W., & Walling, D. E. (1974). Local variation in background water quality. *Science of the Total Environment,* **3**, 141–53.

Webster, J. R., Benfield, E. F., & Cairns, J., jr (1979). Model predictions of effects of impoundment of particulate organic matter transport in a river system. Pp. 339–64 in *The Ecology of Regulated Streams* (Ed. J. V. Ward, & J. A. Stanford). Plenum Press, New York, USA: xi + 398 pp., illustr.

Welch, E. B. (1980). *Ecological Effects of Waste Water.* Cambridge University Press, Cambridge, UK: xii + 337 pp., illustr.

Welcomme, R. L. (1979). *Fisheries Ecology of Floodplain Rivers.* Longman, London & New York, USA: vii + 317 pp., illustr.

Wesche, T. A., & Rechard, P. A. (1980). *A Summary of Instream Flow Methods for Fisheries and Related Research Needs.* Eisenhower Consortium Bulletin 9, Water Research Institute, University of Wyoming, Laramie, Wyoming, USA: 122 pp.

Westlake, D. F. (1975). Macrophytes. Pp. 106–28. in *River Ecology* (Ed. B. A. Whitton). Blackwell Scientific Publications, Oxford, England, UK: x + 725 pp., illustr.

Westman, W. E. (1978). Measuring the inertia and resilience of ecosystems. *BioScience,* **28**, 705–10.

Wetzel, R. G. (1973). Primary production. Pp. 230–47 in *River Ecology* (Ed. B. A. Whitton). Blackwell Scientific Publications, Oxford, England, UK: x + 725 pp., illustr.

Wetzel, R. G. (1975). *Limnology*. W. B. Saunders Co., Philadelphia, USA: 743pp.

White, G. F. (Ed.) (1977*a*). *Environmental Effects of Complex River Development*. Westview, Boulder, Colorado, USA: xi + 172 pp., illustr.

White, G. F. (1977*b*). Comparative analysis of complex river development. Pp. 1–21 in *Environmental Effects of Complex River Development* (Ed. G. F. White). Westview, Boulder, Colorado, USA: xvi + 172 pp., illustr.

Whiteside, B. G., & McNutt, R. (1972). Fish species diversity in relation to stream order. *American Midland Naturalist*, **88** (1), 90–101.

Whitley, J. R., & Campbell, R. S. (1974). Some aspects of water quality and biology of the Missouri River. *Transactions of the Missouri Academy of Science*, **8**, 60–72.

Whitton, B. A. (Ed.) (1975), *River Ecology*. Blackwell Scientific Publications, Oxford, England, UK: x + 725 pp., illustr.

Wiebe, A. H. (1938). Limnological observations on Norris Reservoir with special reference to dissolved oxygen and temperature. *Transactions of the North American Wildlife Conference*, **3**, 440–57.

Williams, D. D. (1980). Some relationship between stream benthos and substrate heterogeneity. *Limnology and Oceanography*, **25**, 166–72.

Williams, D. D., & Hynes, H. B. N. (1974). The occurrence of benthos deep in the substatum of a stream. *Freshwater Biology*, **4**, 233–56.

Williams, R. D., & Winget, R. H. (1979). Macroinvertebrate response to flow manipulation in the Strawberry River, Utah (USA). Pp. 365–77 in *The Ecology of Regulated Streams* (Ed. J. V. Ward, & Stanford). Plenum Press, New York, USA: xi + 398 pp., illustr.

Williams, W. D. (1967). The changing limnological environment in Victoria. Pp. 240–51 in *Australian Inland Waters and Their Fauna* (Ed. A. H. Weatherley). Australian National University Press, Canberra: 287 pp.

Wilson, R. S., Sleigh, M. A., Maxwell, T. R. A., Mance, G., & Milne, R. A. (1975). Physical and chemical aspects of Chew Valley and Blagdon Lakes, two eutrophic reservoirs in North Somerset, England. *Freshwater Biology*, **5**, 357–77.

Winkle, W. van, Christensen, S., & Mattice, J. S. (1976). Two roles of ecologists in defining and determining the acceptability of environmental impacts. *International Journal of Environmental Studies*, **9**, 247–54.

Wolfe, R. S. (1960). Microbial concentration of iron and manganese in water with low concentrations of these elements. *Journal of the American Water Works Association*, **52**, 1335–7

Wolman, M. G. (1967). Two problems involving river channels and their background observations: Quantitative Geography Part II. *Northwestern University Studies in Geography*, **14**, 67–107.

Wolman, M. G., & Gerson, R. (1978). Relative scales of time and effectiveness in watershed geomorphology. *Earth Surface Processes*, **3**, 189–208.

Wunderlich, W. O. (1971). The dynamics of density—stratified reservoirs. Pp. 219–32 in *Reservoir Fisheries and Limnology* (Ed. G. E. Hall). Special Publication No. 8, American Fisheries Society, Washington, DC, USA: x + 511 pp., illustr.

Wunderlich, W. O., & Elder, R. A. (1973). Mechanics of flow through man-made lakes. Pp. 300–10 in *Man-Made Lakes: Their Problems and Environmental Effects* (Ed. W. C. Ackermann, G. F. White, & E. R. Worthington). Geophysical Monograph 17 American Geophysical Union, Washington, DC, USA: xx + 847 pp., illustr.

Yakovleva, L. V. (1965). Characteristics of silting of small reservoirs of the Central Chernozem provinces and computation of density of bottom deposits. In *Water Balance and Silting of Small Reservoirs in the Central Chernozem of the Russian Soviet Federal Socialist Republic, 1965*. (Transl. by D. V. Krimgold, US Dept of Agriculture Research Service, Washington, DC, USA: 1967).*

Young, W. C., Kent, D. H., & Whiteside, B. G. (1976). The influence of a deep storage reservoir on the species diversity of benthic macroinvertebrate communities of the Guadalupe River, Texas. *Texas Journal of Science*, **27**, 213–24.

Zakharyan, G. B. (1972). The natural reproduction of sturgeons in the Kura River following its regulation. *Journal of Ichthyology*, **12** (2), 249–59.

Zalumi, S.G. (1970). The fish fauna of the lower reaches of the Dnieper: its present composition and some features of its formation under conditions of regulated and reduced river discharge. *Journal of Ichthyology*, **10**, 587–96.

Zhadin, V. I., & Gerd, S. V. (1963). *Fauna and Flora of the Rivers, Lakes and Reservoirs of the USSR*. Israel Program for Scientific Translations, Jerusalem: 625 pp.

Zumberge, J. H., & Ayers, J. C. (1964). Hydrology of lakes and swamps. Ch 23, pp. 1–33, in *Handbook of Applied Hydrology* (Ed. V. T. Chow). McGraw-Hill, New York: xiv + 29 sec., pp. var., illustr.

Zwerner, G. A., Johnson, J. W., & Flaxman, E. M. (1942). *Advance Report on the Sedimentation Survey and Suspended Matter Observations in Lake Issaquaeena, Clemson, S. C., 1940–1*. US Soil Conservation Service, Washington, DC: November, SCS-SS-37.

Author Index

308

310

Poole, W. C., 176, 292
Pople, W., 161, 181
Por, F. D., 28, 290
Posey, F. H., 84, 294
Potter, H. R., 93, 274
Powell, G. C., 183, 292
Pratts, J. D., 122, 142, 144, 251, 292
Pritchard, A. L., 216, 274, 292
Pritchard, D. W., 78, 267
Pritchard, R. J., 288
Prowse, G. A., 10, 104–105, 292, 294
Purcell, L. T., 78, 293
Pyefinch, K. A., 229, 293

Quinn, J. J., 89, 285

Rabeni, C. F., 203, 293
Radford, D. S., 178, 183–185, 200, 218, 293
Radoane, M., 130, 283
Rantz, S. E., 43, 296
Ras, K. L., 160, 293
Rasid, H., 88, 99, 120, 123, 293
Rausch, D. L., 96–97, 100, 281, 293, 295
Raymond, H. L., 209, 226–230, 233, 269, 275, 293
Raynes, J. J., 84, 294
Rechard, P. A., 253–255, 300
Rees, W. A., 169, 172, 293
Rahacková, V., 219, 271
Reiger, W. A., 148, 292
Reif, C. B., 112
Reiners, W. A., 16, 299
Reteyum, A. Yu., 36, 106, 108, 272
Reynolds, F. R., 226, 293
Rhoades, E. D., 40, 281
Rhodes, D. D., 271
Rice, C. E., 85–86, 278
Richards, K. S., 21, 38, 130, 141, 293
Riddick, T. M., 84, 293
Ridley, J. E., 10, 293
Rigley, F. H., 67, 293
Ritter, J. R., 258, 293
Robeck, G. G., 85, 284, 297
Roche, M., 36, 106–108, 143, 296
Robinson, W. L., 226, 293
Rodhe, W., 63, 293
Roehl, J. W., 95, 294
Roff, J. B., 71, 288
Rogers, H. H., 84, 294
Rogers, K. H., 163, 284
Rogers, P. M., 24, 298
Romamenko, V. I., 36, 106, 108, 272
Rooke, J. B., 71, 288

Rosenberg, R., 238, 244, 268, 270, 289–290
Ross, L. E., 153–154, 294
Rouse, H., 271
Ruane, R. J., 81, 294
Ruffner, G. A., 168, 294
Ruland, W. E., 84, 294
Rushforth, S. R., 153–154, 286
Russell, E. W., 290, 297
Rutter, E. J., 29, 32–33, 37, 294
Ruttner, F., 61, 64, 77, 294
Rzóska, J., 10, 104–106, 111, 114, 271, 276, 294, 297

Sabol, K. J., 164, 294
Sahasrabudhe, S. R., 28, 284
Saltveit, S. J., 180, 210, 287
Sandison, E. E., 243, 294
Sasaki, E. N., 257–258, 260–261, 294
Sauer, S. P., 39, 53, 278, 294
Saunders, J. W., 222, 294
Savory, C. R., 170, 294
Schiewe, M., 227–228, 233, 274
Schlosser, I. J., 21, 294
Schmitz, W. R., 84, 294
Schneider, N. J., 274
Schoeneman, D. E., 228, 230, 294–295
Schoof, R. R., 40, 281
Schreiber, J. D., 67, 98–99, 295
Schumm, S. A., 14, 18, 21, 23, 132, 134, 138, 143, 144, 199, 244, 249, 280, 286, 295
Schwartz, H. I., 99, 295
Scope (1972), 10, 12, 263, 295
Scott, G. R., 53, 295
Scullion, J., 177, 193–195, 200, 202–203, 206, 295
Seavy, L. M., 100, 295
Sedell, J. R., 20–21, 23, 281, 299
Seim, W. K., 255, 269, 296
Senuya, C., 61, 66, 295
Senuya, S., 61, 295
Serr, E. F., 258, 295
Severn–Trent Water Authority (UK), 192
Shalash, S., 124, 295
Shaw, H. B., 91, 98, 283
Shaw, T. L., 97, 100, 272
Sharonov, I. V., 212, 295
Shelford, V. E., 21, 295
Shenouda, W. E., 23, 93, 121, 166, 285
Sherman, B. J., 153, 296
Shiel, R. J., 102, 108, 111, 165, 296

312

Geographical Index

'R.' before proper noun indicates 'River'.

313

Subject Index

Allochthonous organic matter, 15, 21, 89, 150, 246–247
Algae, 21, 23, 50, 64–66, 78, 101–116, 151, 217
  Bacillariophyceae, 103–116, 152–158, 241
    Acanthes spp., 109
    Asterionella spp., 103–104, 114
      formosa, 108
    Calothrix spp., 155
      papietrina, 157
    Cocconeis spp., 109
    Cyclotella spp., 109, 114, 153
      meneghiniana, 153
    Cymbella spp., 155, 157–158
    Diatoma elongatum, 157–158
    Diatomella spp., 154
    Frustulia spp, 154
    Gomphonema spp., 154
    Melosira spp., 103, 111
      granulata, 107–108
      varians, 155, 157
    Navicula spp., 104, 109
      avenacea, 157
    Nitzschia spp., 109
    Pinnularia spp., 153
    Synedra spp., 104
      acus, 107
    Tabellaria spp., 153
  Charophyceae, 112, 151, 159
  Chlorophyceae, 109–110, 152–159, 173, 241
    Ankistrodesmus falcatus, 105
    Cladophora spp., 153, 158–159, 174, 193
      glomerata, 153, 157–158
    Clamydomonas spp., 153
    Closterium spp., 104, 158
    Desmids, 158
    Enteromorphea spp., 163
    Hydra vulgaris, 111
    Miprospora amoena, 157–158
    Micrasterias spp., 158

    Mougeotia spp., 158
    Oedogonium spp., 155, 157–158
    Pediastrum spp., 105
    Scenedesmus spp., 105, 153
    Spirogyra spp., 104, 155, 158
    Stigeoclonium tenue, 157
    Ulothrix spp., 153, 158–159
      zonata, 153, 157–158, 174, 241
    Volvocales, 116
    Volvox aureus, 105
    Zygnema spp., 153, 158–159
  Chrysophyceae, 104, 115–116
    Hydrurus spp., 174
      foetidus, 158
  Cryptophyceae, 116
  Cyanophyceae, 102–116, 152, 154, 241
    Anabaena spp., 103, 115
      flos-aquae, 105–106
      spiroides, 107
    Anabaenopsis raciborskii, 107
    Aphanizomenon spp., 103, 115
    Homeothrix varians, 155, 157–158
    Lyngbya sp., 157, 158
      limnetica, 106, 115
    Microcystis spp., 103, 111, 115
      flos-aquae, 105
    Oscillatoria spp., 111
      Phormidium spp., 157–158
      Tolypothrix sp., 155, 157
  Phaeophyceae
    Heribaudiella fluviatilis, 158
  Rhodophyceae, 157
    Lemanea spp., 157–158
  Xanthophyceae, 157
    Vaucheria spp., 157–158
Algal blooms, 115
Arkansas River Project 118
Attached algae, *see* Algae *and* Periphyton
Autochthonous primary production, 21, 23, 64, 89, 150–164
Aves, *see* Birds

319